Books are to be returned on or before
the last date below.

STRESS

STRESS
Neuroendocrine and Molecular Approaches

Volume 1

Proceedings of the Fifth International Symposium on
Catecholamines and Other Neurotransmitters in Stress
Smolenice Castle, Czechoslovakia
24–29 June 1991

Edited by

RICHARD KVETŇANSKÝ
Institute of Experimental Endocrinology
Slovak Academy of Sciences
Bratislava, Czechoslovakia

RICHARD McCARTY
Department of Psychology
University of Virginia
Charlottesville, USA

JULIUS AXELROD
Laboratory of Cell Biology
National Institute of Mental Health
Bethesda, Maryland, USA

Gordon and Breach Science Publishers
Philadelphia • Reading • Paris • Montreux • Tokyo • Melbourne

Gordon and Breach Science Publishers

5301 Tacony Street, Drawer 330
Philadelphia, Pennsylvania 19137
United States of America

Post Office Box 161
1820 Montreux 2
Switzerland

Post Office Box 90
Reading, Berkshire RG1 8JL
United Kingdom

3-14-9, Okubo
Shinjuku-ku, Tokyo 169
Japan

58, rue Lhomond
75005 Paris
France

Private Bag 8
Camberwell, Victoria 3124
Australia

Library of Congress Cataloging-in-Publication Data

International Symposium on Catecholamines and Other Neurotransmitters
in Stress (5th : 1991 : Smolenice, Czechoslovakia)
   Stress : neuroendocrine and molecular approaches : proceedings of
the Fifth Symposium on Catecholamines and Other
Neurotransmitters in Stress, Smolenice Castle, Czechoslovakia, June
24–29, 1991 / edited by Richard Kvetňanský, Richard McCarty, Julius
Axelrod.
      p.   cm.
   Includes bibliographical references and index.
   ISBN 2-88124-506-4 (set : hardcover)
   1. Stress (Physiology)--Congresses.   2. Neuroendocrinology-
-Congresses.   3. Catecholamines--Physiological effect--Congresses.
4. Neuropeptides--Physiological effect--Congresses.   I. Kvetňanský,
Richard.   II. McCarty, Richard, 1947–   .   III. Axelrod, Julius,
1912–   .   IV. Title.
   [DNLM: 1. Neuroregulators--physiology--congresses.
2. Neurosecretory Systems--physiopathology--congresses.   3. Stress-
-physiopathology--congresses.   QZ 160 I61s 1991]
QP82.2.S8I58   1991
599'.0188--dc20
DNLM/DLC
                                                              92-1426

CONTENTS

VOLUME 2

PART 6. SYMPATHETIC RESPONSES TO STRESS

PART 7. PSYCHONEUROIMMUNE RESPONSES TO STRESS

PREFACE

Stress research continues as a vital and expanding field of scientific inquiry. Its strong interdisciplinary character is reflected in the diversity of experimental approaches taken by investigators involved in contemporary stress research. This diversity of research strategies was also reflected in the presentations at the Fifth International Symposium on Catecholamines and Other Neurotransmitters in Stress held at Smolenice Castle, Czechoslovakia, 24–29 June 1991. Distinguished basic and clinical researchers in many disciplines gathered at this important conference to highlight recent empirical and conceptual advances relating to the field of stress research. Given the quality of the oral and poster presentations and the intensity of the scientific interchange among the participants, the symposium was a complete success.

The fifth in this series of international conferences maintained a focus on peripheral and central catecholamine systems and stress. In addition, stress-related alterations in other neurotransmitter and neuropeptide systems received considerable attention. The presentations included a wide range of experimental approaches, including anatomical, physiological, neurochemical, pharmacological, molecular, immunobiological, endocrine and behavioral techniques. The strong interdisciplinary flavor of the conference provided for stimulating discussions and a number of new ways to look at continuing problems in the field of stress research.

Significant advances have been made in the field of stress research since the previous symposium, held in 1987. Highlights from the latest symposium included presentations on the central anatomy of catecholamine and neuropeptide pathways subserving the response to stressful stimulation. Measurements of neurotransmitters and their metabolites as well as hormones in stressed animals were the focus of several presentations. A substantial contingent of molecular biologists attended the symposium and provided a fascinating picture of molecular responses to stress. This volume provides a glimpse of an important future direction for research in stress.

Research in psychoneuroimmunology was also well represented. A number of presenters emphasized the dynamic interactions that occur between the neural, endocrine and immune systems, especially during stressful stimulation. Studies in this area have redefined many concepts in biology and medicine, and advances continue at a very rapid pace.

The countries of Eastern Europe underwent a tumultuous period of political, economic and social changes in the two years prior to this symposium. In spite of this period of intense upheaval, one factor remained constant. The people of Slovakia, especially the scientists from the Institute of Experimental Endocrinology, maintained their very high standards for scientific research and their equally high standards for hospitality. Since this series of symposia began in 1975, many people from East and West have come together in Slovakia united by their common interests in stress research. These symposia have also been fertile ground for the formation of lasting friendships and productive scientific collaborations. Indeed, the fruits of some of these interactions are represented in the pages of this volume.

On a personal level, my life has been enriched immeasurably by my four visits to Slovakia over the past twelve years. I look forward with great anticipation to a return visit to Slovakia in 1995.

Richard McCarty

ACKNOWLEDGEMENTS

Many individuals contributed to the overwhelming success of the Fifth International Symposium on Catecholamines and Other Neurotransmitters in Stress. Since 1975, Richard Kvetňanský has marshalled considerable support and enthusiasm around the world for these outstanding meetings. The most recent conference was again a scientific and cultural *tour de force*. The local organizing committee, headed by Dr. I. Vietor, attended to every need of the participants. Academician Ladislav Macho, chairman of the Slovak Academy of Sciences, and Dr. Milan Vigas, head of the Institute of Experimental Endocrinology, provided support and encouragement for this important international conference. Mrs. Darina Kvetnanská organized an excellent program for spouses of attendees. Finally, the executive secretaries and the international organizing committee, especially Dr. Errol B. De Souza, assisted Dr. Kvetňanský in program planning and enlisted financial support from a number of organizations. These included:

Nippon Zoki Pharmaceutical Company, Ltd.
Servier International Research Institute
UCB–Pharmaceutical Sector
The Upjohn Company
Merrell Dow Research Institute of Marion Merrell Dow
DuPont Medical Products
Reid–Rowell, Inc.
Center for Brain Sciences, Cambridge
Berlex Laboratories
Verla Pharm
ICI Pharmaceutical Group

The generous support of these organizations ensured the success of the meeting and is gratefully acknowledged. We are especially grateful to Nippon Zoki Pharmaceutical Company, Ltd. for providing a complimentary copy of the two-volume proceedings to each attendee.

Over the past sixteen years, these symposia have provided an important opportunity for scientists from East and West to interact on personal and professional levels. In addition, many friendships and scientific collaborations have developed as a result of these gatherings. On behalf of all attendees, we extend our warmest good wishes to the Slovak scientists and staff members who welcomed us to their magnificent homeland. A special note of thanks is extended to Mrs. Olga Kratka and the staff of Smolenice Castle for their warm hospitality.

The preparation of this camera-ready manuscript required a heroic effort by a dedicated group of editorial assistants. The editors extend their sincere thanks to Ms. Karen Webster, Ms. Lynn Webster and Ms. Kim Jenner for their exceptional efforts and dedication to this most demanding project. Without them, this volume would not have been possible. In addition, Dr. Magda Kourilova provided invaluable assistance to the local organizing committee by receiving completed manuscripts and preparing the author and subject indices.

We look forward with great enthusiasm to the Sixth International Symposium on Catecholamines and Other Neurotransmitters in Stress, to be held again at Smolenice Castle, Czechoslovakia, in 1995.

FIRST AUTHORS

E.D. Abercrombie, Center for Molecular and Behavioral Neuroscience, Rutgers University, 197 University Avenue, Newark, New Jersey 17102, USA

L.F. Agnati, Department of Human Physiology, University of Modena, Modena, Italy

G. Aguilera, Section on Endocrine Physiology, Developmental Endocrinology Branch, NICHD, National Institutes of Health, Building 10, Room 10N262, Bethesda, Maryland 20892, USA

S. Al-Damluji, Endocrinology and Reproduction Research Branch, National Institutes of Health, Building 10, Room B1L400, Bethesda, Maryland 20892, USA

I. Armando, CNB, NINDS, National Institutes of Health, Building 10, Room 5N214, Bethesda, Maryland 20892, USA

I. Assenmacher, Endocrinological Neurobiology Laboratory, URA 1197-CNRS, Department of Physiology, University of Montpellier-2, 34095 Montpellier, France

F. Berkenbosch, Department of Pharmacology, Medical Faculty, Free University, Van der Boechorststraat 7, 1081 BT Amsterdam, The Netherlands

M.C. Bohn, National Science Foundation, Room 320, 1800 G Street, NW, Washington, DC 20550, USA

B. Bohus, Department of Animal Physiology, University of Groningen, PO Box 14, 9750 AA Haren, The Netherlands

M.R. Brown, Peptide Biology Laboratory (0817), University of California, San Diego, Gilman Drive, La Jolla, California 92093, USA

J. Bugajski, Institute of Pharmacology, Polish Academy of Sciences, 12 Smetna Str., 31-343 Kraków, Poland

V.M. Chesnokova, Institute of Cytology and Genetics, Siberian Branch of the Russian Academy of Sciences, 630090 Novosibirsk, Russia

L. Chomicka, Institute of Animal Physiology and Nutrition, Polish Academy of Sciences, 05-110 Jablonna, Poland

G. Cizza, Clinical Neuroendocrinology Branch, National Institute of Mental Health, Building 10, Room 3S-235, Bethesda, Maryland 20892, USA

S.F. de Boer, Department of Psychopharmacology, Faculty of Pharmacy, University of Utrecht, Sorbonnelaan 16, 3584 CA Utrecht, The Netherlands

D.C.E. de Goeij, Department of Pharmacology, Free University, Van der Boechorststraat 7, 1081 BT Amsterdam, The Netherlands

E.R. de Kloet, Division of Medical Pharmacology, Center for Bio-Pharmaceutical Sciences, PO Box 9503, 2300 RA Leiden, The Netherlands

E.B. De Souza, The DuPont Merck Pharmaceutical Company, Experimental Station, E400/4352, PO Box 80400, Wilmington, Delaware 19880-0400, USA

M. Dobrakovova, Institute of Experimental Endocrinology, Slovak Academy of Sciences, Vlarska 3, 833 06 Bratislava, Czechoslovakia

W. Emsenhuber, Institute of Functional Pathology, Endocrinological Research Unit, University of Graz, Graz, Austria

K. Fukuhara, CNB, NINDS, National Institutes of Health, Building 10, Room 5N214, Bethesda, Maryland 20892, USA

W.F. Ganong, Department of Physiology, University of California, San Francisco, PO Box 0444, San Francisco, California 94143-0444, USA

D.S. Goldstein, NINDS, National Institutes of Health, Building 10, Room 5N214, Bethesda, Maryland 20892, USA

F.K. Goodwin, Office of the Director, Alcohol, Drug Abuse, and Mental Health Administration, Rockville, Maryland 20857, USA

D.E. Grigoriadis, The DuPont Merck Pharmaceutical Company, Experimental Station, E400/4442, PO Box 80400, Wilmington, Delaware 19880-0400, USA

M. Grino, Laboratoire de Neuroendocrinologie Experimentale, INSERM U 297, Marseille 13326, France

M. Haass, Department of Cardiology, University of Heidelberg, Bergheimerstr. 58, D-6900 Heidelberg, Germany

N.R.S. Hall, Psychoimmunology Division, Department of Psychiatry and Behavioral Medicine, University of South Florida College of Medicine, 3515 East Fletcher Avenue, Tampa, Florida 33613, USA

T. Hata, Department of Pharmacology, Faculty of Pharmacy, Kinki University, Kowakae, Higashi-Osaka 577, Japan

N. Hubalik, Department of Physical Education, University of Alaska, Fairbanks, Alaska 99775, USA

D. Jezova, Institute of Experimental Endocrinology, Slovak Academy of Sciences, Vlarska 3, 833 06 Bratislava, Czechoslovakia

N. Kalita, Institute for Standardization and Control of Drugs, Russian Ministry of Health, 117246 Moscow, Russia

D.T. Kiem, Institute of Experimental Medicine, Hungary Academy of Sciences, H-1450 Budapest POB 67, Hungary

B.K. Kiran, Department of Pharmacology, Uludag University Medical School, 16059 Bursa, Turkey

Z.M. Kiselyova, Clinical Biochemistry Laboratory, All-Union Cardiology Research Center, 3rd Cherepkovskaya Str., 15-a, 121552 Moscow, Russia

I. Kitayama, Department of Psychiatry, Mie University School of Medicine, 2-174, Edobashi, Tsu, Mie-ken 514, Japan

Y. Kurimoto, Institute of Bio-Active Science, Nippon Zoki Pharmaceutical Company, Ltd., Hyogo 673-14, Japan

R. Kvetnansky, Institute of Experimental Endocrinology, Slovak Academy of Sciences, Vlarska 3, 833 06 Bratislava, Czechoslovakia

J. Lavicky, Institute of Pharmacology, 2nd Medical Faculty, Charles University, 128 00 Prague 2, Albertov 4, Czechoslovakia

S.L. Lightman, Neuroendocrinology Unit, Charing Cross and Westminster Medical School, Charing Cross Hospital, Fulham Palace Road, London W6 8RF, UK

L. Macho, Institute of Experimental Endocrinology, Slovak Academy of Sciences, Vlarska 3, 833 06 Bratislava, Czechoslovakia

E. Mamalaki, Department of Endocrinology, University Hospital of Crete, Stavrakia, Heraklion, Crete 71702, Greece

A. Matsui, Department of Physiology, School of Dentistry, Showa University, Tokyo, Japan

N. Matsuki, Department of Chemical Pharmacology, Faculty of Pharmaceutical Sciences, University of Tokyo, Tokyo 113, Japan

R. McCarty, University of Virginia, Psychology Department, 102 Gilmer Hall, Charlottesville, Virginia 22903, USA

F. Menzaghi, Laboratory of Cellular Biology, INSERM U.308, 38, rue Lionnois, 54000 Nancy, France

L.L. Miner, c/o Barry B. Kaplan, Molecular Neurobiology and Genetics Program, Western Psychiatric Institute and Clinic, 3811 O'Hara Street, Pittsburgh, Pennsylvania 15213, USA

K. Murgas, Institute of Experimental Endocrinology, Vlarska 3, 833 06 Bratislava, Czechoslovakia

T. Nagatsu, Department of Biochemistry, Nagoya University School of Medicine, Nagoya 466, Japan

E.V. Naumenko, Institute of Cytology and Genetics, Siberian Branch of the Russian Academy of Sciences, 630090 Novosibirsk, Russia

J. Nikolic, Institute of Nuclear Sciences "Boris Kidrich," Vincha, Yugoslavia

H. Ohara, Institute of Bio-Active Science, Nippon Zoki Pharmaceutical Company, Ltd., Hyogo 673-14, Japan

Y. Oomura, Institute of Bio-Active Science, Nippon Zoki Pharmaceutical Company, Ltd. Hyogo 673-14, Japan

H. Osada, Division of Cardiology, Showa University, Fujigaoka Hospital, Yokohama, Japan

K. Pacak, National Institutes of Health, Building 10, Room 5N214, Bethesda, Maryland 20892, USA

M. Palkovits, First Department of Anatomy, Semmelweis University Medical School, Tüzoltó u. 58, H-1450 Budapest, Hungary

C. Pihoker, Department of Psychiatry, PO Box 3859, Duke University Medical Center, Durham, North Carolina 27710, USA

J. Polkowska, Institute of Animal Physiology and Nutrition, Polish Academy of Sciences, 05-110 Jablonna, Poland

S. Porta, Institute of Functional Pathology, Endocrinological Research Unit, University of Graz, Mozartgasse 14, A-8010 Graz, Austria

J.C. Porter, Department of Obstetrics and Gynecology, University of Texas, Southwestern Medical Center, 5323 Harry Hines Boulevard, Dallas, Texas 75235-9032, USA

J. Rauter, Institute of Functional Pathology, Endocrinological Research Unit, University of Graz, Graz, Austria

R. Richter, Institute of Drug Research, Alfred-Kowalkestr. 4, 0-1136 Berlin, Germany

I. Rinner, Institute of Functional Pathology, University of Graz, Mozartgasse 14, A-8010 Graz, Austria

E.L. Sabban, Department of Biochemistry and Molecular Biology, New York Medical College, Valhalla, New York 10595, USA

H. Saito, Department of Chemical Pharmacology, Faculty of Pharmaceutical Sciences, University of Tokyo, Tokyo 113, Japan

D. Saphier, Department of Pharmacology and Therapeutics, Louisiana State University Medical Center, 1501 Kings Highway, Shreveport, Louisiana 71130-3932, USA

P.E. Sawchenko, The Salk Institute, PO Box 85800, San Diego, California 92186, USA

K. Schauenstein, Institute of Functional Pathology, Medical School, University of Graz, Mozartgasse 14, A-8010 Graz, Austria

L.I. Serova, Institute of Cytology and Genetics, Siberian Branch of the Russian Academy of Sciences, 630090 Novosibirsk, Russia

M.S. Sothmann, Department of Human Kinetics, Enderis Hall, Room 419, University of Wisconsin-Milwaukee, PO Box 413, Milwaukee, Wisconsin 53201, USA

V. Stich, 4th Department of Internal Medicine, University Hospital, U Nemocnice, 128 00 Praha 2, Czechoslovakia

K.V. Sudakov, P.K. Anokhin Institute of Normal Physiology, Russian Academy of Medical Sciences, 6 Hertzen Street, Moscow 103009, Russia

R. Tamura, Department of Physiology, Faculty of Medicine, Toyama Medical and Pharmaceutical University, Sugitani, Toyama 930-01, Japan

R. Tigranian, Institute for Standardization and Control of Drugs, Russian Ministry of Health, 117246 Moscow, Russia

F.J.H. Tilders, Department of Pharmacology, Free University, Van der Boechorststraat 7, 1081 BT Amsterdam, The Netherlands

T. Torda, Institute of Experimental Endocrinology, Slovak Academy of Sciences, Vlarska 3, 833 06 Bratislava, Czechoslovakia

G.M. Tyce, Physiology Department, Mayo Clinic and Foundation, Rochester, Minnesota 55905, USA

H.H. Van Dijken, Department of Pharmacology, Medical Faculty, Free University, Van der Boechorststraat 7, 1081 BT Amsterdam, The Netherlands

J. Vernikos, Life Science Division, Mail Stop 239-11, NASA Ames Research Center, Moffett Field, California 94035, USA

T.C. Wessel, Laboratory of Molecular Neurobiology, The Burke Medical Research Institute, Cornell University Medical College, 785 Mamaroneck Avenue, White Plains, New York 10605, USA

M.H. Whitnall, Department of Physiology, Armed Forces Radiobiology Research Institute, Bethesda, Maryland 20889-5145, USA

W. Wuttke, Division of Clinical and Experimental Endocrinology, Department of Obstetrics and Gynecology, University of Göttingen, Göttingen, Germany

H. Yago, Institute of Bio-Active Science, Nippon Zoki Pharmaceutical Company, Ltd., Hyogo 673-14, Japan

R. Yoneda, Institute of Bio-Active Science, Nippon Zoki Pharmaceutical Company, Ltd., Hyogo, 673-14, Japan

W.S. Young III, Laboratory of Cell Biology, National Institute of Mental Health, Building 36, Room 2D10, Bethesda, Maryland 20892, USA

Z. Zukowska-Grojec, Department of Physiology and Biophysics, Georgetown University Medical Center, 3900 Reservoir Road NW, Washington, DC 20007, USA

ABBREVIATIONS

ACh = acetylcholine
ADX = adrenalectomy
ALD = aldosterone
ANF = atrial natriuretic factor
AVVP = arginine vasopressin
BZD = benzodiazepine
CA = catecholamine(s)
CRH = corticotropin releasing hormone
DA = dopamine
DBH = dopamine β-hydroxylase
DHPG = dihydroxyphenylglycol
DOPA = dihydroxyphenylalanine
DOPAC = dihydroxyphenylacetic acid
DR = dorsal raphe
DSIP = delta sleep inducing peptide
ECD = electrochemical detection
EPI = epinephrine
GH = growth hormone
GR = glucocorticoid receptor
5-HIAA = 5-hydroxyindoleacetic acid
HPAA = hypothalamo-pituitary-adrenal axis
HPLC = high performance liquid chromatography
HT = hypothalamus
5-HT = serotonin

HVA = homovanillic acid
ICV = intracerebroventricular
IL = interleukin
IMO = immobilization stress
IR = immunoreactive
LH = luteinizing hormone
LHRH = luteinizing hormone releasing hormone
MHPG = 3-methoxy-4-hydroxyphenylglycol
NE = norepinephrine
PNMT = phenylethanolamine N-methyltransferase
POMC = proopiomelanocortin
PRA = plasma renin activity
PRL = prolactin
PVN = paraventricular nucleus
SART = specific alteration of rhythm in temperature
SC = spinal cord
SHR = spontaneously hypertensive rats
SON = supraoptic nucleus
TH = tyrosine hydroxylase
TRH = thyroliberin
VP = vasopressin
vs = versus
→ = effect on
← = effect of

PART ONE

CENTRAL MONOAMINE AND NEUROPEPTIDE PATHWAYS AND STRESS

Stress: Neuroendocrine and Molecular Approaches
Edited by R. Kvetnansky, R. McCarty and J. Axelrod

1992 Gordon and Breach Science
Publishers S.A., New York, USA.
Photocopying permitted by license only.

CHEMICAL NEUROANATOMY OF BRAIN STRUCTURES INVOLVED IN THE STRESS RESPONSE WITH SPECIAL REFERENCE TO CORTICOTROPIN RELEASING FACTOR

M. Palkovits[1], E. Mezey[1], A. Csiffary[1], F. A. Antoni[1]
W. Vale[2] and R. L. Eskay[3]

[1]Laboratory of Cell Biology, National Institute of Mental Health
Bethesda, MD, USA and First Department of Anatomy, Semmelweis
University Medical School, Budapest, Hungary
[2]Salk Institute, La Jolla, CA
3Laboratory of Clinical Studies, National Institute on Alcohol Abuse
and Alcoholism, Bethesda, MD, USA

NEUROANATOMICAL AND NEUROCHEMICAL ORGANIZATION OF THE STRESS RESPONSE

Neuronal pathways involved in stress responses may be divided into two major groups: 1) Short loop reflexes, and 2) long loop pathways with ascending afferents carrying stress signals to hypothalamic, limbic and cortical areas, from where descending efferents run to the median eminence, as well as to medullary and spinal cord preganglionic efferents.

1) The short loop consist of 3 (sometimes only 2) neurons: a) Stressful stimuli from the periphery are carried by spinal and cranial nerve fibers to the dorsal horn, as well as to primary sensory nuclei in the lower brainstem. Besides amino acids, neuropeptides such as substance P, calcitonin gene-related peptide (CGRP), somatostatin,

cholecystokinin (CCK) and vasoactive intestinal polypeptide (VIP) are present as putative neurotransmitters in these afferent neurons. b) From primary sensory neurons, stress signals may be relayed by interneurons to preganglionic efferent neurons located in the medulla oblongata (in and around the dorsal vagal and ambiguous nuclei) and in the spinal cord (in the thoracolumbar intermedio-lateral cell column). The chemical nature of these interneurons is unknown; catecholamines, amino acids and neuropeptides (including opioids) are potential candidates. c) Spinal and medullary preganglionic efferents are cholinergic. Increasing immunohistochemical evidence suggests that certain neuropeptides may co-localize in these cholinergic neurons, and in addition some catecholaminergic neurons (co-localized with neuropeptide Y and neurotensin) also project to the periphery. Both spinal and medullary efferents terminate in peripheral ganglionic cells which innervate the adrenal gland.

2) In the long-loop system, primary (peripheral) afferents and preganglionic efferents are the same as those in the short-loop reflex arc. Instead of involving interneurons, however, stress signals ascend to higher forebrain regions where stress responses are organized, and from where neuronal or neurohormonal efferents arise and exert their actions on the adrenal.

a) Stress-conducting afferent fibers (ascending fibers in the long-loop). Signals from primary sensory nuclei in the spinal cord and the medulla oblongata ascend in multisynaptic pathways. Somato- and viscerosensory signals, including pain, mechanical (tactile, pressure, etc.), thermal and proprioceptive-kinesthetic signals in certain conditions may prove to be stressful stimuli for various organisms. These signals ascend in the spinothalamic, the spino-cervico-thalamic and the spinoreticular tracts, and in the dorsal column-medial lemniscus system to thalamic and cortical centers (for details and references, see Palkovits, 1987, 1989). Collaterals of these fibers may reach hypothalamic and certain limbic nuclei directly or through the brainstem reticular formation, or via relay neurons in the medial forebrain bundle and through lower brainstem biogenic amine-containing neurons. Some of the direct ascending projections, such as fibers from the nucleus of the solitary tract (NTS) to the paraventricular (PVN) or central amygdaloid nuclei are peptidergic. Recently, enkephalin, somatostatin, neuropeptide Y, neurotensin, dynorphin, bombesin and inhibin-ß have been demonstrated in these fibers by immunohistochemical and tract-tracing techniques (Riche *et al*, 1990; Sawchenko *et al*, 1990).

The second group of ascending stress-conducting fibers run in special sensory pathways. Photic, olfactory, acoustic, vestibular and taste signals may serve as stressful stimuli. These fibers may reach the hypothalamus either by direct projections or be relayed indirectly by neurons in subcortical centers (for details, see Palkovits, 1987). These stressful stimuli may reach the paraventricular nucleus through the medial forebrain bundle (Feldman, 1985). Direct connections between medial forebrain bundle neurons and CRF-immunoreactive cells in the PVN have been verified by a combination of tract-tracing and immunohistochemical techniques (Kiss, Gorcs and Palkovits, to be published).

b) Neuronal efferents in the stress response. Numerous neuronal pathways leading to the hypothalamus convey stress-induced signals. The hypothalamic-pituitary-adrenal axis is activated at three different levels: 1) actions on hypothalamic neurons (mainly in the PVN); 2) actions on nerve terminals in the median eminence, and 3) direct actions on anterior pituitary cells.

Among several neurogenic substances, corticotropin releasing factor (CRF) represents the predominant component in the release of ACTH from anterior pituitary cells (Makara, 1985; Rivier and Vale, 1985). CRF neurons in the PVN are a major target of neural stressors. Several peptidergic and aminergic axons from various brain regions (refer to Palkovits, 1986; Swanson *et al*, 1986; Sawchenko and Swanson, 1990) terminate on PVN CRF neurons. They may participate in the organization of stress responses by influencing CRF synthesis in and axonal transport from paraventricular neurons (Makara, 1985; Rivier and Vale, 1985; Tilders *et al*, 1985; Mezey *et al*, 1987; Mezey and Palkovits, 1991a).

Paraventricular CRF neurons are involved in 2 different pathways in stress responses: a) neurohumoral pathway via median eminence - portal blood - anterior pituitary to release ACTH, which mobilizes corticosterone from the adrenal cortex and influences PNMT activity in the adrenal medulla, b) a neuronal pathway via descending CRF fibers to the spinal cord and medullary preganglionic afferents (Swanson and Kuypers, 1980; Sawchenko and Swanson, 1982,1990). Fourteen days after bilateral paraventricular lesions concentrations of CRF were depleted in the medial subdivision of the NTS and disappeared from the ventrolateral medulla in and around the A1 catecholaminergic cell groups (Table 1). Bilateral transection of the

TABLE 1.　Corticotropin releasing factor (ng CRF/mg protein) in the median eminence (ME) and lower brainstem nuclei 14 days after bilateral lesions of the paraventricular nucleus. (PVN).

	SHAM	PVN-lesion
ME	52.85 ± 3.45 (8)	10.00 ± 3.03** (7)
NTSm	0.74 ± 0.07 (4)	0.23 ± 0.10* (6)
NTSc	0.64 + 0.05 (6)	0.67 ± 0.11 (6)
A1	0.60 ± 0.06 (4)	N.D. **

mean ± S.E.M.; ( ) = number of animals. ** - $p < 0.001$; * - $p < 0.05$; N.D. - non-detectable. NTSm = nucleus of the solitary tract, medial part. NTSc = nucleus of the solitary tract, commissural part. A1 = A1-norepinephrine- containing cell group.

ventral tegmental bundle at the pontomedullary junction resulted in similar alterations. By immunohistochemistry, a total disappearance of CRF-ir fibers and varicosities was observed in the ventrolateral medulla, including the descending spinal path, after PVN-lesions or brainstem transections. CRF immunoreactivity remained almost unchanged in the commissural parts of the NTS of these animals while it was markedly reduced in the medial part of the nucleus. Due to the retrograde accumulation of CRF immunoreactivity, CRF-immunostained cells became visible in the PVN 3 days after lesioning the rostral part of the nucleus of the solitary tract (NTS), or after brainstem transections. Cells in the PVN were retrogradely labeled after medullary (NTS and its vicinity) injection of True Blue (Sawchenko, 1987). Neither PVN lesions (Table 1) nor brainstem hemisections could alter CRF concentrations in the commissural part of the NTS (subdivisions which belong to the vagus nerve). This part of the nucleus is innervated by local CRF neurons (Figures 1 and 2).

CRF-containing cells are located just rostro-medial to the NTS, in the nucleus prepositus. PVN lesions or brainstem hemisections did not affect CRF immunoreactivity of these cells. After a medullary parasagittal cut, which transects the axons of these cells, a rapid and significant increase in CRF mRNA was observed in the nucleus, ipsilateral to the knife cut (Figure 3). This alteration seems to be a consequence of axotomy, since similar changes were found in CRF neurons of the inferior olive after transections of the olivocerebellar tract (Palkovits *et al*, 1987; Mezey and Palkovits, 1991b).

In addition to CRF, other peptidergic neurons such as

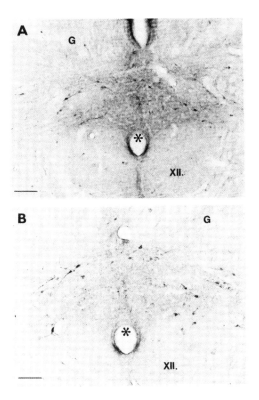

FIGURE 1. CRF-immunoreactive neurons in the commissural part of the nucleus of the solitary tract in rats (48 hours after 60 μg/10 μl colchicine i.c.v.). A - 0.5 mm, B - 1.0 mm caudal to the obex. Abbr.: G - nucleus gracilis, XII - motor hypoglossal nucleus. The central canal is indicated by stars. Bar scales = 100 μm.

enkephalin, substance P, neurotensin and somatostatin project from the PVN to medullary and spinal cord preganglionic efferents (Sawchenko and Swanson, 1982, 1990; Swanson *et al*, 1986). Descending fibers arise also in the hypothalamic arcuate nucleus, central amygdaloid nucleus, bed nucleus of the stria terminalis, insular cortex and parabrachial nuclei. Vasopressin, oxytocin, opioids (enkephalins, ß-endorphin, dynorphins), galanin, substance P, neurotensin, and CCK have been characterized in these descending projections (Veening *et al*, 1984; Gray and Magnuson, 1987). Furthermore, TRH, substance P, somatostatin, enkephalin and NPY immunoreactive fibers from the

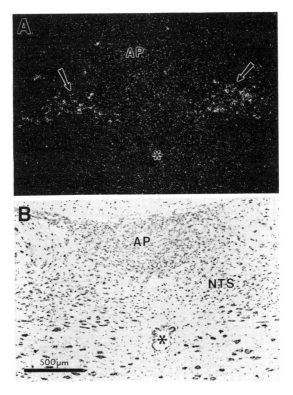

FIGURE 2.   CRF mRNA in neurons (indicated by arrows) of the nucleus of the solitary tract (commissural part-NTS).   Figure 2A is a dark field microphotograph of an area shown in Figure B.   The central canal is indicated by stars.   AP-area postrema.

medulla descend to the intermediolateral cell column in the spinal cord (Strack *et al*, 1989).

The median eminence receives intra- and extrahypothalamic fibers.   Intrahypothalamic fibers are mainly peptidergic while the extrahypothalamic ones derive from lower brainstem biogenic amine-containing cell groups.   The aminergic nuclei are innervated by axons or axon collaterals of spinal and medullary somato- and viscerosensory neurons.   Therefore, these ascending aminergic fibers to the median eminence may constitute the major route for nociceptive stressful stimuli to influence CRF (and/or additional peptide) release from nerve terminals to the portal circulation in the external zone of the median eminence.

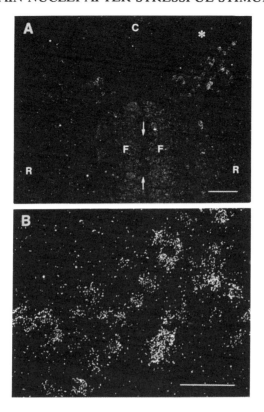

FIGURE 3. CRF mRNA in cells of the nucleus prepositus hypoglossi in rats with a unilateral parasagittal knife cut in the medulla oblongata. Coronal section, 1 mm rostral to the obex. A strong accumulation of labeling can be seen in neurons ipsilateral to the lesion (indicated by a star on Figure 3A, and shown with high magnification on Figure 3B). C - choroid plexus in the fourth ventricle, F - medial longitudinal fascicle, R - parvicellular reticular nucleus, the midline (medullary raphe) is indicated by arrows. Bar scales: A = 200 μm, B = 100 μm.

CRF in Brain Nuclei after Various Types of Stressful Stimuli

Stressful stimuli alter CRF levels in the hypothalamic-pituitary system. Depending on the intensity and duration of stressful stimuli and the time following stress, alterations in CRF neurons and terminals become visible by immunohistochemical or *in situ* hybridization techniques or by radioimmunoassay. In the present experiment, the acute effects of three different stressors on hypothalamic and CRF-rich extrahypothalamic nuclei were determined by radioimmunoassay (for technical details, see Palkovits *et al*, 1985). Stressful stimuli were as follows: 1) ether stress-5 minutes of ether anesthesia; 2) surgical stress-under short-term (2 minutes) and superficial ether anesthesia, the head of each animal was fixed in a stereotaxic device, the skin was cut along the sagittal suture and a one mm long hole placed in the skull with a drill, then the skin was closed; and 3) restraint stress-animals were kept in a small plastic box for 2 minutes. Animals in groups 1 and 2 were sacrificed by decapitation 1 or 6 hours after stress. In group 3, animals were sacrificed 30 seconds or 30 minutes after stress.

Depletion of CRF concentrations in the median eminence became significant by 6 hours after ether and surgical stress, and 30 minutes after restraint stress. Likewise, a decrease in CRF concentrations in the whole hypothalamus was found throughout the first 30 minutes of restraint stress, which was followed by an increase in CRF levels (Moldow *et al*, 1987). In the present study, no alterations were found in PVN CRF levels after ether and surgical stress but it was elevated after restraint stress. By using acute immobilization stress with different time courses, reduced CRF levels have been reported in the median eminence by quantitative immunohistochemistry (de Goeij *et al*, 1991) or radioimmunoassay (Chappell *et al*, 1986). This reduction appears to be the consequence of a rapid release of CRF from nerve terminals in the external layer of the median eminence into the portal circulation. Indeed, significant elevations of CRF levels were measured in portal blood after hemorrhage (Plotsky and Vale, 1984). Median eminence CRF is mainly of PVN origin. Depletion of CRF in nerve terminals may induce increased synthesis in cell bodies, which was indicated by increased CRF levels in the PVN following restraint stress.

None of the stressful stimuli altered CRF concentrations in the bed nucleus of the stria terminalis, central amygdaloid nucleus, parabrachial nuclei or nucleus of the solitary tract. Similar findings have been reported by Beyer *et al* (1988). In response to adrenalectomy, which increased CRF mRNA levels in the PVN, there was no change in the NIST or the central amygdaloid nucleus. In contrast to these findings, marked elevations in the concentration of CRF were found in the locus coeruleus of rats exposed to acute and chronic stress (Chappell *et al*, 1986). This finding also indicates a close association of the noradrenergic system with CRF in responses to stress.

Among extrapyramidal brain regions, the inferior olive showed altered CRF concentrations after acute stress. Significantly depleted CRF concentrations were measured here 6 hours after ether and surgical stress, and 30 minutes after restraint stress. The inferior olive is known to be a component of the extrapyramidal motor system. Neurons in the inferior olive contain CRF and project as climbing fibers to the cerebellum (Palkovits et al, 1987). Descending neuronal inputs from higher brain regions appear to be important in CRF synthetic activity of olivary neurons since brainstem hemisection inhibited CRF mRNA levels in the ipsilateral inferior olive several hours after surgery (unpublished observations). Although these findings may not provide convincing evidence for a specific hypothesis, it is possible that CRF in that part of the extrapyramidal system may be influenced by stressful stimuli.

REFERENCES

Beyer, H.S., Matta, S.G., and Sharp, B.M. (1988). Regulation of the messenger ribonucleic acid for corticotropin-releasing factor in the paraventricular nucleus and other brain sites of the rat. *Endocrinology* **123**, 2117-2123.

Chappell, P.B., Smith, M.A., Kilts, C.D., Bissett, G., Ritchie, J., Anderson, C., and Nemeroff, C.B. (1986). Alterations in corticotropin-releasing factor-like immunoreactivity in discrete rat brain regions after acute and chronic stress. *Journal of Neuroscience* **6**, 2908-2914.

de Goeij, D.C.E., Kvetnansky, R., Whitnall, M.H., Jezova, D., Berkenbosch, F., and Tilders, F.J.H. (1991). Repeated stress-induced activation of corticotropin-releasing factor neurons enhances vasopressin stores and colocalization with corticotropin-releasing factor in the median eminence of rats. *Neuroendocrinology* **53**, 150-159.

Feldman, S. (1985). Neural pathways mediating adrenocortical responses. *Federation Proceedings* **44**, 169-175.

Gray, T.S., and Magnuson, D.J. (1987). Neuropeptide neuronal efferents from the bed nucleus of the stria terminalis and central amygdaloid nucleus to the dorsal vagal complex in the rat. *Journal of Comparative Neurology* **262**, 365-374.

Makara, G.B. (1985). Mechanisms by which stressful stimuli activate the pituitary-adrenal system. *Federal Proceedings* **44**, 149-153.

Mezey, E., and Palkovits, M. (1991a). CRF-containing neurons in the hypothalamic paraventricular nucleus: Regulation, especially by catecholamines. *Frontiers in Neuroendocrinology* **12**, 23-37.

Mezey, E., and Palkovits, M. (1991b). Time dependent changes in CRF and its mRNA in the neurons of the inferior olive following surgical transection of the olivocerebellar tract in the rat. *Molecular Brain Research* **10**, 55-59.

Mezey, E, Young, W.S.III., Siegel, R.E., and Kovacs, K. (1987). Neuropeptides and neurotransmitters involved in regulation of corticotropin-releasing factor-containing neurons in the rat. *Progress in Brain Research* 72, 119-127.

Moldow, R.L., Kastin, A.J., Graf, M., and Fischman, A.J. (1987). Stress mediated changes in hypothalamic corticotropin-releasing factor-like immunoreactivity. *Life Sciences* 40, 413-418.

Palkovits, M. (1986). Afferents onto neuroendocrine cells. In: D. Ganten and D. Pfaff (Eds.), "Current Topics in Neuroendocrinology, Vol. 7: Morphology of Hypothalamus and Its Connections," pp. 197-222. Berlin: Springer-Verlag.

Palkovits, M. (1987). Organization of the stress response at the anatomical level. *Progress in Brain Research* 72, 47-55.

Palkovits, M. (1989). Neuroanatomical overview of brain neurotransmitters in stress. In: G.R. Van Loon, R. Kvetnansky, R. McCarty and J. Axelrod (Eds.), "Stress: Neurochemical and Humoral Mechanisms," Vol. 1, pp. 31-42. New York: Gordon & Breach.

Palkovits, M., Brownstein, M.J., and Vale, W. (1985). Distribution of corticotropin-releasing factor in rat brain. *Federation Proceedings* 44, 215-219.

Palkovits, M., Leranth, Cs., Gorcs, T., and Young, W.S. III. (1987). Corticotropin-releasing factor in the olivocerebellar tract of rats: Demonstration by light-and electron microscopic immunohistochemistry and *in situ* hybridization histochemistry. *Proceedings of the National Academy of Sciences* (USA) 84, 3911-3915.

Plotsky, P.M., and Vale, W. (1984). Hemorrhage-induced secretion of corticotropin-releasing factor-like immunoreactivity into the rat hypophysial portal circulation and its inhibition by glucocorticoids. *Endocrinology* 114, 164-169.

Riche, D., De Pommery, J., and Menetrey, D. (1990). Neuropeptides and catecholamines in efferent projections of the nuclei of the solitary tract in the rat. *Journal of Comparative Neurology* 293, 399-424.

Rivier, C., and Vale, W. (1985). Effects of corticotropin-releasing factor, neurohypophyseal peptides, and catecholamines on pituitary function. *Federation Proceedings* 44, 189-195.

Sawchenko, P.E. (1987). Evidence for differential regulation of corticotropin-releasing factor and vasopressin immunoreactivities in parvocellular neurosecretory and autonomic-related projections of the paraventricular nucleus. *Brain Research* 437, 253-263.

Sawchenko, P.E., Arias, C., and Bittencourt, J.C. (1990). Inhibin-ß, somatostatin, and enkephalin immunoreactivities coexist in caudal medullary neurons that project to the paraventricular nucleus of the hypothalamus. *Journal of Comparative Neurology* 291, 269-280.

Sawchenko, P.E., and Swanson, L.W. (1982). Immunohistochemical identification of neurons in the paraventricular nucleus of the hypothalamus that project to the medulla or to the spinal cord of the rat. *Journal of Comparative Neurology* 205, 260-272.

Sawchenko, P.E., and Swanson, L.W. (1990). Organization of CRF immunoreactive cells and fibers in the rat brain: Immunohistochemical studies. In: E.B. de Souza and C.B. Nemeroff (Eds.), "CRF: Basic and Clinical Studies of a Neuropeptide," pp. 29-51. Boca Raton, FL: CRC, Inc.

Strack, A.M., Sawyer, W.B., Platt, K.B., and Loewy, A.D. (1989). CNS cell groups regulating the sympathetic outflow to adrenal gland as revealed by transneuronal cell body labeling with pseudorabies virus. *Brain Research* **491**, 274-296.

Swanson, L.W., and Kuypers, H.G.J.M. (1980). The paraventricular nucleus of the hypothalamus: Cytoarchitectonic subdivisions and organization of projections to the pituitary, dorsal vagal complex, and spinal cord as demonstrated by retrograde fluorescence double-labeling methods. *Journal of Comparative Neurology* **194**, 555-570.

Swanson, L.W., Sawchenko, P.E., and Lind, R.W. (1986). Regulation of multiple peptides in CRF parvocellular neurosecretory neurons: implications for the stress response. *Progress in Brain Research* **68**, 169-190.

Tilders, F.J.H., Berkenbosch, F., Vermes, I., Linton, E.A., and Smelik, P.G. (1985). Role of epinephrine and vasopressin in the control of the pituitary-adrenal response to stress. *Federation Proceedings* **44**, 155-160.

Veening, J.G., Swanson, L.W., and Sawchenko, P.E. (1984). The organization of projections from the central nucleus of the amygdala to brainstem sites involved in central autonomic regulation: a combined retrograde transport immunohistochemical study. *Brain Research* **303**, 337-357.

Stress: Neuroendocrine and Molecular Approaches
Edited by R. Kvetnansky, R. McCarty and J. Axelrod

1992 Gordon and Breach Science
Publishers S.A., New York, USA.
Photocopying permitted by license only.

AMINERGIC AND PEPTIDERGIC PATHWAYS SUBSERVING THE STRESS RESPONSE

P. E. Sawchenko, E. T. Cunningham, Jr., J. C. Bittencourt and R. K. W. Chan

Laboratory of Neuronal Structure and Function, The Salk Institute for Biological Studies, and The Clayton Foundation for Research - California Division, La Jolla, California USA

INTRODUCTION

The hypothalamo-pituitary-adrenal (HPA) axis represents a major avenue through which adaptive, homeoregulatory responses to challenges posed from an organism's internal or external environment may be achieved. In its simplest form, this cascade involves the successive release of corticotropin-releasing factors by hypothalamic neurosecretory neurons into the hypophyseal-portal vasculature, which serve to stimulate the secretion of pituitary adrenocorticotrophic hormone (ACTH) into the systemic circulation, which, in turn, effects the release of glucocorticoids from the adrenal cortex. Regulation of the axis is accomplished by both blood-borne and neurogenic factors. Negative feedback effects of circulating glucocorticoids, exerted at the levels of the hypothalamus and pituitary, provide a major regulatory influence on hormone synthesis and secretion within the system. Neuronal circuits play critical roles in conveying to the hypothalamus inputs from the myriad of sensory systems that are capable of supporting a stress response. In addition to this information-bearing capacity, neurogenic mechanisms have also been implicated in more regulatory (i.e., biosynthetic) functions, including as transsynaptic

15

mediators of corticosteroid feedback effects on the HPA axis, and in providing the positive drive on the increased output of the central limb of the axis seen in response to chronic stress.

Here we summarize results from our laboratory bearing on the organization and function of projections arising from the caudal portion of the medulla. These are in a position to be pivotally involved in conveying interoceptive information to the neuroendocrine hypothalamus, are the most fully characterized set of inputs to the neurosecretory system, and seem likely to comprise the arena in which a detailed anatomy of substrates underlying modality-specific influences on the HPA axis will first be achieved.

Effector Neuron Organization

There is now general agreement that the 41-residue peptide, corticotropin-releasing factor (CRF), produced by parvocellular neurosecretory neurons localized within the paraventricular nucleus of the hypothalamus (PVH), provides the principle stimulatory drive for pituitary ACTH secretion (Antoni, 1986). It is also clear, however, that other ACTH secretagogues and other cell types play important roles in sculpting the pituitary-adrenal response to stress (Figure 1). Two nonapeptide hormones, oxytocin (OT) and vasopressin (AVP), are the principal secretory products of the magnocellular neurosecretory system, whose cells of origin are centered in distinct regions of the PVH and in the supraoptic (SO) nucleus. Both OT and AVP are present, and differentially regulated, in hypophyseal portal plasma (Plotsky, 1985) and each is capable of interacting with CRF, synergistically or additively, to promote ACTH secretion. Adrenal steroid-dependent expression of AVP in parvocellular neurosecretory CRF-producing neurons of the PVH has been established (e.g., Sawchenko *et al*, 1984a; Sawchenko, 1987) suggesting one structural basis for interactions among secretagogues, but OT is yet to be definitively localized in a substantial number of hypophysiotropic neurons under any condition (Sawchenko *et al*, 1984b). Magnocellular neurons would seem the only viable candidate sources for such an influence (Figure 1). Consistent with this view, calcium-dependent, potassium-stimulated release of AVP from magnocellular axons has been demonstrated *in vitro* (Holmes *et al*, 1986) and exocytotic release of neurosecretory granule contents has been imaged ultrastructurally from magnocellular "axons-of-passage" coursing through the internal

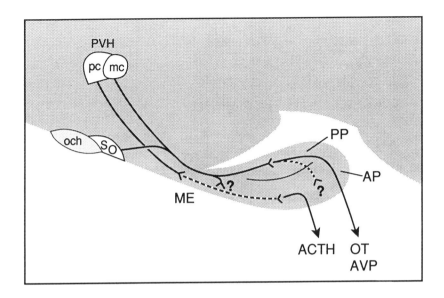

FIGURE 1. Schematic sagittal view of the hypothalamus and pituitary to show the organization of magnocellular (mc) and parvocellular (pc) neurosecretory systems, highlighting potential sites at which interactions relevant to the control of pituitary ACTH secretion may occur. CRF is produced by cells in the parvocellular division of the PVH, which project to the external zone of the median eminence (ME), ultimately reaching the anterior pituitary (AP) via the portal vasculature (dashed line). Magnocellular neurosecretory projections from the PVH and the supraoptic nucleus (SO) pass through the internal lamina of the median eminence en route to the posterior lobe (PP), where oxytocin (OT) and vasopressin (AVP) are released into the systemic circulation. OT and AVP of magnocellular origin may gain access to the portal vasculature by way of exocytotic release at the median eminence, or via vascular links between the posterior and anterior lobes (question marks).

layer of the median eminence (Buma and Nieuwenhuys, 1987). In addition, vascular links permitting blood flow from the neural to the anterior lobes have been suggested (Page, 1986). Thus, the potential exists for OT and AVP of magnocellular origin to gain access to anterior pituitary corticotropes via two distinct vascular routes.

There exists strong physiological evidence that CRF serves primarily to set the stimulatory tone on corticotropes, with the situational response determined largely by the relative abundance of co-secretagogues (Plotsky *et al*, 1985). The implication that magnocellular and parvocellular neurosecretory neurons, may, in a manner yet to be fully appreciated, jointly comprise the central limb of the HPA axis significantly broadens the array of neuronal pathways that must be considered as conveying stress-related information to the hypothalamus.

Ascending Catecholaminergic Projections

That the PVH and SO comprise two of the more prominent catecholaminergic terminal fields in the mammalian brain has been recognized since the advent of catecholamine histofluorescence methods. A massive literature has accumulated to support an involvement of these pathways in the control of HPA axis output (Plotsky *et al*, 1989). Modern immunohistochemical and axonal transport techniques have been applied to decipher the organization of these projections (Figure 2). This circuitry revolves about the nucleus of the solitary tract (NTS), the principal recipient of primary vagal and glossopharyngeal afferents. Distinct rostral and caudal territories of the NTS carry adrenergic and noradrenergic signatures, respectively, and project preferentially to the parvocellular neurosecretory zone of the PVH, as does a major projection field of the NTS, the C1 adrenergic cell group (Cunningham and Sawchenko, 1989; Cunningham *et al*, 1990; Sawchenko and Swanson, 1981, 1982). By contrast, the A1 noradrenergic cell group, which also appears to receive an NTS input, projects strongly and preferentially to regions of the magnocellular neurosecretory system in which AVP-containing cells are sequestered (Cunningham and Sawchenko, 1988).

   While we refer to these projections as being catecholaminergic, it is important to emphasize that each of them has the capacity to express an impressive array of additional neuroactive substances. For example, virtually all adrenergic neurons that project to the PVH also express neuropeptide Y, while subsets of A1 noradrenergic neurons that give rise to ascending projections display neuropeptide Y-, galanin- and substance P-immunoreactivity (Bittencourt *et al*, 1991; Levin *et al*, 1987; Sawchenko *et al*, 1985). Although each of these peptides has been shown capable of acting alone or in synergy with catecholamines in modifying the electrical and/or secretory activities of hypothalamic neurons (e.g., Day, 1989), the paucity of information bearing on the questions of whether and how these co-localized principles may be differentially regulated in a situation-dependent

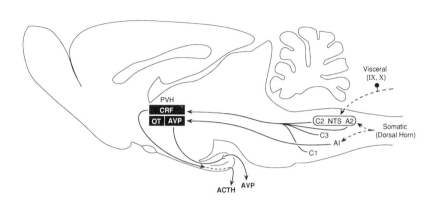

FIGURE 2.   Organization of brainstem catecholaminergic projections to the neurosecretory hypothalamus. Schematic drawing of a sagittal section through the rat brain to show the origins and targets of adrenergic and noradrenergic projections to the PVH. Adrenergic projections arising from each of the adrenergic cell groups (C1-3) are joined by noradrenergic projections from the A2 cell group in projecting preferentially to the parvocellular division of the PVH.   The A1 noradrenergic group, by contrast, projects quite specifically to magnocellular neurosecretory AVP neurons. These pathways are in a position to relay visceral, and perhaps somatic, sensory information to the neurosecretory hypothalamus. Other abbreviations: NTS, nucleus of the solitary tract; IX, glossopharyngeal nerve; X, vagus nerve.

manner has limited understanding of the significance of these co-localization phenomena. Recent indications that at least some of the medullary aminergic cell groups also contain excitatory amino acid-like moieties (Nicholas *et al*, 1991) raise the possibility, yet to be critically explored, that fast-acting neurotransmitters like glutamate and its congeners may play important roles in conveying information concerning moment-to-moment alterations in the internal milieu from brainstem to hypothalamus (Van Den Pol *et al*, 1990), with co-stored amines and neuropeptides relegated, perhaps, to serve slower, more modulatory, functions.

    Lesions that involve various components of these pathways can disrupt pituitary-adrenal responses to such diverse challenges as exposure to ether, hemorrhage and circadian cues (Darlington *et al*,

1986; Szafarczyk *et al*, 1985). To gain some insight into the relative importance of medullary catecholaminergic neurons in orchestrating responses to systemic stress, we have followed the time course of *c-fos* mRNA and protein expression in immunohistochemically identified medullary neurons following acute hypotensive hemorrhage (Chan and Sawchenko, 1991), a challenge known to impact CRF, OT and AVP secretion by hypothalamic neurons. The *c-fos* gene encodes an inducible transcription factor, *Fos*, whose appearance in neuronal nuclei appears capable of serving as a widespread, though not necessarily universal, index of functional activation (Morgan and Curran, 1991). Removal of 10 ml/kg blood, which reduced mean arterial pressure by some 50%, resulted in a highly preferential induction of *Fos* protein in each of the medullary catecholamine cell groups that projects to the PVH. This response was detectable within 30 minutes of the challenge, peaked at roughly 2 hours, and was virtually absent at 4 hours, the longest interval examined. With the exception of a late-appearing induction of *Fos* protein in the area postrema, which may have been due to activation of the renin-angiotensin system, catecholaminergic neurons constituted a substantial majority of all medullary neurons displaying *Fos* induction, suggesting that they may play a privileged role in organizing hypothalamic and other adaptive responses to a hemorrhage challenge, at least.

In addition to their capacity as "relays" for visceral sensory control of stress-related circuitry, evidence exists to suggest that the integrity of catecholaminergic pathways may to be required for corticosteroid feedback influences to be exerted in at least some models (e.g., Smythe *et al*, 1983). This, coupled with the fact that the cells of origin of these projections express the type II glucocorticoid receptor (Sawchenko and Bohn, 1989) raised the possibility that some form or fraction of feedback may be mediated remotely, via the aminergic projections. Transections of these pathways in intact and steroid-manipulated rats, however, resulted in effects on CRF and AVP immunostaining that were opposite in sign, and independent, from those that resulted from manipulations in steroid titers (Sawchenko, 1988). Moreover, these lesions fail to exert effects on CRF or AVP mRNA levels (Swanson and Simmons, 1989), which are known to be regulated by glucocorticoids. Thus, while the aminergic pathways appear capable of modifying peptide dynamics in CRF neurons, the effect appears to be exerted post-transcriptionally. The mechanism(s) through which such effects may be exerted, and how they might be integrated with the synaptic signaling properties of these pathways remain to be clarified.

Ascending Peptidergic Projections

Despite the fact that at least some interoceptive stimuli that modify CRF and/or AVP secretion exert correspondingly potent effects on OT release into the portal and/or systemic vasculatures, until quite recently little was known of the routes by which such stimuli might reach magnocellular OT neurons. Moreover, literally no prominent set of ascending projections to OT neurons had been described that might provide a basis for understanding the manner in which the prototypic stimuli for OT secretion, suckling or vaginal distension, might be conveyed to the hypothalamus. Recent anatomical work has identified a cell group that provides a potential substrate for controlling OT secretion in response to both visceral and somatic sensory cues (Figure 3).

These neurons are centered in the caudomedial part of the NTS, and are topographically and biochemically distinct from catecholaminergic cells, expressing instead at least three neuropeptides (Sawchenko *et al*, 1988ab, 1990). One which serves as a convenient marker for the system is the ß subunit of inhibin and activin, which are

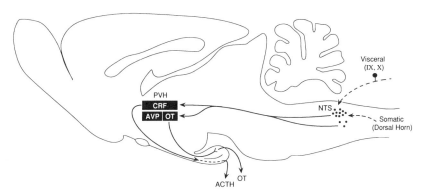

FIGURE 3.   Organization of medullary peptidergic projections to the neurosecretory hypothalamus. Neurons situated principally within the caudal NTS, distinct from catecholaminergic cells, and expressing multiple neuropeptides, project to both the CRF-rich zone of the parvocellular division of the PVH and to OT territories of the magnocellular neurosecretory system. These, too, are in a position to relay visceral and/or somatic sensory information.

best known as dimeric gonadal glycoproteins, though their tissue distributions and functions are now known to be far more expansive (see Vale *et al*, 1988). Peptides derived from both prepro-somatostatin and prepro-enkephalin have been co-localized within sizeable complements of inhibin ß-immunoreactive cells in the NTS (Sawchenko *et al*, 1990). The projection pattern of these caudal medullary neurons provides interesting parallels and contrasts with the catecholamine-containing cells that innervate the hypothalamus, in that their distribution suggests interactions with both CRF-containing parvocellular neurosecretory neurons and magnocellular neurosecretory OT neurons. Preferential synaptic interactions of inhibin ß- and somatostatin-containing terminals with OT neurons in the PVH and SO have been demonstrated (Sawchenko *et al*, 1988bc).

Plotsky and colleagues have carried out pharmacologic studies to evaluate the contextual roles suggested by the anatomic observations. It has been shown that centrally administered activin (inhibin ß homodimer) promotes the release of CRF into portal plasma and increases circulating ACTH levels (Plotsky *et al*, 1991). Similarly, activin administered directly to the PVH stimulates OT secretion, and, intriguingly, local administration of inhibin ß antiserum to lactating dams attenuates OT secretion seen in response to suckling (Plotsky *et al*, 1988). The area of the NTS in which this inhibin ß cell group resides falls within the projection field of vagal afferents, and this pathway would thus seem ideally situated to mediate the OT and/or CRF secretory responses to nausea or emetic agents (Verbalis *et al*, 1986). In addition, however, the caudal NTS receives direct inputs from the dorsal horn of the spinal cord (Menetrey and Basbaum, 1987), raising the possibility that caudal medullary peptidergic neurons could provide the long-elusive substrate for the milk ejection reflex. Clearly, knowledge of the modalities of visceral and/or somatic sensory inputs that may impinge upon these peptidergic cells within the NTS will be required to establish fully and rigorously the functional associations of this cell group.

DISCUSSION

The results summarized above have served to define an elaborately organized set of projections originating in the caudal medulla that provides for a rich, and differential, ascending afferent control of three neurosecretory cell types that are critically involved in determining pituitary-adrenal output in response to a variety of stressors (Figure 4). The high degree of target specificity exhibited by these pathways contrasts with their biochemical makeup, which is complex and multiply determined. As noted above, one major task remaining is to follow these circuits one step backward, into the periphery, to determine the particular kinds of sensory information that may be gated through these medullary neurons. Is it possible, for example, that some might represent sites of convergence of disparate afferent modalities? Such a scenario is particularly intriguing for the caudal medullary inhibin ß-containing cell group, which could conceivably play roles in modulating OT secretion in response to stimuli as diverse as suckling and nausea. Addressing issues such as this may also provide insight into the nature of the participation of magnocellular neurosecretory neurons in controlling the HPA axis. These cells produce physiologically important circulating hormones, and their controls must presumably be independent of those that govern the delivery of OT and AVP to the portal vasculature. Do neuronal inputs play a role in such differential regulation, or must we look to local

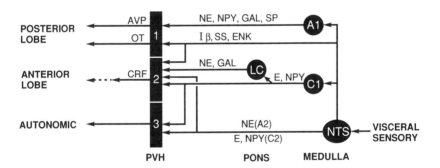

FIGURE 4. Schematic summary of the organization and putative transmitter specificity of medullary projections to the PVH. Visceral sensory information impinging upon the NTS may be distributed to particular visceromotor targets in the PVH via direct or indirect routes that are anatomically and biochemically differentiated. Other abbreviations: ENK, enkephalin; GAL, galanin; NPY, Iß, inhibin ß; LC, locus coeruleus; NPY, neuropeptide Y; SS, somatostatin; SP, substance P.

mechanisms in the hypothalamus or pituitary to explain the bifunctionality of the nonapeptides? Finally, yet to be addressed systematically is the question of whether medullary afferents participate in the regulation of relevant hypothalamic gene expression, apart from their documented influences on peptide secretion. Although the medullary afferents considered here do not appear to play a role in mediating corticosteroid feedback effects on hypothalamic gene expression, many laboratories have now described elevated hypothalamic levels of mRNAs encoding CRF and/or AVP and/or OT in various stress paradigms (e.g., Imaki *et al*, 1991), suggesting that neural mechanisms must provide the heightened drive on the HPA axis seen in chronic stress. That such elevations occur despite high circulating corticosteroids might indicate a role for neural inputs in regulating the biosynthesis, as well as the activity, of stress-related neuropeptides.

ACKNOWLEDGEMENTS

The work from our laboratory that is summarized here was supported by NIH grants NS-21182 and HL-35137, and was conducted in part by the Clayton Foundation for Research - California Division. P.E.S. is a Clayton Foundation Investigator.

REFERENCES

Antoni, F.A. (1986). Hypothalamic control of adrenocorticotropin secretion: Advances since the discovery of 41-residue corticotropin-releasing factor. *Endocrine Reviews* **7**, 351-378.

Bittencourt, J.C., Benoit, R., and Sawchenko, P.E. (1991). Distribution and origins of substance P-immunoreactive projections to the paraventricular and supraoptic nuclei: Partial overlap with ascending catecholaminergic projections. *Journal of Chemical Neuroanatomy* **4**, 63-78.

Buma, P., and Nieuwenhuys, R. (1987). Ultrastructural demonstration of oxytocin and vasopressin release sites in the neural lobe and median eminence of the rat by tannic acid and immunogold methods. *Neuroscience Letters* **74**, 151-157.

Chan, R.K.W., and Sawchenko, P.E. (1991). Time course of hemorrhage-induced *c-fos* expression in brain stem cardiovascular nuclei. *Neuroscience Abstracts* **17**, 614.

Cunningham, Jr., E.T., Bohn, M.C., and Sawchenko, P.E. (1990). The organization of adrenergic projections to the paraventricular and supraoptic nuclei of the rat hypothalamus. *Journal of Comparative Neurology* **292**, 651-667.

Cunningham, Jr., E.T., and Sawchenko, P.E. (1988). Anatomical specificity of noradrenergic inputs to the paraventricular and supraoptic nuclei of the rat hypothalamus. *Journal of Comparative Neurology* **274**, 60-76.

Darlington, D.N., Shinsako, J., and Dallman, M.F. (1986). Medullary lesions eliminate ACTH responses to hypotensive hemorrhage. *American Journal of Physiology* **251**, R106-R115.

Day, T.A. (1989). Control of neurosecretory vasopressin cells by noradrenergic projections of the caudal ventrolateral medulla. *Progress in Brain Research* **81**, 303-315.

Holmes, M.C., Antoni, F.A., Aguilera, G., and Catt, K.J. (1986). Magnocellular axons in passage through the median eminence release vasopressin. *Nature* **319**, 326-329.

Imaki, T., Nahon, J.-L., Rivier, C., Sawchenko, P.E., and Vale, W. (1991). Differential regulation of corticotropin-releasing factor mRNA in rat brain cell types by glucocorticoids and stress. *Journal of Neuroscience* **11**, 585-599.

Levin, M.C., Sawchenko, P.E., Howe, P.R.C., Bloom, S.R., and Polak, J.M. (1987). The organization of galanin- immunoreactive inputs to the paraventricular nucleus with special reference to their relationship to catecholaminergic afferents. *Journal of Comparative Neurology* **261**, 562-582.

Menetrey, D., and Basbaum, A.I. (1987). Spinal and trigeminal projections to the nucleus of the solitary tract: a possible substrate for somatovisceral and viscerovisceral reflex activation. *Journal of Comparative Neurology* **255**, 439-450.

Morgan, J.I., and Curran, T. (1991). Stimulus-transcription coupling in the nervous system: Involvement of the inducible proto-oncogenes *fos* and *jun*. *Annual Review of Neuroscience* **14**, 421-451.

Nicholas, A.P., Cuello, A, Goldstein, M., and Hokfelt, T. (1990). Glutamate-like immunoreactivity in medulla oblongata catecholamine/substance P neurons. *Neuroreports* **1**, 235-238.

Page, R.B. (1986). The pituitary portal system. In: D. Ganten and D. Pfaff (Eds.), "Current Topics in Neuroendocrinology, Vol. 7. Morphology of the Hypothalamus and its Connections," pp. 1-47. Berlin: Springer-Verlag.

Plotsky, P.M. (1985). Hypophyseotropic regulation of adenohypophyseal adrenocorticotropin secretion. *Federation Proceedings* **44**, 207-213.

Plotsky, P.M., Bruhn, T.O., and Vale, W. (1985). Hypophysiotropic regulation of adrenocorticotropin secretion in response to insulin-induced hypoglycemia. *Endocrinology* **117**, 323-329.

Plotsky, P.M., Cunningham Jr., E.T., and Widmaier, E.P. (1989). Catecholaminergic modulation of corticotropin-releasing factor and adrenocorticotropin secretion. *Endocrine Reviews* **10**, 437-458.

Plotsky, P.M., Kjaer, A., Sutton, S., Sawchenko, P.E., and Vale, W. (1991). Central activin administration modulates CRF and ACTH secretion. *Endocrinology* **128**, 2520-2525.

Plotsky, P.M., Sawchenko, P.E., and Vale, W. (1988). Evidence for inhibin ß-chain like peptide mediation of suckling-induced oxytocin secretion. *Neuroscience Abstracts* **14**, 627.

Sawchenko, P.E. (1988). The effects of catecholamine-depleting medullary knife cuts on CRF- and vasopressin-immunoreactivity in the hypothalamus of normal and steroid-manipulated rats. *Neuroendocrinology* **48**, 459-470.

Sawchenko, P.E., Arias, C., and Bittencourt, J.C. (1990). Inhibin ß-, somatostatin- and enkephalin-immunoreactivities coexist in caudal medullary neurons that project to the paraventricular nucleus of the hypothalamus. *Journal of Comparative Neurology* **291**, 269-280.

Sawchenko, P.E., Benoit, R., and Brown, M.R. (1988a). Somatostatin 28-immunoreactive inputs to the paraventricular nucleus: Origin from non-aminergic neurons in the nucleus of the solitary tract. *Journal of Chemical Neuroanatomy* **1**, 81-94.

Sawchenko, P.E., and Bohn, M.C. (1989). Glucocorticoid receptor immunoreactivity in C1, C2 and C3 adrenergic neurons that project to the hypothalamus or to the spinal cord in the rat. *Journal of Comparative Neurology* **205**, 107-116.

Sawchenko, P.E., Pfeiffer, S., Roberts, V.J., Cunningham, Jr., E.T., Benoit, R., Brown, M.R., and Vale, W. (1988c). Inhibin ß- and somatostatin-28-immunoreactive projections from the nucleus of the solitary tract to oxytocinergic cell groups. *Neuroscience Abstracts* **14**, 442.

Sawchenko, P.E., Plotsky, P.M., Cunningham, Jr., E.T., Vaughan, J., Rivier, J., and Vale, W. (1988b). Inhibin ß-immunoreactivity in a visceral sensory system controlling oxytocin secretion in the rat brain. *Nature* **344**, 315-317.

Sawchenko, P.E., and Swanson, L.W. (1981). Central noradrenergic pathways for the integration of hypothalamic neuroendocrine and autonomic responses. *Science* **214**, 685-687.

Sawchenko, P.E., and Swanson, L.W. (1982). The organization of noradrenergic projections from the brainstem to the paraventricular and supraoptic nuclei in the rat. *Brain Research Reviews* **4**, 275-325.

Sawchenko, P.E., Swanson, L.W., Grzanna, R., Howe, P.R.C., Polack, J., and Bloom, S.R. (1985). Co-localization of neuropeptide Y-immunoreactivity in brainstem catecholaminergic neurons that project to the paraventricular nucleus of the hypothalamus. *Journal of Comparative Neurology* **241**, 138-153.

Sawchenko, P.E., Swanson, L.W., and Vale, W.W. (1984b). Corticotropin releasing factor: Co-expression within distinct subsets of oxytocin-, vasopressin-, and neurotensin-immunoreactive neurons in the hypothalamus of the adult male rat. *Journal of Neuroscience* **4**, 1118-1129.

Sawchenko, P.E., Swanson, L.W., and Vale, W.W. (1984a). Co-expression of CRF- and vasopressin-immunoreactivity in parvocellular neurosecretory neurons in the adrenalectomized rat. *Proceedings of the National Academy of Sciences (USA)* **81**, 1883-1887.

Smythe, G.A., Bradshaw J.E., and Vining, R.F. (1983). Hypothalamic monoamine control of stress-induced adrenocorticotropin release in the rat. *Endocrinology* **113**, 1062-1071.

Swanson. L.W., and Simmons, D.M. (1989). Differential steroid hormone and neural influences on peptide mRNA levels in CRH cells of the paraventricular nucleus: A hybridization histochemical study in the rat. *Journal of Comparative Neurology* **285**, 413-435.

Szafarczyk, A., Alonso, G., Ixart, G., Malaval, F., and Assenmacher, I. (1985). Diurnal-stimulated and stress-induced ACTH release is mediated by ventral noradrenergic bundle. *American Journal of Physiology* **249**, E219-E226.

Vale, W., Rivier, C., Hsueh, A., Campen, C., Meuniere, H., Bicsak, T., Vaughan, J., Corrigan, A., Bardin, W., Sawchenko, P., Petraglia, F., Yu, J., Plotsky, P., Speiss, J., and Rivier, J. (1988). Chemical and biological characterization of the inhibin family of protein hormones. *Recent Progress in Hormone Research* **44**, 1-34.

Van Den Pol, A.N., Wuarin, J.P., and Dudek, F.E. (1990). Glutamate, the dominant excitatory transmitter in neuroendocrine regulation. *Science* **250**, 1276-1278.

Verbalis, J.G., McCann, M.J., McHale, C.M., and Stricker, E.M. (1986). Oxytocin secretion in response to cholecystokinin and food: differentiation of nausea from satiety. *Science* **232**, 1417-1419.

Stress: Neuroendocrine and Molecular Approaches
Edited by R. Kvetnansky, R. McCarty and J. Axelrod

IMPACT OF ACUTE AND CHRONIC STRESS ON THE RELEASE AND SYNTHESIS OF NOREPINEPHRINE IN BRAIN: MICRODIALYSIS STUDIES IN BEHAVING ANIMALS

E. D. Abercrombie[1], L. K. Nisenbaum[2] and M. J. Zigmond[2]

[1]Center for Molecular and Behavioral Neuroscience, Rutgers University, Newark NJ USA and [2]Department of Behavioral Neuroscience, University of Pittsburgh, Pittsburgh PA USA

LOCUS COERULEUS ACTIVITY AND STRESS

Stress Increases Activity in LC Neurons

A wide variety of acutely presented stressful stimuli, including footshock, cold environment, and immobilization, have been shown to affect indices of activity in the system of norepinephrine (NE)-containing neurons originating in the pontine nucleus locus coeruleus (LC) (see Anisman and Zacharko, 1990; Stone, 1975). Among the changes that have been demonstrated to occur as a result of acute stress are increases in the firing rate of NE neurons in LC (Abercrombie and Jacobs, 1987a), decreases in brain NE content (Bliss *et al*, 1968; Kvetnansky *et al*, 1977), increases in NE turnover (Korf *et al*, 1973; Tanaka *et al*, 1983; Thierry *et al*, 1968), and increases in the extracellular level of NE (Abercrombie *et al*, 1988; Finlay *et al*, 1990; Kalen *et al*, 1989). The stress-induced increases in these indices of LC activation take place in multiple brain areas that receive afferent inputs from this structure, suggesting a concerted activation of this central NE system during stress. This is consistent with both anatomical (Lindvall and Bjorklund, 1974; Moore and Bloom, 1979; Segal *et al*, 1973) and biochemical (Crawley *et al*, 1980) data which emphasize the widespread distribution of LC-NE terminals throughout the brain.

Following chronic exposure to stress, the reductions in brain NE levels that occur after acute stress no longer are observed (Ritter and Ritter, 1977; Zigmond and Harvey, 1970). Indeed, brain NE levels may actually be increased in response to chronic stress exposure (Adell *et al*, 1988; Irwin *et al*, 1986; Kvetnansky *et al*, 1977; Roth *et al*, 1982; Thierry *et al*, 1968). For example, following prolonged cold exposure the rate of NE utilization was reported to be increased; however, brain NE concentrations were increased as well (Bhagat, 1969; Simmonds, 1969). In another set of studies, repeated immobilizations of short duration or a single prolonged immobilization period were not associated with the NE depletions observed in response to acute immobilization. Indeed, NE levels were increased in several hypothalamic nuclei (Kvetnansky *et al*, 1977). Based on results such as these, we hypothesize that chronic stress leads to a compensatory increase in the biosynthesis of NE associated with increased TH activity. This increased synthesis would serve to support a sustained increase in NE release such that NE content does not decline and may even increase. These changes, then, represent adaptations of LC-NE neurons to conditions of increased demand.

Increased Activity in LC Neurons Leads to Increased TH Activity

It is well established that the activity of TH, the rate-limiting enzyme in catecholamine synthesis, is regulated in response to the demand for catecholamines. The increase in TH activity that occurs during stress presumably serves to maintain an adequate supply of neurotransmitter in the face of increased demand. Early studies of the sympathetic nervous system demonstrated that the enhancement of TH activity that accompanied an acute increase in sympathetic activity was related to an increase in the activity of existing enzyme molecules. Based on this observation, it was proposed that NE synthesis may be controlled by negative feedback inhibition of TH (Udenfriend 1966; Weiner, 1970). In this model, during periods of increased impulse flow when more transmitter is released and metabolized, a strategic regulatory pool of NE normally accessible to TH is depleted, end-product inhibition is removed, and TH activity is increased.

In more recent years, however, it has become apparent that the regulation of TH is very complex and other processes, linked to neuronal activity, also are involved in the control of NE synthesis. Electrical stimulation of both central and peripheral catecholamine neurons results in an apparent activation of TH which persists following the termination of the stimulation period (Morgenroth *et al*, 1974; Salzman and Roth, 1980). Furthermore, an activation of TH has been shown to occur in response to various stimuli which are known to increase the activity of central NE neurons, including administration of catecholamine depleting drugs, partial lesion of central catecholamine systems, and various forms of stress (Acheson and Zigmond, 1981; Iuvone and Dunn, 1986; Salzman and Roth, 1980; Stone *et al*, 1978). This effect has been noted both in the region of the LC as well as in brain areas receiving NE innervation from this structure.

In addition to the increases in TH activity discussed above, which represent immediate adaptation to increased transmitter utilization, a second mechanism may come into play after prolonged increases in the activity of NE neurons. This latter mechanism is reflected by an increase in the maximal activity of TH measured *in vitro* under saturating concentrations of substrate and cofactor. This increase appears to be due to an actual increase in the number of active enzyme molecules (Acheson and Zigmond, 1981; Chuang and Costa, 1974; Fluharty *et al*, 1985; Thoenen, 1970).

Examining the Relation Between NE Efflux and TH Activity

A recent series of experiments conducted in our laboratory has been aimed at examining the functional significance of elevations in TH activity in relation to release of transmitter. Since both transmitter release and synthesis are regulated in a complex manner by a variety of factors, the nature of the relationship between NE release and TH activity is difficult to ascertain. Indeed, it has been suggested that under some conditions these two variables are inversely related. For example, it has been demonstrated that adaptation to a number of stressors results in reduced NE turnover in response to subsequent exposure to the same stressor during a period in which the maximal activity of TH is elevated (Kvetnansky *et al*, 1983; Stone and McCarty, 1983; Stone *et al*, 1978). We propose a two-process model to explain this apparent paradox. First, habituation occurs to the stressor, thus reducing the need for and the secretion of brain catecholamines. In support of this idea, we have observed that the increased firing rate of LC neurons that occurs in response to acute stress declines with repeated or prolonged exposure to the stressor (Abercrombie and Jacobs, 1987b). A second process simultaneously prepares the organism for a greater maximum release of NE during emergencies and this involves biochemical changes such as alterations in TH activity. This model predicts that animals in which stress-induced changes in biosynthetic capacity have occurred will show a greater maximum capacity for NE release than control animals in response to an emergency situation. We have tested this model by comparing the release and synthesis of NE in the LC system of chronically stressed rats and of naive control rats in response to challenge with a novel stressor, tail-shock (Nisenbaum and Abercrombie, 1991; Nisenbaum *et al*, 1991). In most experiments, we have utilized *in vivo* microdialysis to monitor the release and synthesis of NE.

CHRONIC STRESS EFFECTS ON NE RELEASE AND SYNTHESIS: *IN VIVO* MICRODIALYSIS STUDIES

Methodological Aspects of *In Vivo* Microdialysis

In vivo microdialysis is a method that permits one to monitor changes in the extracellular level of various neuroactive substances in brain of unanesthetized rats and to make repeated measurements over time in the same animal. The method involves implantation of a small piece of dialysis membrane containing numerous small pores into the brain area of interest. This dialysis "probe" is continuously perfused at a very slow rate with an artificial cerebrospinal fluid solution. Small molecules in the extracellular space diffuse through the pores in the membrane and the solution is collected and analyzed for the content of specific neuroactive compounds using high pressure liquid chromatography with electrochemical detection (HPLC-EC). In the present series of experiments, we have used this approach to measure the extracellular concentration of NE and DOPAC in the dentate gyrus region of the dorsal hippocampus (Figure 1A). Detailed discussions of the methods used to collect the data presented in this chapter can be found elsewhere (Abercrombie and Finlay, 1991; Nisenbaum and

Abercrombie, 1991).

NE Release After Chronic Stress Exposure

Extracellular NE in the dorsal hippocampus was measured under resting conditions and in response to 30 minutes of intermittent tail-shock. Two groups of rats were studied, a naive control group and a group that previously had been exposed to 3-4 weeks of chronic cold stress. The chronic stress procedure consisted of shaving the animals and placing them into a cold room (5°C), where they were housed singly with free access to food and water. Cold exposure was chosen as the chronic stressor because it has been shown that this stimulus produces an increase in maximal TH activity within the cell bodies of the LC (Richard *et al*, 1988; Thoenen, 1970; Zigmond *et al*, 1974). On the day of removal from the cold room, the dialysis probe was implanted into hippocampus. Testing was conducted the following day. The basal extracellular concentration of NE in hippocampus was found to be the same in naive and chronically cold-stressed rats. However, 30 minutes of intermittent tail shock produced a significantly greater elevation of extracellular NE in the cold stressed animals compared to the naive control animals (Figure 2). Thus, to the extent that changes in NE measured in the dialysate are a reflection of changes in released neurotransmitter, these results suggest that stress-induced NE release is increased to a greater degree in hippocampus of chronically cold-stressed rats than in controls.

NE Synthesis After Chronic Stress Exposure

Maximal TH activity. Since a prolonged increase in the V_{max} for TH has been demonstrated in LC cell bodies in response to various chronic stressors, including chronic cold (see above), we thought it important to measure this variable in relation to possible changes in maximal TH activity at the level of the noradrenergic nerve terminal. TH activity was assayed *in vitro* in a cell-free homogenate in the presence of saturating concentrations of cofactor and at the pH optimum for the enzyme. Whereas chronic cold exposure produced the expected increase in maximal TH activity in the cell body region of the LC, no significant change in this variable could be detected in hippocampus of chronically cold-stressed rats. Although these data suggest that no increase in maximal TH activity occurred in hippocampus following chronic cold stress, such assays are relatively insensitive to changes of 25% or less. Small changes in TH activity might have occurred that were physiologically significant but could not be detected (Nisenbaum *et al*, 1991).

Extracellular DOPAC. As discussed above, much evidence exists to suggest a coupling between neurotransmitter release and neurotransmitter synthesis (see Fillenz, 1990; Zigmond *et al*, 1989). We have proposed that DOPAC, which appears in our hippocampal dialysis samples, may provide an indirect measure of tyrosine hydroxylation *in vivo* (Abercrombie and Zigmond, 1989; Nisenbaum *et al*, 1991). Because no significant dopaminergic innervation of the dorsal dentate gyrus region of the hippocampus appears to exist (Verney *et al*, 1985), this DOPAC probably is produced in NE terminals by metabolism of the NE precursor dopamine within NE terminals (Figure 1A). Extracellular DOPAC in the dentate gyrus may therefore reflect the balance between dopamine synthesis and transport into NE storage vesicles within the NE terminals.

The resting extracellular concentration of DOPAC in hippocampus was not significantly different in the rats that had been chronically cold stressed when compared to naive control rats. These data suggest that under resting conditions the synthesis of NE in the two experimental groups was the same, and are compatible with the finding that the basal extracellular NE concentration in control and chronically stressed rats did not differ. The increase in extracellular DOPAC in response to the tail-shock stress suggests an increase in tyrosine hydroxylation in response to this manipulation. This result is consistent with previous studies (see above) showing that increased neuronal activity in LC-NE neurons produces a short-term activation of TH in NE neurons. Furthermore, the stress-induced increase in DOPAC observed in our experiments was greater in rats exposed to chronic cold than in control rats, suggesting that a larger increase in NE synthesis occurred in the chronically stressed rats (Figure 3).

Extracellular DOPA Accumulation After AADC Inhibition. Another approach that we have employed to examine changes in NE biosynthesis after chronic stress involves the administration of an inhibitor of AADC, NSD-1015, locally via the dialysis probe and the measurement of the resulting accumulation of DOPA in the extracellular fluid of hippocampus (Figure 1B; Nisenbaum and

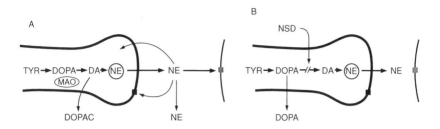

FIGURE 1.   The use of *in vivo* microdialysis to monitor the release and synthesis of NE in hippocampus.   *Panel A:*   Extracellular NE was monitored in the hippocampus as an index of the release of this neurotransmitter.   In addition, DOPAC in extracellular fluid of this structure was monitored as an index of NE synthesis.   The DOPAC in hippocampal dialysates is thought to be derived during the process of conversion of dopamine to NE in the noradrenergic nerve terminal. *Panel B:*   Another means by which the synthesis of NE was assessed was by measuring the accumulation of DOPA in extracellular fluid during local application of the AADC inhibitor NSD-1015.

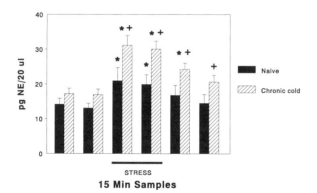

FIGURE 2.   The effect of acute tail shock on extracellular NE in naive and chronically cold-stressed rats.   Thirty minutes of intermittent tail shock (line) was administered after obtaining at least four stable baseline samples.   Basal NE levels did not differ between the two groups.   In naive rats (solid bars), tail shock produced a 54% increase in extracellular NE (n=9), while in chronically cold-stressed rats (hatched bars), an 82% increase above baseline occurred (n=9). Results are expressed as means $\pm$ SEM; *, $p < 0.05$ versus respective baseline; +, $p < 0.05$ chronically cold-stressed versus naive rats.   Figure adapted from Nisenbaum *et al*, 1991.

Abercrombie, 1991; Westerink *et al*, 1990). As a validation of this method, we have observed that administration of the TH inhibitor, alpha-methyl-p-tyrosine, decreases extracellular DOPA to undetectable levels. In rats previously exposed to chronic cold stress, the basal accumulation of extracellular DOPA did not differ from that observed in naive controls. However, acute tail shock produced a significantly greater and more prolonged elevation of extracellular DOPA in hippocampus of chronically stressed rats (Figure 4). We also have utilized the more traditional method of measuring the post-mortem accumulation of DOPA in tissue after inhibition of AADC to examine this issue (Carlsson *et al*, 1972). In agreement with our *in vivo* microdialysis data, the results revealed that this variable was elevated to a greater extent in hippocampus of chronically stressed rats exposed to acute tail shock as compared to naive controls (Nisenbaum *et al*, 1991). Taken together, these data provide further support for the conclusion that under basal conditions NE synthesis is not altered by chronic stress but that in response to a novel stressor a greater elevation of NE biosynthesis is elicited in chronically stressed rats.

THE NATURE OF THE RELATION BETWEEN RELEASE AND SYNTHESIS OF NE

Pharmacological Manipulations and the Release and Synthesis of NE

An important question that emerges from the above data focusses on the nature of the relation between the enhanced release of NE in response to a novel stressor in chronically stressed rats and the enhanced synthesis of NE observed under these conditions. Is the enhanced release of NE a consequence of increased synthesis of a releasable pool of transmitter or is the enhanced level of synthesis a secondary response to the increased release of NE? We have begun to examine this question by studying changes in extracellular NE and in extracellular DOPAC as indices of release and synthesis, respectively, in response to inhibition of TH activity with alpha-methyl-p-tyrosine and in response to stimulation of alpha$_2$ adrenergic receptors with clonidine.

Alpha-methyl-p-tyrosine administration. In preliminary experiments, inhibition of tyrosine hydroxylase by administration of alpha-methyl-p-

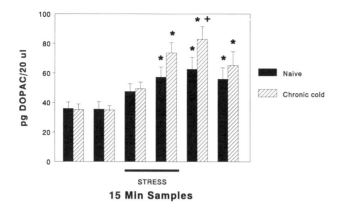

15 Min Samples

FIGURE 3. The effect of acute tail shock on extracellular DOPAC in naive and chronically cold-stressed rats. Experimental procedure was as in Figure 1. Basal DOPAC levels did not differ between the two groups. In naive rats (solid bars), tail shock produced a 75% increase in extracellular DOPAC (n=9), while in chronically cold-stressed rats (hatched bars), a 136% increase above baseline occurred (n=8). Results are expressed as means $\pm$ SEM; *, $p<0.05$ versus respective baseline; +, $p<0.05$ chronically cold-stressed versus naive rats. Figure adapted from Nisenbaum et al, 1991.

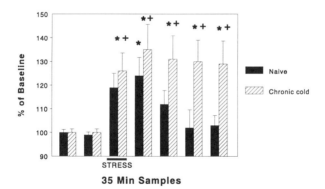

35 Min Samples

FIGURE 4. The effect of acute tail shock on extracellular DOPA in naive and chronically stressed rats during local infusion of NSD-1015, an inhibitor of AADC. Basal DOPA levels did not differ significantly between the two groups (naive: 278 ± 28 pg/50 μl; chronic cold: 372 ± 65 pg/50 μl). The increase in extracellular DOPA induced by tail shock (line) was significantly greater and more prolonged in the chronically stressed rats (hatched bars) than in naive controls (solid bars). Results are expressed as mean percent of baseline $\pm$ SEM; *, $p<0.05$ versus respective baseline; +, $p<0.05$ chronically cold-stressed versus naive rats. Figure from Nisenbaum and Abercrombie, 1991.

tyrosine (400 mg/kg, i.p.) rapidly decreased extracellular DOPAC to undetectable levels. This finding supports the idea that extracellular DOPAC provides a reliable measure of NE biosynthesis. In contrast, the extracellular concentration of NE in hippocampus decreased more slowly and was still detectable in the absence of measurable DOPAC concentrations (Figure 5A). These data suggest that NE release can occur in the absence of ongoing NE synthesis.

Clonidine administration. Stimulation of alpha$_2$ adrenergic receptors is known to produce a decrease in NE release from various noradrenergic terminal regions (Abercrombie *et al*, 1988; L'Heureux *et al*, 1986). Administration of clonidine (0.1 mg/kg, i.p.), an alpha$_2$ adrenergic receptor agonist, rapidly reduced extracellular NE in hippocampus to undetectable levels. In contrast, clonidine produced a maximal decrease of only 40% in extracellular DOPAC (Figure 5B). Thus, alpha$_2$ adrenergic receptor stimulation produces a cessation of NE release while synthesis of NE appears to be only mildly reduced.

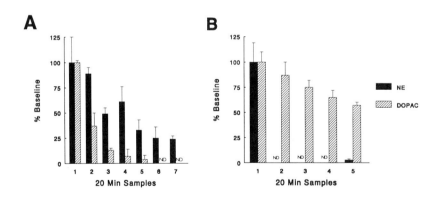

FIGURE 5. Preliminary data showing the effect of 400 mg/kg i.p. alpha-methyl-p-tyrosine (Panel A; n=2), a TH inhibitor, and 0.1 mg/kg i.p. clonidine (Panel B; n=4), an agonist of alpha$_2$ adrenergic receptors, on extracellular NE (solid bars) and DOPAC (hatched bars) in hippocampus. The drugs were administered between sample 1 and sample 2.

SUMMARY AND CONCLUSIONS

The present data suggest that prior exposure to chronic stress results in enhanced hippocampal NE release in response to a novel stressor. There was no apparent increase in the maximal activity of TH or in the resting level of tyrosine hydroxylation in hippocampus of chronically stressed animals. The increase in tyrosine hydroxylation produced by acute tail-shock stress, however, was significantly greater in the hippocampus of the chronically stressed rats compared to naive controls. The preliminary pharmacological data presented suggest that NE release persists for some time in the absence of ongoing synthesis and that inhibition of NE release is accompanied by only modest changes in synthesis. We conclude from these latter data that ongoing synthesis is not an absolute prerequisite for release of NE and that there exist conditions under which release and synthesis are not tightly coupled. We speculate, therefore, that the enhanced biosynthetic response to stress observed in chronically stressed animals represents a replenishment secondary to the release of transmitter rather than an enabling phenomenon that provides a mechanism for the enhanced transmitter release.

REFERENCES

Abercrombie, E.D., and Jacobs, B.L. (1987a). Single-unit response of noradrenergic neurons in the locus coeruleus of freely moving cats. I. Acutely presented stressful and nonstressful stimuli. *Journal of Neuroscience* 7, 2837-2843.

Abercrombie, E.D., and Jacobs, B.L. (1987b). Single-unit response of noradrenergic neurons in the locus coeruleus of freely moving cats. II. Adaptation to chronically presented stressful stimuli. *Journal of Neuroscience* 7, 2844-2848.

Abercrombie, E.D., Keller, R.W., and Zigmond, M.J. (1988). Characterization of hippocampal norepinephrine release as measured by microdialysis perfusion: Pharmacological and behavioral studies. *Neuroscience* 27, 897-904.

Abercrombie, E.D., and Zigmond, M.J. (1989). Partial injury to central noradrenergic neurons: Reduction of tissue norepinephrine content is greater than reduction of extracellular norepinephrine measured by microdialysis. *Journal of Neuroscience* 9, 4062-4067.

Abercrombie, E.D., and Finlay, J.M. (1991). Monitoring extracellular norepinephrine in brain using *in vivo* microdialysis and HPLC-EC. In: T. Robinson and J. Justice Jr. (Eds.), "Microdialysis in the Neurosciences," in press. Amsterdam: Elsevier Scientific.

Acheson, A.L., and Zigmond, M.J. (1981). Short and long term changes in tyrosine hydroxylase activity in rat brain after subtotal destruction of central noradrenergic neurons. *Journal of Neuroscience* **1**, 493-504.

Adell, A., Garcia-Marquez, C., Armario, A., and Gelpi, E. (1988). Chronic stress increases serotonin and noradrenaline in rat brain and sensitizes their responses to a further acute stress. *Journal of Neurochemistry* **50**, 1678-1681.

Anisman, H., and Zacharko, R.M. (1990). Multiple neurochemical and behavioral consequences of stressors: Implications for depression. *Pharmacology and Therapeutics* **46**, 119-136.

Bhagat, B. (1969). Effect of chronic cold stress on catecholamine levels in rat brain. *Psychopharmacologia* **16**, 1-8.

Bliss, E., Ailion, J., and Zwanziger, J. (1968). Metabolism of norepinephrine, serotonin and dopamine in rat brain with stress. *Journal of Pharmacology and Experimental Therapeutics* **164**, 122-134.

Carlsson, A., Davis, J.N., Kehr, W., Lindqvist, M., and Atack, C.V. (1972). Simultaneous measurement of tyrosine and tryptophan activities in brain *in vivo* using an inhibitor of the aromatic amino acid decarboxylase. *Naunyn-Schmiedeberg's Archives of Pharmacology* **275**, 153-168.

Chuang, D., and Costa E. (1974). Biosynthesis of tyrosine hydroxylase in rat adrenal medulla after exposure to cold. *Proceedings of the National Academy of Sciences* (USA) **71**, 4570-4574.

Crawley, J.N., Maas, J.W., and Roth, R.H. (1980). Biochemical evidence for simultaneous activation of multiple locus coeruleus efferents. *Life Sciences* **26**, 1373-1378.

Fillenz, M. (1990). Regulation of catecholamine synthesis: Multiple mechanisms and their significance. *Neurochemistry International* **17**, 303-320.

Finlay, J.M., Zigmond, M.J., and Abercrombie, E.D. (1990). Effects of diazepam on the stress-induced increase in extracellular dopamine and norepinephrine in medial prefrontal cortex. *Neuroscience Abstracts* **16**, 1322.

Fluharty, S.J., Snyder, G.L., Stricker, E.M., and Zigmond, M.J. (1985). Tyrosine hydroxylase activity and catecholamine biosynthesis in the adrenal medulla of rats during stress. *Journal of Pharmacology and Experimental Therapeutics* **233**, 32-38.

Irwin, J., Ahluwalia, P., Zacharko, R.M., and Anisman, H. (1986). Central norepinephrine and plasma corticosterone following acute and chronic stressors: Influence of social isolation and handling. *Pharmacology, Biochemistry & Behavior* **24**, 1151-1154.

Iuvone, P.M., and Dunn, A.J. (1986). Tyrosine hydroxylase activation in mesocortical 3,4-dihydroxyphenylethylamine neurons following footshock. *Journal of Neurochemistry* **47**, 837-844.

Kalen, P., Rosegren, E., Lindvall, O., and Bjorklund, A. (1989). Hippocampal noradrenaline and serotonin release over 24 hours as measured by the dialysis technique in freely moving rats: Correlation to behavioural activity state, effect of handling and tail-pinch. *European Journal of Neuroscience* **1**, 181-188.

Korf, J., Aghajanian, G.K., and Roth, R.H. (1973). Increased turnover of norepinephrine in the rat cerebral cortex during stress: Role of the locus coeruleus. *Neuropharmacology* **12**, 933-938.

Kvetnansky, R., Palkovits, M., Mitro, A., Torda, T., and Mikulaj, L. (1977). Catecholamines in individual hypothalamic nuclei of acutely and repeatedly stressed rats. *Neuroendocrinology* **23**, 257-267.

Kvetnansky, R., Nemeth, S., Vigas, M., Oprsalova, Z., and Jurcovicova, J. (1983). Plasma catecholamines in rats during adaptation to intermittent exposure to different stressors. In: E. Usdin, R. Kvetnansky, and J. Axelrod (Eds.), "Stress: The Role of Catecholamines and Other Neurotransmitters," pp. 537-562. New York: Gordon and Breach.

L'Heureux, R., Dennis, T., Curet, O., and Scatton, B. (1986). Measurement of endogenous noradrenaline release in the rat cerebral cortex *in vivo* by transcortical dialysis: Effects of drugs affecting noradrenergic transmission. *Journal of Neurochemistry* **46**, 1794-1801.

Lindvall, O., and Bjorklund, A. (1974). The organization of the ascending catecholamine neuron systems in the rat brain as revealed by the glyoxylic acid fluorescence method. *Acta Physiologica Scandinavia*, Supplement **412**, 1-48.

Moore, R.Y., and Bloom, F.E. (1979). Central catecholamine neuron systems: Anatomy and physiology of the norepinephrine and epinephrine systems. *Annual Review of Neuroscience* **2**, 113-168.

Morgenroth, V.H. III, Boadle-Biber, M.C., and Roth R.H. (1974). Tyrosine hydroxylase: Activation by nerve stimulation. *Proceedings of the National Academy of Sciences* (USA) **71**, 4283-4287.

Nisenbaum, L.K., and Abercrombie, E.D. (1991). Enhanced tyrosine hydroxylation in hippocampus of chronically stressed rats upon exposure to a novel stressor. *Journal of Neurochemistry*, in press.

Nisenbaum, L.K., Zigmond, M.J., Sved, A.F., and Abercrombie, E.D. (1991). Prior exposure to chronic stress results in enhanced synthesis and release of hippocampal norepinephrine in response to a novel stressor. *Journal of Neuroscience* **11**, 1478-1484.

Richard, F., Faucon-Biguet N., Labautu, R., Rollet, D., Mallet, J., and Buda, M. (1988). Modulation of tyrosine hydroxylase gene expression in rat brain and adrenals by exposure to cold. *Journal of Neuroscience Research* **20**, 32-37.

Ritter, S., and Ritter, R.C. (1977). Protection against stress-induced brain norepinephrine depletion after repeated 2-deoxy-D-glucose administration. *Brain Research* **127**, 179-184.

Roth, K.A., Mefford, I.M., and Barchas, J.D. (1982). Epinephrine, norepinephrine, dopamine and serotonin: Differential effects of acute and chronic stress on regional brain amines. *Brain Research* **239**, 417-424.

Salzman, P.M., and Roth, R.H. (1980). Poststimulation catecholamine synthesis and tyrosine hydroxylase activation in central noradrenergic neurons. I. *In vivo* stimulation of the locus coeruleus. *Journal of Pharmacology and Experimental Therapeutics* **212**, 64-73.

Segal, M., Pickel, V., and Bloom, F.E. (1973). The projections of the nucleus locus coeruleus: An autoradiographic study. *Life Sciences* **13**, 817-821.

Simmonds, M.A. (1969). Effect of environmental temperature on the turnover of noradrenaline in hypothalamus and other areas of rat brain. *Journal of Physiology* **203**, 199-204.

Stone, E.A. (1975). Stress and catecholamines. In: A.J. Friedhoff (Ed.), "Catecholamines and Behavior, Vol. 2," pp.31-72. New York: Plenum.

Stone, E.A., and McCarty, R. (1983). Adaptation to Stress: Tyrosine hydroxylase activity and catecholamine release. *Neuroscience & Biobehavioral Reviews* **7**, 29-34.

Stone, E.A., Freedman, L.S., and Morgano, L.E. (1978). Brain and adrenal tyrosine hydroxylase activity after chronic footshock stress. *Pharmacology, Biochemistry & Behavior* **9**, 551-553.

Tanaka, M., Kohno, Y., Nakagawa, R., Ida, Y., Takeda, S., Nagasaki, N., and Noda, N. (1983). Regional characteristics of stress-induced increases in brain noradrenaline release in rats. *Pharmacology, Biochemistry & Behavior* **19**, 543-547.

Thierry, A M., Javoy, J., Glowinski, J., and Kety, S.S. (1968). Effects of stress on the metabolism of norepinephrine, dopamine and serotonin in the central nervous system of the rat. I. Modifications of norepinephrine turnover. *Journal of Pharmacology and Experimental Therapeutics* **163**, 163-171.

Thoenen, H. (1970). Induction of tyrosine hydroxylase in peripheral and central adrenergic neurones by cold-exposure of rats. *Nature* **228**, 861-862.

Udenfriend, S. (1966). Tyrosine hydroxylase. *Pharmacological Review* **18**, 43-51.

Verney, C., Baulac, M., Berger, B., Alvarez, C., Vigny, A., and Helle, K.B. (1985). Morphological evidence for a dopaminergic terminal field in the hippocampal formation of young and adult rat. *Neuroscience* **14**, 1039-1052.

Weiner, N. (1970). Regulation of norepinephrine biosynthesis. *Annual Review of Pharmacology* **10**, 273-290.

Westerink, B.H.C., DeVries, J.B., and Duran, R. (1990). The use of microdialysis for monitoring tyrosine hydroxylase activity in the brain of conscious rats. *Journal of Neurochemistry* **54**, 381-387.

Zigmond, M.J., and Harvey, J.A. (1970). Resistance to central norepinephrine depletion and decreased mortality in rats chronically exposed to electric foot shock. *Journal of Neuro-Visceral Relations* **31**, 373-381.

Zigmond, R.E., Schon, F., and Iversen, L.L. (1974). Increased tyrosine hydroxylase activity in the locus coeruleus of rat brain stem after reserpine treatment and cold stress. *Brain Research* **70**, 547-552.

Zigmond, R.E., Schwarzschild, M.A., and Rittenhouse, A.R. (1989). Acute regulation of tyrosine hydroxylase by nerve activity and by neurotransmitters via phosphorylation. *Annual Review of Neuroscience* **12**, 415-461.

Stress: Neuroendocrine and Molecular Approaches
Edited by R. Kvetnansky, R. McCarty and J. Axelrod

1992 Gordon and Breach Science
Publishers S.A., New York, USA.
Photocopying permitted by license only.

SEROTONERGIC INVOLVEMENT IN FEEDING SUPPRESSION AFTER IMMOBILIZATION STRESS

Y. Oomura,[1] N. Shimizu[2] and Y. Kai[2]

[1]Institute of Bio-Active Science, Nippon Zoki Pharmaceutical Co., Ltd., Yashiro, Hyogo, and Toyama Medical and Pharmaceutical University, Toyama, Japan
[2]Department of Physiology, Faculty of Medicine, Kyushu University, Japan

INTRODUCTION

Many studies have considered possible roles of neurotransmitters, peptides and endogenous chemical substances in feeding behavior (Oomura, 1989), and have reached various conclusions. Brain serotonin (5-HT) may be involved in feeding control and facilitation of serotonergic functions may cause anxiety (Iversen, 1984; Leysen, 1984). Administration of p-chlorophenylalanine into the cerebral ventricles produced marked depletion of brain 5-HT and induced hyperphagia and obesity while destruction of the 5-HT fiber system could lead to obesity (Blundell, 1984). Chemical lesions of 5-HT nerve terminals by intraventricular injection of 5,6-dihydroxytryptamine produced increases in food consumption (Diaz *et al*, 1974). The dorsal raphe nucleus (DR) contains the highest density of serotonergic neurons in brain and 5-HT fibers extend to the LHA. Based upon immunocytochemical studies, 5-HT-positive fibers and DR varicosities have been found in the hypothalamus (Kawata *et al*, 1984). The concentration of serotonin is greater in the rat LHA than in other regions (Saavedra *et al*, 1974). Biochemical and electrochemical studies demonstrated that various stressful conditions caused serotonergic and dopaminergic activities in brain to increase significantly (Curzon et al, 1972). Immobilization stress is also accompanied by elevated brain tryptophan (Knott et al, 1973). Exposure to stress increases 5-hydroxyindoleacetic acid (5-HIAA) in brain that reflects increased release and utilization of 5-HT.

The present study was undertaken to clarify the contribution of acute immobilization stress to anorexia by measurements of changes in 5-HIAA, dopamine metabolites, neuronal activity in the LHA and synaptic connections from the DR to the LHA.

METHODS

Rats were immobilized for two hours by strapping their paws to a restraining board with adhesive tape.

Voltammetric Measurements

Carbon fiber (5~7) electrodes (Torayca, Type M-40, 7 μm o.d.) were supported in pulled glass capillaries. The reference and auxiliary electrodes were an Ag/AgCl and a silver wire, respectively. Differential pulse voltammetry (BAS, DPV-5) was used for *in vitro* calibration and *in vivo* calibration and *in vivo* measurements (Shimizu et al, 1989a). The parameters used were: potential range -0.2 V to +0.5 V; scan rate 50 mV/sec; modulation amplitude 50 mV, pulse frequency 10 Hz. Working electrodes were calibrated in artificial cerebrospinal fluid (CSF) containing 5×10^{-6} to 7×10^{-5} M 3,4-dihydroxyphenylacetic acid (DOPAC, Sigma) and 5-HIAA (Sigma). To sensitize the electrode to 5-HIAA and DOPAC it was necessary to pretreat the carbon fiber electrode electrically prior to its use *in vivo*. The electrode was treated with a triangular wave (0-3 V vs. Pt, 70 Hz for 30 seconds) in 0.1 M H_2SO_4 solution. After pretreatment, distinct peaks were evident at -0.09V (P1, ascorbic acid, 5×10^{-4} M), +0.04 V (P2, DOPAC, 5×10^{-5} M) and +0.23 V (P3, 5-HIAA, 5×10^{-5} M). Because the oxidation potentials of ascorbic acid (P1) and DOPAC (P2) were similar, P1 and P2 could not be completely separated. For the separation, the oxidation current of P1 could be reduced by holding the working electrode at an oxidation potential of -0.05 V versus the reference electrode for 5 minutes before scanning. The oxidation current measured with DOPAC and 5-HIAA varied linearly in the concentration range of 5×10^{-6} to 7×10^{-5} M. The sensitivity of the electrode used in this study was 7 times higher to 5-HIAA than to DOPAC. The electrode was inserted into the LHA of male Wistar rats (A = 4.6, L = 1.5, H = -2.7) according to the atlas of König and Klippel (1963), and the reference and auxiliary electrodes were placed on the dural surface of the frontal cortex.

Cerebrospinal fluid electrochemical detector of high-performance liquid chromatography (HPLC-ECD). CSF (30 μl) before, during and after immobilization was withdrawn through stainless steel cannulae implanted chronically in the third cerebral ventricle or cisterna magna and catechol- and indoleamines and their metabolites in CSF were analyzed (Bioanalytical System, LC-48).

Neuronal activity in the LHA. Under ketamine (100 mg/kg, i.p.) anesthesia, a bundle of recording electrodes (eight flexible teflon-coated platinum-iridium wires (Medwire, New York, 25 μm in diameter), was chronically implanted in the LHA. A dual-channel FET (2SK18, Toshiba) mounted directly on the animal's head was driven differentially from two of the eight implanted electrodes, one for recording single neuronal activity and the other as an indifferent electrode.

DR stimulation. Concentric bipolar stainless steel electrodes (i.d., 0.1 mm; o.d., 0.4 mm) were used (DR, A, 0.3 ± 0.3; L, 0.5 ± 0.3; H 0.6 ± 0.3). Rats were anesthetized with i.p. administration of urethane plus α-chloralose. Rectal temperature was kept at 38 ± 0.5 °C by an electric blanket. Single neuron discharges were recorded from the LHA. An extracellular recording electrode, filled with a solution of pontamine sky blue dissolved in 0.5 M sodium acetate (DC resistance, 5-10 MΩ) for marking the recording site, was glued to a 7-barrel pipette with its tip extending about 30 μm beyond the pipette tip. Each barrel was filled with one of the following chemicals: 0.5 M glucose (pH 6.0), 0.5 M sucrose (pH 6.0), 50 mM lisuride hydrogen maleate (Sherring, pH 5.0), 50 mM (-)-propranolol (Sigma, pH 5.0), 5 mM methysergide (Sandoz, pH 5.0), or 25 mM 5-HT creatinine sulfate (Sigma pH 5.0 in 1% ascorbic acid), each dissolved in 0.15 M NaCl; 0.5 M monosodium l-glutamate or 0.15 M NaCl for current-balancing (DC resistance, 40-100 MΩ). Monosodium-l-glutamate, a non-specific neural excitant, was applied to confirm that the assembly was intact and the chemicals could reach their target. Since methysergide and (-) -propranolol sometimes lightly suppress neural activity, the excitatory effect of glutamate was confirmed. If glutamate did not excite neural activity, the data were excluded. Intracellular recording was through glass micropipettes filled with 3 M potassium acetate (DC resistance, 40-100 MΩ). The input membrane resistance of an impaled cell was measured by voltage deflection from the resting membrane potential while passing 0.1-0.5 nA, 100 ms constant current pulses through a conventional bridge circuit.

Histology. After all experiments, electrolytic lesions were made by a 200 μA, 20 second cathodal current passed through the working electrode. Under deep anesthesia, the rats were transcardially perfused with physiological saline followed by 10% neutral formaldehyde, the brain was cut into 100 μm serial sections in a freezing microtome and the sections were stained with neutral red.

RESULTS

Food Intake and Immobilization Stress

Food intake and body weight. On the first three days, after two hours of immobilization, food intake of the immobilized group was significantly lower than basal intake. Basal levels of food intake before and during immobilization were 24.3 ± 0.3 g (n = 10), and 15.9 ± 0.7 g, respectively, each day. There was no adaptation to immobilization. Immediate recovery to basal levels occurred on day 4 (22.6 ± 0.5 g), and there was no compensatory overeating. Water intake was also suppressed during the test period, probably due to reduction of prandial drinking. Body weight loss was significant on the experimental days. The body weight change of the control group was + 18.0 ± 1.5 g (+ 7.1 ± 0.6%) by day 3, and that of the immobilized group was -13.4 ± 0.8 g (-5.5 ± 0.3%).

Anorexia caused by immobilization stress may be mediated through the serotonergic or the opioid system or both, so their antagonist effects were studied. Either physiological saline, methysergide (5 mg/kg) or naloxone (3 mg/kg) was injected (i.p.) one hour before immobilization. Controls received the same drug injection one day before immobilization. Food intake for the 3 hours after immobilization and for the succeeding 9 hours was measured beginning at 2000 hours. Immobilization significantly decreased food intake for both the 3 and 9 hour periods compared to controls. Food intake decreased to 74% of the control amount during the 3 hour period (physiological saline, 6.2 ± 0.3 g; physiological saline plus immobilization, 4.6 ± 0.2 g; n = 8) and to 68% of control during the next 9 hours (physiological saline, 17.3 ± 0.2 g; physiological saline plus immobilization, 11.7 ± 0.7 g). Methysergide antagonized the immobilization-induced anorexia for 3 hours (methysergide, 6.5 ± 0.4 g; methysergide plus immobilization, 5.9 ± 0.3 g; n = 8). The difference in 3 hour food intake between physiological saline and methysergide was statistically significant (one-way ANOVA). The following 9 hour food intake effect, however, was not antagonized by methysergide; there was no significant difference between physiological saline and methysergide (11.7 ± 0.7 g and 11.1 ± 0.6 g, respectively). Intraperitoneal administration of naloxone did not affect the anorexia caused by immobilization stress in either the 3 or 9 hour measurements.

Voltammetry. Voltamograms in the LHA had only one peak, P3, which appeared at an oxidation potential of about +0.25 V. We tested whether this peak reflected the extracellular level of 5-HIAA by pharmacological manipulations in anesthetized rats. As shown in Figure 1, 5-hydroxytryptophan (5-HTP) (30 mg/kg, i.p.), a precursor of 5-HT, augmented the amplitude of P3, to about a 250% increase 90 minutes after the application, and then a gradual return to basal level. Pargyline, (50 mg/kg, i.p.) a monoamine oxidase inhibitor, decreased the P3 amplitude, to about a 60% decrease in P3 amplitude 2 hours after the application. Injection of pargyline during the increase of P3 amplitude, 30 minutes after injection of 5-HTP, decreased the P3 amplitude rapidly after a latency of 15 minutes. The oxidation potential of P3 recorded in the LHA was reproducible in size and shape for the 2 hour period of the experiment. The results indicate that P3 reflects extracellular levels of 5-HIAA.

During immobilization stress the amplitude of P3 increased rapidly in the LHA. Figure 2A shows results from the LHA before, during and 120 minutes after cessation of immobilization stress. The P3 amplitude increased 46% in the 45 minutes after starting immobilization (Figure 2B). The electrochemical signals gradually

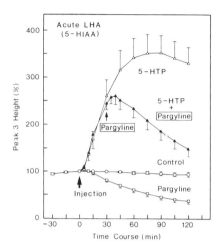

FIGURE 1.   *In vivo* voltammetric measurements in anesthetized rats and pharmacological manipulations of 5-HIAA levels in LHA.   Upward arrowhead, injections of drugs.   Changes in peak 3 (P3) height, plotted as % of peak height at 0 time.   Mean ± S.E., n = 5.   Controls were injected with physiological saline. During the course of increasing P3 height after injection of 5-HTP (30 mg/kg), injection of pargyline (50 mg/kg, time at 30 minutes) blocked the rise of P3 and caused a rapid decrease with a 15-minute latency.

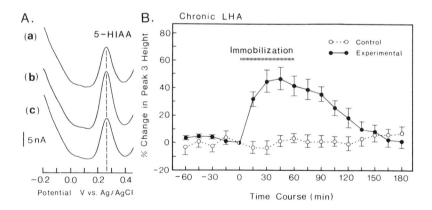

FIGURE 2.   *In vivo* voltammetric measurement in LHA.   Unanesthetized rat. A: voltammograms recorded before (a), 60 minutes after starting (b) and 2 hours after termination of immobilization (c).   Only one peak corresponding to 5-HIAA was detectable. B: changes in 5-HIAA concentrations affected the peak amplitude of P3 relative to 5-HIAA concentrations in the LHA during and after immobilization.

decayed immediately after the end of immobilization and returned to pre-immobilization levels within 180 minutes. During a 4 hour test of controls (freely moving), the amplitude of P3 was relatively stable. Statistical analysis revealed that the two groups were significantly different during immobilization and for 60 minutes after stopping immobilization (two-way ANOVA, after angular transformation).

HPLC-ECD measurement.   The concentrations of catecholamines, indoleamines and their metabolites in CSF were measured by HPLC-ECD before and 90 minutes after the beginning immobilization. The peaks of 5-HIAA and DOPAC increased significantly after immobilization. Norepinephrine (NE) and homovanillic acid also increased. After 2 hours of immobilization, 5-HIAA and DOPAC levels gradually increased to $82.3 \pm 10.1\%$ and $116.3 \pm 16.5\%$, respectively. The time course of change in 5-HIAA levels was slower than for DOPAC levels. After cessation of immobilization, levels of both 5-HIAA and DOPAC began to decrease, and returned to basal levels within 3 hours.

Neuronal activity in the LHA under immobilization stress.   Single neuron activity in the LHA was recorded during immobilization stress by chronically implanted electrodes. The activity of 18 of 25 LHA neurons decreased significantly during immobilization stress. The mean firing rate (impulses/second) decreased from 27 ± 2.4 before to 8 ± 1.0 during immobilization. The firing rate of the other 7 neurons did not change significantly and there was no neuron whose firing rate increased during immobilization. Of the 18 neurons, the suppressed neural activity of 7 recovered to baseline within 30 minutes after the end of immobilization (Figure 3A). Among the 18 neurons that decreased firing during immobilization stress, long-term suppression of spontaneous firing rate was observed in 9. As shown in the upper part of Figure 3B, the low rate of firing continued for 3 to 12 hours after immobilization and irregular bursts appeared in most cases. The suppression of neuronal activity was antagonized by methysergide (5 mg/kg, i.p.), 50 minutes before immobilization. As shown in the lower part of Figure 3B, firing recovered to the baseline rate 22 hours after the end of immobilization. The methysergide treatment almost completely attenuated suppression by immobilization stress. In 5 neurons tested for the effect of methysergide, all responses were antagonized by this treatment.

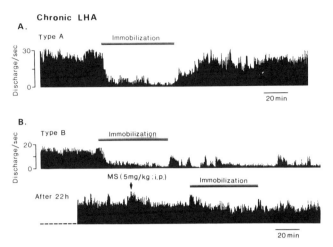

FIGURE 3. Single neuron activity in the LHA during immobilization stress. Immobilization stress suppressed the spontaneous activity of LHA neurons. A: suppression of neural activity recovered soon after release from immobilization. B: upper and lower graphs, same neuron. Immobilization suppressed neuron activity for a prolonged period. Downward arrow, intraperitoneal application of methysergide (MS) at 5 mg/kg. MS, attenuated suppression of neuron activity and blocked immobilization-induced anorexia.

Extracellular and intracellular recording of LHA neuron responses to DR stimulation. Extracellular recordings were made from 287 neurons in the LHA, and 230 responded to DR stimulation. Among the responsive neurons, 137 (60%) were inhibited, 20 (9%) were excited after inhibition, 31 (13%) were excited and 42 (18%) were inhibited after excitation.

Primary inhibitory response neurons. This pattern was observed in 157 neurons (55%). Intracellular recordings were made from 49 neurons. The resting membrane potential was -45 ± 5.0 mV and the membrane resistance was 40 ± 5.5 MΩ. As shown in Figure 4A and B, amplitude and duration of inhibitory postsynaptic potential (IPSP) increased with an increase in stimulus intensity. As shown in Figure 4C, the amplitude of hyperpolarization of IPSP was decreased and reversed by application of hyperpolarizing current. The reversal potential of the IPSP was estimated to be about -94 mV. The IPSP appeared to be monosynaptic since the latency was invariant with a change in stimulus intensity. The mean latency was 4.2 ± 2.6 ms (n = 26).

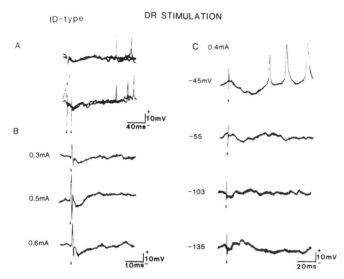

FIGURE 4.   Intracellular recordings of primary inhibitory response type neuron to dorsal raphe nucleus (DR) stimulation.▲, stimulation. Resting membrane potential, -50mV. A: IPSP in LHA neuron (two superimposed sweeps). (Stimulus intensity, 0.5 mA). B: increase in amplitude and duration of IPSP with increase in stimulus intensity (mA). Latency, no change. C: reversal of IPSP by change in membrane potential due to hyperpolarizing current. Reversal potential for IPSP, -94mV.

To identify a possible transmitter of the primary inhibitory response, 5-HT receptor antagonists were electrophoretically administered on extracellulary recorded neurons.   The neuronal responses were inhibited to 10% (control) of the basal firing rate; application of lisuiride for 3 minutes attenuated the inhibitory response to 32% from 9% (control), and these responses recovered to 6% after the end of the application (Figure 5 left).   The attenuation by lisuride was significant (paired sample t test after angular transformation, $p < 0.01$).   The 5-HT$_1$ receptor antagonist, (-)-propranolol (Middlemiss et al, 1977), applied electrophoretically for 3 minutes, attenuated the inhibition, and recovery was achieved after cessation of the application (Figure 5 right).   This attenuation by (-)-propranolol was significant (paired sample t test after angular transformation, $p < 0.05$). Electrophoretic applications of 5-HT and its antagonists were studied for their effects on spontaneous activity of primary inhibitory response type neurons.   As shown in Figure 6, glucose and 5-HT applications inhibited spontaneous neuronal activity.   During lisuride (upper) and

DR stimulation

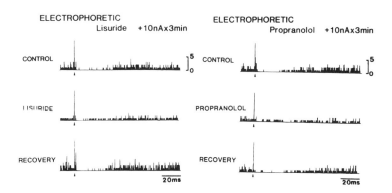

FIGURE 5. Effects of serotonin antagonists on LHA neuron responses to DR stimulation. PSTH of 100 sweeps. Bin size, 0.2 msec. Left: control, inhibition by DR stimulation. Primary inhibitory response type neuron. After 3 minutes electrophoretic application of lisuride, inhibition was attenuated. Recovery 3 minutes after cessation of lisuride application. Right: same type neuron. Control, inhibition by DR stimulation. After 3 minutes electrophoretic application of (-)-propranolol, inhibition was attenuated. Recovery 3 minutes after cessation of propranolol application.

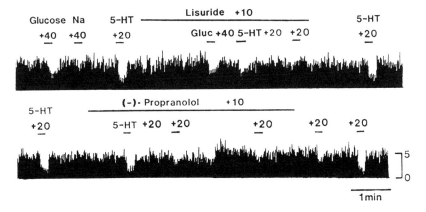

FIGURE 6. Effects of glucose, serotonin (5-HT) and serotonergic antagonists on spontaneous activity of primary inhibitory response type neurons. Same neuron. Upper: responses of glucose-sensitive neuron to 5-HT. Responses to glucose not attenuated; but those to 5-HT, attenuated by lisuride (10 nA). Lower: responses to 5-HT attenuated by (-)-propranolol.

(-)-propranolol (lower) applications, the 5-HT effect was attenuated, but the glucose response was not changed. Of 49 primary inhibitory response type neurons tested with 5-HT, 43 (88%) were inhibited and three (6%) were excited. Of 41 neurons tested with glucose, 21 (51%) were inhibited and 20 were not affected. The characteristic primary inhibition to glucose was significant (X^2 test, $p < 0.01$) (Kai *et al*, 1988).

Primary excitatory response neurons. Primary excitatory responses were observed in 73 neurons (25%). Intracellular recordings were made from 49 neurons. The resting membrane potential and membrane resistance were -48 ± 8.4 mV and 41 ± 3.0 MΩ, respectively. As shown in Figures 7A and B, excitatory postsynaptic potentials (EPSPs) in response to stimulation were observed in 13 (27%) of the 49 neurons; the amplitude and duration increased with increases in stimulus intensity although no changes in latency of the EPSP were noted. The mean latency was 4.8 ± 2.9 msec (n = 13) and the response was monosynaptic. The EPSP increased upon application of hyper-polarizing current (Figure 7C); the reversal potential for the EPSP was estimated to be approximately -17 mV.

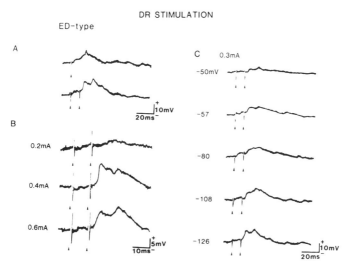

FIGURE 7. Intracellular recording of primary excitatory responses in LHA neurons. ▲, DR stimulation. Resting membrane potential, -40 mV. A and B: EPSP. Increase in amplitude and duration of EPSP with increase in stimulus intensity (mA). Latency, no change. C: increase of EPSP amplitude by changes of membrane potential by hyperpolarizing current. Reversal potential for EPSP, -17 mV.

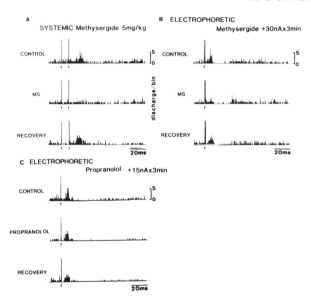

FIGURE 8. Effects of serotonin antagonists on primary excitatory responses to DR stimulation in LHA. PSTH of 100 sweeps. Bin size, 0.2 ms. A: control, excitation followed by inhibition. Methysergide (IP) attenuated both excitation and inhibition. B: control, excitation. After 3 minutes electrophoretic application of methysergide, excitation was attenuated. Recovery 3 minutes after end of methysergide application. C: control, excitation, (-)-propranolol application, no attenuation. Recovery 3 minutes after cessation of propranolol application.

As shown in Figure 8A, intraperitoneal administration of methysergide attenuated both excitation and inhibition after the excitation. Recovery occurred 40 minutes after drug administration. Electrophoretic methysergide applied for 3 minutes attenuated excitation, which recovered 20 minutes after the end of the application. The inhibition after the excitation was not affected by this treatment (Figure 8B). Electrophoretically applied (-)-propranolol did not affect the response (Figure 8C). Attenuation of the response by systemic and electrophoretic methysergide was significant (paired sample t test after angular transformation, $p < 0.05$). The effects of electrophoretic glucose and 5-HT on spontaneous activity of primary excitatory response type neurons were investigated. As shown in Figure 9, glucose did not affect the firing, and 5-HT dose-dependently inhibited the discharge frequency. During electrophoretic application of methysergide, the inhibitory effect of 5-HT was attenuated, and this

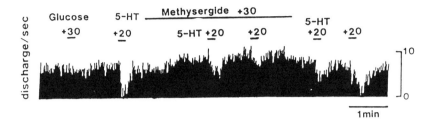

FIGURE 9. Effects of glucose, 5-HT and methysergide on spontaneous activity of primary excitatory response type neurons. Glucose, no effect; 5-HT, suppression. Methysergide attenuated glucose-insensitive neuronal responses to 5-HT. Recovery after methysergide application.

recovered after stopping the methysergide. Of 29 primary excitatory response type neurons tested with 5-HT, 21 (72%) were inhibited and one was excited. Of 14 neurons tested with glucose, 3 (21%) were inhibited, 1 was excited, and 10 (72%) were not changed (Kai *et al*, 1988).

DISCUSSION

Stable monitoring of the 5-HIAA peak was possible in the LHA of unanesthetized rats. The carbon fiber electrode could detect 5 μM DOPAC *in vitro*, but we could not identify any clear sign of DOPAC in the LHA, even after immobilization stress. Changes in the concentrations of DOPAC and 5-HIAA in the CSF were clear when analyzed by HPLC-ECD, and both 5-HIAA and DOPAC increased significantly after immobilization stress to 182% and 216% of baseline, respectively. This increase of DOPAC concentration in CSF might be attributed to the activation of the striatum, an area known to receive dense dopaminergic projections from the substantia nigra pars compacta. Stress-induced increases in DOPAC have been observed *in vivo* by voltammetry in the striatum (Fadda *et al*, 1978). Dopaminergic innervation of the hypothalamus has been reported and insufficient sensitivity of the carbon fiber electrode might explain why we could not detect the peak of DOPAC if its concentration was less than 5×10^{-6}M in the LHA.

In the feeding behavior experiment, methysergide significantly antagonized immobilization-induced anorexia while naloxone did not. Continuous recording of single neuronal activity in the LHA indicated that many neurons were inhibited by immobilization and the inhibition was significantly antagonized by preinjection of methysergide (Figure 3). Further, most LHA neurons were inhibited by electrophoretic application of 5-HT and this effect was attenuated by methysergide, which is known to be a potent 5-HT receptor blocking agent (Figure 9). These results suggest that immobilization-induced anorexia was mediated, at least in part, through serotonergic mechanisms in the LHA, which is known to be important in the regulation of feeding behavior.

Neuronal primary inhibition (55%) and primary excitation (25%) in response to DR stimulation were shown to be due mostly to monosynaptic IPSPs and EPSPs, since the latencies of the responses did not change with change in stimulus intensity. Electrophoretic 5-HT usually inhibited both primary inhibition and primary excitation response type neurons. Two questions arise: one, what factors caused the difference between these responses? The other, why did electrophoretically applied 5-HT inhibit the spontaneous activity of both neuronal types? One possibility is that the reactions were mediated by different receptor subtypes. It has been reported that 5-HT_1 receptors are labeled by [^{3}H] 5-HT and 5-HT_2 receptors are labeled by [^{3}H] spiroperidol (Peroutka and Snyder, 1979). Methysergide, a general 5-HT antagonist, is reported to be 40-400 times as potent at 5-HT_2 receptors than it is at 5-HT_1 receptors, and lisuride, which is 15-200 times as potent at the 5-HT_1 receptor (Peroutka and Snyder, 1979), attenuated the responses of primary inhibitory response type neurons to electrophoretic applications. This response type was also attenuated by (-)- propranolol (Middlemiss et al, 1977). The results support the idea that primary inhibitory response type neurons might function through 5-HT_1 receptors, and primary excitatory response type neurons might involve 5-HT_2 receptors. Since this latter type of response was attenuated by methysergide, but not by (-)-propranolol, its monosynaptic nature suggests that it might be serotonergic, and the inhibitory effect of 5-HT on this type of neuron was antagonized by methysergide. Although electrophoretically applied 5-HT was not excitatory in the present experiment, the following possibility must be considered (Roberts and Stravghan, 1967). It has been reported that NE caused opposing dose-dependent actions: low concentrations were excitatory on supraoptic vasopressin neurons, and high concentrations were inhibitory (Day et al, 1985). It is possible that there are two kinds of 5-HT receptors on the same LHA neuron, and the excitatory component might be masked by the inhibitory component that is mediated by the 5-HT$_1$ receptor (Fukuda et al, 1987).

Serotonergic functions in the LHA were investigated by identifying glucose-sensitive neurons. The glucose-sensitive neurons, which are influenced by endogenous chemical information and affect feeding behavior (Oomura, 1989) received significant primary inhibitory response type inputs from the DR that responded through 5-HT$_1$ receptors. Glucose-insensitive neurons received primary excitatory and primary inhibitory response type inputs from the DR that responded through 5-HT$_2$ and 5-HT$_1$ receptors, respectively. Although the physiological correlates of our results are speculative, it is possible that 5-HT$_1$ receptors are associated with suppression of feeding by 5-HT application (Blundell, 1984). There results strongly suggest that an increase in 5-HT release in the LHA during immobilization stress is the main cause of anorexia. The delayed suppression of food intake after stress should be clarified in the near future.

ACKNOWLEDGEMENTS

We thank Professor A. Simpson for help in preparing this manuscript. This study was supported in part by Grants-in-Aid for Scientific Research 58870118, 60440097, 61870102, 63480470 and 62304031 from the Ministry of Education, Science and Culture of Japan.

REFERENCES

Blundell, J.E. (1984). Serotonin and appetite. Neuropharmacology 23, 1537-1551.

Curzon, G., Joseph, M.H., and Knott, P.J. (1972). Effects of immobilization and food deprivation on rat brain tryptophan metabolism. Journal of Neurochemistry 19, 1967-1974.

Day, T.A., Randle, J.R., and Renaud, L.P. (1987). Opposing α- and ß-adrenergic mechanisms mediate dose-dependent actions of noradrenalin on supraoptic vasopressin neurons in vivo. Brain Research 358, 171-179.

Diaz, J., Ellison, G., and Masuoka, D. (1974). Opposed behavior syndromes in rats with partial and some complete central serotonergic lesions made with 5,6-

58   Y. OOMURA et al

dihydroxytryptamine. *Psychopharmacologia* **37**, 67-79.

Fadda, F., Argiolas, A., Melis, M.R., Tissari, A.H., Onali, P.L., and Gessa, G.L. (1978). Stress-induced increase in 3,4-dihydroxyphenylacetic acid (DOPAC) levels in the cerebral cortex and in N. accumbens: Reversal by diazepam. *Life Sciences* **23**, 2219-2224.

Fukuda, A., Minami, T., Nabekura, J., and Oomura, Y. (1987). The effects of noradrenaline on neurons in the rat dorsal motor nucleus of the vagus, *in vitro*. *Journal of Physiology* **393**, 213-231.

Iversen, S.D. (1984). 5-HT and anxiety. *Neuropharmacology* **23**, 1553-1560.

Kai, Y., Oomura, Y., and Shimuzu, N. (1988). Responses of rat lateral hypothalamic neuron activity to dorsal raphe nuclei stimulation. *Journal of Neurophysiology* **60**, 524-535.

Kawata, M., Takeuchi, Y., Ueda, S., Matsuura, T., and Sano, Y. (1984). Immunohistochemical demonstration of serotonin-containing nerve fibers in the hypothalamus of the monkey, *Macaca fuscata*. *Cell and Tissue Research* **236**, 495-503.

Knott, P.J., Joseph, M.H., and Curzon, G. (1973). Effects of food deprivation and immobilization on tryptophan and other amino acids in rat brain. *Journal of Neurochemistry* **20**, 249-251.

König, J., and Klippel, R.A. (1963). "The Rat Brain: A Stereotaxic Atlas of the Forebrain and Lower Parts of the Brain Stem." Baltimore: Williams and Wilkins.

Leysen, J. (1984). Problem in *in vitro* receptor binding studies and identification and role of serotonin receptor sites. *Neuropharmacology* **23**, 247-254.

Middlemiss, D.N., Blakeborough, L., and Leather, S.R. (1977). Direct evidence for an interaction of ß-adrenergic blockers with the 5-HT receptor. *Nature* **267**, 289-290.

Oomura, Y. (1989). Sensing of endogenous chemicals in control of feeding. In: H. Autrum, D. Ottoson, E.R. Perl, R.F. Schmidt, H.C. Shimazu, and W.D. Willis (Eds.), "Progress in Sensory Physiology 9" pp. 171-191. Berlin: Springer-Verlag.

Peroutka, S.J., and Snyder, S.H. (1979). Multiple serotonin receptors: Differential binding of [³H]5-hydroxytryptamine and [³H]-lysergic acid diethylamine and [³H]-spiroperidol. *Molecular Pharmacology* **16**, 687-699.

Peroutka, S.J., Lebovitz, R.M., and Snyder, S.H. (1981). Two distinct central serotonin receptors with different physiological functions. *Science* **212**, 827-829.

Roberts, M.H.T., and Stravghan, D.W. (1967). Excitation and depression of cortical neurons by 5-hydroxytryptamine. *Journal of Physiology* **193**, 269-294.

Saavedra, J.M., Palkovits, M., Brownstein, M.J., and Axelrod, J. (1974). Serotonin distribution in the nuclei of the rat hypothalamus and preoptic region. *Brain Research* **77**, 157-162.

Shimizu, N., Oomura, Y., and Aoyagi, K. (1989a). Electrochemical analysis of hypothalamic serotonin metabolism accompanied by immobilization stress in rats. *Physiology and Behavior* **46**, 829-834.

Shimizu, N., Oomura, Y., Aoyagi, K., and Kai, Y. (1989b). Stress-induced anorexia in rats mediated by serotonergic mechanisms in the hypothalamus. *Physiology and Behavior* **46**, 835-841.

Stress: Neuroendocrine and Molecular Approaches
Edited by R. Kvetnansky, R. McCarty and J. Axelrod

1992 Gordon and Breach Science
Publishers S.A., New York, USA.

STRESS COPING EFFECTS OF DELTA SLEEP INDUCING PEPTIDE

K. V. Sudakov

P. K. Anokhin Institute of Normal Physiology
USSR Academy of Medical Sciences, Moscow 103009 USSR

INTRODUCTION

Delta sleep inducing peptide (DSIP) has been isolated from venous blood dialyzate outflowing from brains of rabbits in a state of sleep induced by electrical stimulation of the intralaminar nuclei of the thalamus (Monnier *et al*, 1977). EEG studies have reported slow high-amplitude delta activity in response to DSIP injected into awake animals.

DSIP has been widely discussed in the scientific literature (Bjartell, 1990; Monnier and Schoenenberger, 1976; Mejerson *et al*, 1984; Schoenenberger and Schneider-Helmert, 1983; Graf and Kastin, 1986; Kovalson, 1986; Lubbers and Strossek, 1970; Medvedjev *et al*, 1983). However, evidence for a somnogenic effect of DSIP is contradictory. Some researchers (Kovalson and Tsybulski, 1980; Tobler and Borbely, 1980) believe that DSIP has nothing to do with mechanisms of natural sleep. It is possible that electrical stimulation of brain produces a state of electrosleep which is usually characterized by slow, high-amplitude EEG patterns (Sudakov *et al*, 1972).

As some data show (Ivashchenko *et al*, 1976) electrosleep has stress coping effects. This led us to hypothesize that DSIP produced by electrostimulation of brain can have similar effects when administered exogenously.

METHODS

Experiments were conducted on Chinchilla male rabbits, and Wistar and August rats. Rabbits and rats with bipolar nichrome electrodes implanted into the ventromedial hypothalamic nuclei (VMH) under local procaine anesthesia were kept in a stereotaxic device to study behavioral and autonomic responses and development of emotional stress. The diameter of implanted electrodes was 0.2mm. VMII stimulation (current of 100-350μA, frequency of 100Hz, pulse duration of 1.4ms) lasting 3 seconds induced an escape reaction, heart and respiration rate increases, urination and defecation.

DSIP, synthetized at the M.M. Shemiakin Institute of Bio-Organic Chemistry of the USSR Academy of Sciences and produced by Serva and Ciba Companies, was injected (10μl) into the lateral brain ventricles via implanted cannulas of 0.8mm diameter in doses of 0.15-1.0μg. In another series of experiments a dose of 60 nmole/kg of DSIP was injected intra-arterially to rats and intravenously to rabbits.

RESULTS

Behavioral and Autonomic Manifestations of DSIP Induced Escape Reaction

These experiments were conducted on 20 rabbits with previously implanted VMH electrodes. Animals were immobilized on a board. In this inescapable conflict situation engendering emotional stress, animals were electrostimulated. A catheter was implanted into a femoral artery under local procaine anesthesia for arterial blood pressure (BP) recording. Latent periods of escape response and changes in ECG (2nd standard lead), BP and respiration rate (RR) were recorded during VMH stimulation.

The results show that VMH stimulation using threshold values of electrical current induced RR increases and escape responses after 1.5-2 seconds. This response was accompanied by BP elevations and heart rate (HR) decreases in 30% of rabbits. Sixty percent of animals responded with a two-phase pressor-depressor reaction or with a depressor reaction alone. No vascular changes were observed in 10% of test animals. At 20-30 minutes after i.c.v. DSIP injection in a dose of 100-150ng, latent period (LP) of escape reaction in VMH-stimulated rabbits increased and autonomic components became less marked. Somatoautonomic manifestations of escape reaction disappeared in 9 rabbits though they continued receiving electrical stimulation of the same parameters (Figure 1). Only a 2-3 fold increase in stimulation parameters restored escape reactions. Inhibitory effects of DSIP continued for 2 additional days.

These results prove that DSIP inhibits somatoautonomic symptoms of escape reactions in VMH-stimulated animals. The excitability of brain structures responsible for escape responses reduced significantly.

DSIP Effects During Acute Emotional Stress

Emotional stress was produced in 50 male Wistar rats immobilized in a stereotaxic device by means of a metal plate fixed to the cranium. Animals were aperiodically given VMH stimulation alternating stochastically with electrocutaneous stimulation. Rats had been stimulated for 3 hours using AC threshold values (1.4-6V, 50Hz, 1ms). Each stimulation lasted 30 seconds. BP was recorded using a catheter implanted via the tail artery into the descending aorta.

By the end of the 3rd hour of the conflict situation, only 10 (35.7%) of 28 control animals survived. Five rats died during the 1st hour, 7 rats during the 2nd hour and 6 rats during the 3rd hour.

DSIP in a dose of 60 nmole/kg was injected into the descending aorta of 22 rats. Of these animals, 17 (77.3%) survived under the same acute conflict situation. Two animals died within the 2nd hour and 3 died during the 3rd hour.

Our previous studies revealed that a general population consists of resistant, adapted and stress-susceptible animals. This differentiation was based on their cardiovascular indices (Sudakov *et al*, 1981). DSIP injections decreased by 3-fold the number of animals predisposed to changes in cardiovascular responses under acute emotional stress. Compared to controls, the number of DSIP-treated animals exposed to emotional stress but maintaining BP at a constant level increased by 45%. Simultaneously, vascular responses to episodic VMH stimulation changed significantly. Stress-exposed rats exhibited pressor, pressor-depressor, or depressor responses. Only a few rats had no vascular responses under these conditions. Most of the DSIP-treated rats exposed to 3 hours of emotional stress had no

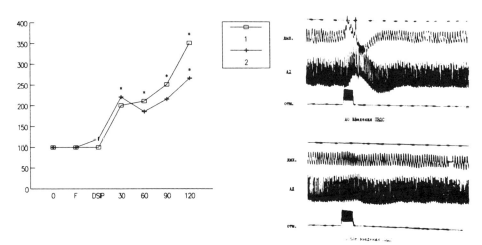

FIGURE 1.   Behavioral and autonomic responses to VMH stimulation in DSIP-treated rabbits.   Site of injection - the lateral brain ventricle.   Left, DSIP-induced changes in the latent period of escape reaction to VMH electrical stimulation (arrow). 1) 176 pmole; 2) 117 pmole; ordinate, LP value in percent of control; abscissa, time in minutes. *, $p < 0.05$ relative to control; Right, autonomic reactions to VMH stimulation before (top) and 30 minutes after (bottom) DSIP injection; 1) respiration; 2) BP; 3) time mark, 1 second,and VMH stimulation mark.

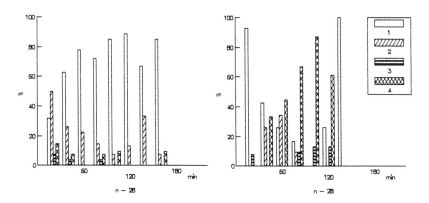

FIGURE 2.   Dynamics of vascular responses in immobilized rats predisposed to emotional stress and exposed to episodic VMH electrostimulation.   Left, before; right, after i.a. DSIP injection, every 20 minutes for 3 hours 1, pressor; 2, pressor-depressor; 3, depressor responses; 4, absence of vascular responses. Ordinate, percent of vascular responses every 20 minutes of the test; abscissa, time in minutes.

vascular responses to VMH stimulation (Figure 2).

These results permitted us to suggest that DSIP increased survival rates in animals exposed to acute emotional stress by decreasing the number of pressor responses capable of summation (Sudakov, 1981).

DSIP Effects on Cortical-Subcortical Relationships

We have tested DSIP effects on neural responses in cerebral cortex of rabbits following electrostimulation of VMH. These results showed that background EEG desynchronization totally disappeared in 6 of 8 DSIP-treated rabbits. Using doses of 0.6-1.0μg, we observed shortening of EEG desynchronization responses and eventually its total inhibition in 17 of 23 VMH stimulated animals 10-60 minutes after DSIP was injected into the lateral brain ventricles.

Coefficients of EEG correlation between the symmetric points in the sensorimotor and occipital cortex were brought to normal values following DSIP treatment. In 7 rabbits with high coefficients of inter-hemispheric correlation initially (C = 0.234±0.06) and in response to VMH stimulation (C = 0.285±0.09), a decrease in these values to 0.129±0.02 and 0.157±0.03, respectively, was observed 30 minutes after DSIP injection. In 3 rabbits with an initial coefficient of 0.156±0.01 and a coefficient of 0.167±0.05 in response to VMH stimulation, administration of DSIP resulted in increases in the coefficients to 0.343±0.05 and 0.317±0.08, respectively. We found that the decrease in coefficients' values was accompanied by inhibition of VMH ascending activating influences on the cortex. The normalizing effect of DSIP on correlations between cortical and subcortical brain structures has been observed by other researchers (Malyshenko *et al*, 1986).

This indicates that decreasing the excitability of VMH defense centers by DSIP also inhibits their ascending activating influences on the cortex apparently normalizing cortical-subcortical relationships under the impact of emotional stress when these influences are abnormally enhanced.

DSIP Effects on Cerebral Blood Flow

Our colleague E. V. Koplik studied DSIP effects on cerebral microcirculation in cats exposed to VMH electrostimulation.

Experiments were conducted on 18 nembutal-anesthetized animals. BP, ECG, and electrocorticogram of the sensorimotor cortex were analyzed. Arterial blood pO_2, pCO_2 and pH were measured during the experiment. Cerebral microcirculation was recorded by the local hydrogen clearance method (Lubbers and Strossek, 1970). Recording of blood flow was made from 4 points of the cortical surface, 1.6 mm in diameter. VMH stimulation (3-5 V, 50 Hz, 1 ms) lasted 3 seconds. DSIP was injected intravenously in a dose of 60 nmole/kg.

Cerebral microcirculation increased by 48±7% 1 hour after DSIP injection. One hour later it was 66±5% higher and one more hour later it increased to 78±10%. At the same time, BP and ECG did not change. DSIP changed significantly cerebral blood flow responses to VMH stimulation and direct stimulation of the cortex. The peptide increased the duration and the amplitude of cerebral blood flow changes. These experiments demonstrate a facilitating effect of DSIP on cerebral microcirculation. DSIP-induced enhancement of cerebral blood flow is a possible factor underlying the increase in the organism's resistance to acute emotional stress.

DSIP Content in the Blood and the Hypothalamus of Rats with Differences in Resistance to Acute Emotional Stress

The dynamics of changes in DSIP content in blood and hypothalamus of 63 male Wistar and 25 August rats were studied by ELISA test prior to and during the development of acute emotional stress. It was induced by fixing rats to a metal plate in a stereotaxic device. For 3 hours, rats were exposed to electrocutaneous stimulation alternating stochastically with VMH electrostimulation.

Our previous studies (Sudakov et al, 1981) demonstrated that Wistar rats were more resistant to emotional stress than August animals. Resistance to emotional stress was differentiated by latent periods of nociceptive responses in a tail-flick test, by behavioral responses in a Varimex activity monitor and also by changes in ECG, BP, RR, and HR when animals were turned with the abdomen up (Sudakov et al, 1990).

Initial content of DSIP-like substances in blood and in hypothalamus of Wistar rats was higher than in August animals by 40.36% and 51.01%, respectively. DSIP content in blood and hypothalamus of resistant animals of both strains was higher than in predisposed rats (Figure 3). DSIP content in blood and hypothalamus

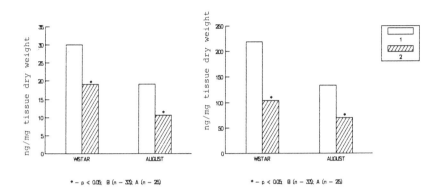

FIGURE 3.   DSIP content in the blood and the hypothalamus of rats with different resistance to emotional stress.   Left, DSIP content in blood; right, DSIP content in hypothalamus; 1) resistant and 2) predisposed to emotional stress.   Ordinate, DSIP content in pg/ml of dried substance; *, significant, p < 0.05, compared with resistant animals of the same strain; n, number of animals.

of Wistar rats resistant to emotional stress exceeded that of resistant August animals by 63.84% and 47.51%, respectively.   These values in predisposed Wistar rats exceeded those in August animals of the same class by 80.9% in blood and 89.06% in hypothalamus.   Therefore, animals resistant to emotional stress have an initially higher content of DSIP-like substances both in blood and hypothalamus.

A 1.5 hour period of emotional stress induced increases in blood and hypothalamic DSIP values both in resistant and predisposed animals.   The increase was more pronounced in blood of resistant rats. Compared to initial levels, DSIP content in blood of resistant rats increased by 171.5% and in hypothalamus by 21.79%, while in predisposed animals the increases were 81.26% and 101.4%, respectively.   Thus, in resistant animals DSIP content increased to a greater extent in blood and in predisposed rats the increase was more marked in the hypothalamus.   By the 3rd hour of stress, DSIP content in blood of resistant animals compared to that after a 1.5-hour stress session decreased by 50.85% and in predisposed rats by 16.24%.

Hypothalamic content continued to increase by 70.18% in resistant animals and by 81.25% in predisposed animals.

Therefore, in the dynamics of acute emotional stress, hypothalamic DSIP content increased. The decrease in its level in blood after an initial rise apparently resulted from a decrease in peripheral DSIP content during the stress protocol.

DSIP-Induced Changes in Content of Cyclic Nucleotides and Neurotransmitters During Acute Emotional Stress

DSIP effects on intracellular mediators (cAMP and cGMP) of the sympathetic and parasympathetic nervous systems were studied in immobilized Wistar rats exposed to aperiodic electrocutaneous stimulation alternating with electrical VMH stimulation. The content of cAMP and cGMP in the VMH, midbrain reticular formation (RF), and the liver was measured by RIA using standard Amersham kits.

In the dynamics of acute emotional stress, unidirectional changes in levels of cyclic nucleotides in cerebral tissues and liver were revealed. By the 30th minute of emotional stress, cAMP levels increased 8-to 10-fold in cerebral structures and 11-fold in the liver. A similar tendency was observed in cGMP values: 6-to 8-fold increases in cerebral structures and a 2-fold rise in liver. By the 60th minute, both cAMP and cGMP levels in cerebral structures and liver decreased slightly but exceeded initial values by 4-to 5-fold in brain and 3-to 4-fold in liver.

The dynamics of changes in the content of biogenic amines and cyclic nucleotides in the VMH and RF under the impact of acute emotional stress was affected by i.a. DSIP injections in a dose of 60nmole/kg. Table 1 shows that by the 30th minute of emotional stress cAMP content in the VMH decreased significantly after DSIP administration (compared to data without DSIP - control). By the 60th minute, histamine content increased while levels of serotonin and norepinephrine decreased. By the 90th minute, levels of histamine and serotonin were elevated and cAMP content was decreased. RF values underwent the following changes: by the 30th minute of emotional stress cGMP levels increased, by the 60th minute levels of serotonin and dopamine fell and by the 90th minute cGMP content increased.

DSIP effects on level of neurotransmitters in brain have been studied further by other researchers (Mendjeritsky et al, 1989) who have found that DSIP increased the level of inhibitory neuro-

TABLE 1.   Changes in biogenic amine and cyclic nucleotide content in the ventromedial hypothalamus (VMH) and reticular formation (RF) in rats after i.p. administration of DSIP.

Substances tested		Control	After DSIP injection (minutes) 30	60	90
Histamine	VMH	0.677±0.08	0.509±0.13*	0.734±0.21*	0.647±0.14
	RF	0.434±0.07	0.281±0.07*	0.329±0.14	0.441±0.09
Serotonin	VMH	0.946±0.12	1.36±0.29*	1.56±0.32*	1.29±0.25*
	RF	0.744±0.14	1.06±0.18*	0.954±0.18*	0.734±0.21
Norepi-	VMH	0.944±0.09	1.37±0.31*	1.21±0.21	0.897±0.23
nephrine	RF	0.469±0.13	0.618±0.14	0.512±0.11	0.578±0.15
Dopamine	VMH	1.27±0.25	1.77±0.56	1.32±0.26	1.04±0.28
	RF	0.792±0.11	1.23±0.21*	0.948±0.12	0.971±0.21
cAMP	VMH	1.45±0.25	2.15±0.45*	1.57±0.41	1.23±0.3
	RF	0.840±0.15	0.72±0.22	2.04±0.52*	1.36±0.25*
cGMP	VMH	0.11±0.06	0.223±0.08*	0.381±0.08*	0.42±0.08*
	RF	0.08±0.16	0.13±0.07*	0.32±0.11*	0.24±0.07*

* differences between test and control groups are significant ($p < 0.05$).

transmitters in the CNS (homocarnosine and GABA) and decreased glutamate content in rats subjected to 6 hours of immobilization.

DSIP-Induced Changes in Substance P Content in the Hypothalamus During Emotional Stress

Hypothalamic substance P (SP) content was determined by RIA in 29 August rats. Emotional stress was induced by immobilizing 19 animals in a group with their tails fixed (Yumatov *et al*, 1988). Of these animals, 8 stressed rats were included in a vehicle-treated group and 11 stressed rats were injected with DSIP in a dose of 60 nmole/kg prior to stress. Another control group not exposed to stress consisted of the 10 remaining animals.

Experimental animals were exposed to a 5 hour period of immobilization stress and at the end of this period SP content in the hypothalamus was 57.8±25.4 pg/mg. Compared to control animals with 89.4±10.0 pg/mg of SP in the hypothalamus, levels decreased by

35.35%. SP levels in the hypothalamus of rats injected with DSIP before immobilization, were 107.9±10.6 pg/mg (Figure 4). Our experiments show that DSIP increases SP levels in the hypothalamus of rats exposed to emotional stress. At the same time single DSIP injections in the above dose prevented stress-induced changes in adrenal and thymus weights (Table 2).

DSIP Effects on Cardiac Function During Acute Emotional Stress

DSIP effects on cardiac stability and development of cardiac dysfunctions under the impact of acute emotional stress were examined in 19 rabbits (Ulyaninsky et al, 1990). Emotional stress in immobilized rabbits was induced by electrocutaneous stimulation alternating stochastically with VMH stimulation. A previously published technique (Lown, 1980) for assessing electrical stability of the heart in rabbits was modified. Animals were implanted with bipolar electrodes into the myocardium of the left ventricle. With discretely increasing intensity of electrical current conveyed to the electrode at the most vulnerable stage of the cardiac cycle, thresholds of ventricular arrhythmias induced by pulse bursts were measured.

DSIP effects on cardiac electrical stability were studied in 7 rabbits exposed to stress. Initial heart rate was 236±3 beats per minute. Initial threshold values of electrical current triggering ventricular arrhythmias were: 17.8±0.7mA for recurrent ventricular extrasystole; 32.2±1.1mA for ventricular tachycardia; and 38.2±0.6mA for ventricular fibrillation.

Emotional stress induced sinus tachycardia and decreased electrical stability of the heart. HR reached 327±7 beats per minute 1 hour after the beginning of stress. Threshold values decreased to 9.7±0.8mA for recurrent ventricular extrasystole, 25.5±0.7mA for ventricular tachycardia, and 31.7±2,0mA for ventricular fibrillation. Two hours after exposure to stress, threshold values for ventricular arrhythmias remained low. Simultaneously, spontaneous ventricular extrasystole developed in 4 of 7 rabbits exposed to stress.

Intravenous DSIP injections (60 nmole/kg) normalized the thresholds of ventricular arrhythmia: 1 hour after injecting DSIP to rabbits with HR of 314±10 beats per minute, threshold values for recurrent ventricular extrasystole were 16.1±0.8mA (p<0.01); for ventricular tachycardia 28.8±0.9mA (p<0.05); and for ventricular fibrillation 37.3±0.7mA (p<0.05). Enhanced cardiac electrical stability

TABLE 2. Adrenal and thymus weights of DSIP-treated rats exposed to acute emotional stress.

Weight	Control	Stress	Stress + DSIP
Adrenals (mg)	40.0±3.50	49.9±2.1	44.2±3.0
% of control		125%	110%
Thymus (mg)	244±41	79.50±18.0	161.1±32.1
% of control		33%	66%

in 4 DSIP-treated animals resulted in a decrease and eventually an elimination of spontaneous ventricular extrasystole. This indicates that DSIP had a stabilizing and antiarrhythmic effect on cardiac function during emotional stress.

Marked sinus tachycardia (to 354±10 beats per minute) and ventricular extrasystoles developed in 12 stress-exposed animals. An increase in T-wave and a shift in ST segment below the isoelectric line were observed in some rabbits. With the development of emotional stress the number of ventricular extrasystoles increased and by the 40–60th minutes became stable equalling 13.3±1.6 (every three minutes).

Intravenous DSIP injection (60 nmole/kg) alleviated and in some cases eliminated ventricular extrasystole. The number of extrasystoles decreased significantly within 10 minutes after DSIP injection. One hour after DSIP administration, ventricular extrasystoles were blocked in 4 rabbits. This antiarrhythmic effect of DSIP persisted for 3-5 days in the majority of animals. These results demonstrate that cardiac dysfunctions induced by acute emotional stress may be normalized by DSIP due to its marked anti-arrhythmic effects.

CONCLUSIONS

These experiments reveal that DSIP has marked stress-coping effects. This was evident in the decrease of excitability of brain structures engendering negative emotional reactions. At the same time, DSIP inhibited activating influences ascending from the VMH to the cortex and blocked somatoautonomic manifestations of the escape reaction to VMH electrical stimulation. Survival rates in DSIP-treated animals exposed to acute conflict situations increased due to a number of

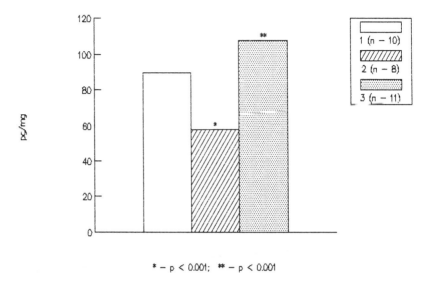

*− p < 0.001; ** − p < 0.001

FIGURE 4. Changes in SP content in the hypothalamus of DSIP-treated animals. 1) controls; 2) animals exposed to conflict; 3) animals treated with DSIP 15 minute before exposure to conflict (single injections of 60 nmole/kg); ordinate, SP content in the hypothalamus (pg/mg); n, number of animals; *, significant, $p < 0.001$, compared with controls; **, significant, $p < 0.001$, compared with control and stress-exposed animals.

factors: decreases in excitability of emotiogenic brain structures, stabilization of cerebral circulation, decreases in the number of pressor responses to VMH stimulation and antiarrhythmic effects.

Our experience have shown that an animal's resistance to emotional stress is determined by a higher initial level of DSIP in blood and hypothalamic tissue. DSIP normalized stress-induced decreases in the content of substance P in hypothalamus and changed the content of cyclic nucleotides and neurotransmitters in various brain structures.

Our results demonstrate that increased levels of DSIP and other neurotransmitters in the hypothalamus are an indication of an animal's resistance to emotional stress. The decrease in hypothalamic DSIP and SP content is associated with decreased resistance of animals to stress and finally in mortality. Additional DSIP injections to stress-exposed animals improved their resistance due to an increase in levels of SP and other neuropeptides in the hypothalamus.

REFERENCES

Bjartell, A. (1990). Delta sleep inducing peptide: a mammalian regulatory peptide. Localization, immunochemical characterization, biosynthesis and functional aspects, p. 217. Lund, Sweden: Grahns Boktryckeri.

Graf, M.V., and Kastin, A.J. (1986). Delta sleep inducing peptide (DSIP): an update. *Peptides* 7, 1165-1187.

Ivatshenko, O.N., Bazhanov, N.N., Bolshakova, T.D., and Sudakov, K.V. (1976). The use of electrosleep to alleviate emotional tension in dentistry. *Stomatology* 2, 50-54.

Kovalson, V.M. (1986). Humoral regulation of sleep. In: "Results of Science and Technology," Series "Physiology of Man and Animal," pp. 3-58, Vol. 31. Moscow: VINITI.

Kovalson, V.M., and Tsybulsky, V.L. (1980). Does synthetic peptide inducing delta sleep have hypnogenic properties? *Zh. Vyssh. Nerv. Deyatelnosty* 30, 1064-1066.

Lown, B. (1977). Role of higher nervous activity in sudden death. In: "Sudden Death", *Proceedings of USA-USSR 1st Joint Symposium*, pp. 387-404. Washington: DHEW Publication No. (NIH) 78-1470 U.S. Government Printing Office.

Lubbers, D.W., and Strossek, K. (1970). Quantitative Bestimmung der localen Durchblutieng durch elekrochemisch im Gewebe erragten Wasserstoff. *Naturwissensch* 57, 31.

Malyshenko, N.M., Kashtanov, S.I., Broshkin, S.V., and Mikhaleva, I.I. (1986). Comparative analysis of changes in cerebral intercentral relationships under DSIP and ACTH effect. *Fiziol. Zh. SSSR* 72, 1, 116-132.

Medvedjev, V.I., Bakharev, V.D., Sargsian, I.S., and Mikhaleva, I.I. (1983). Effects of delta sleep inducing peptide and its analogs on EEG in rabbits in normal conditions and under sleep deprivation and on processes of learning in rats. *Fiziol. Zh. SSSR* 1, 3-9.

Mejerson, F.V., Sukhikh, G.T., Mikhaleva, I.I., Sviriaev, V.I., and Ivanov, V.T. (1984). Prevention of stress-induced decrease in natural killers' activity using sodium oxibutyrate and delta sleep inducing peptide. *Doklady AN SSSR* 274, 482-484.

Mendjeritsky, D.M., Uskova, N.I., and Charayan, I.A. (1989). Effect of delta sleep inducing peptide on cerebral GABA system in rats with hypokinesia. *Neurokhimia* 7, 312-313.

Monnier, M., and Schoenenberger, O.A. (1976). Characterization, sequence, synthesis and specificity of a delta (EEG) sleeping inducing peptide. In: W.P. Koella, and P. Levin (Eds.), "Sleep," pp. 257-263. Basel: Karger.

Monnier, M., Dudler, L., Geachter, H., and Schoenenberger, C.A. (1977). Delta sleep inducing peptide (DSIP); EEG and motor activity in rabbits following intravenous administration. *Neuroscience Letters* 6, 9-13.

Schoenenberger, C.A., and Schneider-Helmert, D. (1983). Psychophysiological functions of DSIP. *Trends in Pharmacology* 4, 307-318.

Sudakov, K.V. (1981). "Systems Mechanisms of Emotional Stress". Moscow: Meditsina.

Sudakov, K.V., Banschikov, B.M., Kulikova, I.L., and Arsentiev, D.A. (1972). Behavior and electroencephalographic reactions in electrosleep. In: *Third International*

Symposium on Electrosleep and Electroanesthesia, pp. 11-12. Varna.

Sudakov, K.V., Yumatov, E.A., and Dushkin, V.A. (1981). Genetic and individual differences of cardiovascular dysfunctions in rats exposed to experimental emotional stress. *Vestnik AMN SSSR* **12**, 32-39.

Sudakov, K.V., Koplik, E.V., Salieva, R.M., and Kamenov, Z.A. (1990). Prognostic criteria of resistance to emotional stress. In: "Emotional Stress. Physiological and Medicosocial Aspects," pp. 12-19. Kharkov: Prapor.

Tobler, J., and Borbely, A.B. (1980). Effects of delta sleep inducing peptide (DSIP) and arginine vasotocin (AVT) on sleep and motor activity in the rat. *Waking and Sleeping* **4**, 139- 153.

Ulyaninsky, L.S., Ivanov, V.T., Mikhaleva, I.I., and Sudakov, K.V. (1990). Delta sleep inducing peptide as a modulator of cardiac activity: theoretical recommendations for practical workers. *Zh. Kosmichesk. i Aviatsionnaya Meditsina* **3**, 23-28.

Yumatov, E.A., Pevtsova, E.I., and Mezentseva, L.N. (1988). Physiologically adequate experimental model of aggression and emotional stress. *Zh. Vyssh. Nervn. Deyatelnosti* **38**, 350-354.

Stress: Neuroendocrine and Molecular Approaches
Edited by R. Kvetnansky, R. McCarty and J. Axelrod

1992 Gordon and Breach Science
Publishers S.A., New York, USA.
Photocopying permitted by license only.

NORADRENERGIC AND NEUROENDOCRINE FUNCTION IN A CHRONIC WALKING STRESS-INDUCED MODEL OF DEPRESSION IN RATS

I. Kitayama[1], S. Kawguchi[1], S. Murase[2], M. Otani[1], M. Takayama[1], T. Nakamura[1], T. Komoiri[1], J. Nomura[1], N. Hatotani[1] and K. Fuxe[3]

[1]Department of Psychiatry and [2]Department of Physiology, Mie University School of Medicine, Tsu, Japan

[3]Department of Histology and Neurobiology, Karolinska Institute, Stockholm, Sweden

INTRODUCTION

Clinically one often encounters many depressed patients in whom a stressful situation precedes the onset of illness. It has been suggested that a chronic intolerable stressor might be an important precipitating factor inducing depressive illness. By exposing female rats to chronic walking stress ("forced running stress"), we produced an animal model which was behaviorally, hormonally and therapeutically analogous to human depression. In our initial studies, we reported a number of central alterations in rats exposed to chronic walking stress, including increased catecholamine (CA) fluorescence, decreased tyrosine hydroxylase (TH) immunofluorescence and decreased turnover rates of CAs in norepinephrine (NE) neurons (Kitayama *et al*, 1984). Additional analyses of noradrenergic and neuroendocrine functions in this model are presented in this chapter.

73

STRESS PROCEDURES AND CHARACTERISTICS
OF THE MODEL

The procedure for producing the depression model was similar to that described before except the rotation of the drum for forced walking was discontinued at a rectal temperature of $34°C$. The sequence of stress until exhaustion and rest for 24 hours was repeated three times, so that the total duration of the stressor was 12 ± 2 days. After stress exposure, approximately 50% of the rats showed a continuous inactivity in spontaneous wheel running and a disappearance of running wheel rhythmicity corresponding to the estrus cycle. These rats were termed "depression model rats". The other half of the animals, however, resumed previous activity and rhythmicity within two weeks. These rats were termed "spontaneous recovery rats". To know the time course of noradrenergic and neuroendocrine alterations, additional rats were exposed to short-term stress for 1 to 4 days.

In addition to inactivity and lack of rhythmicity, the depression model rats revealed a high level of serum 11-hydroxycorticosteroids (11-OHCS) which is compatible with a hypercortisolemia often seen in human depression. Treatment of the depression model rats with the tricyclic antidepressant, imipramine, 8 mg/kg (Figure 1) or the tetracyclic antidepressant, setiptiline, 0.5 mg/kg for 20 days restored activity levels as well as rhythmicity.

Noradrenergic Function

Activity of TH.   TH activity was measured in the locus coeruleus (LC) and hippocampus. The activity in both regions increased within 1-2 days of short-term stress, but decreased in the depression model rats which had been inactive for 2 weeks after long-term stress (Figure 2). TH activity was not altered in the spontaneous recovery rats.

Electrophysiological activity of LC neurons.   Spontaneous firing rate of LC neurons decreased significantly in the depression model rats, but did not change in rats stressed for 3 days and in recovery rats (Figure 3.)

Ultrastructural changes.   Electron microscopic observations of LC neurons of depression model rats revealed a low density area and a destruction of endoplasmic reticulum (ER) membrane in cell bodies

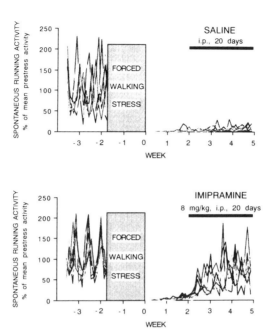

FIGURE 1. Spontaneous running activity of depression model rats before and after long-term forced walking stress as well as the effect of chronic treatment of imipramine and saline on activity.

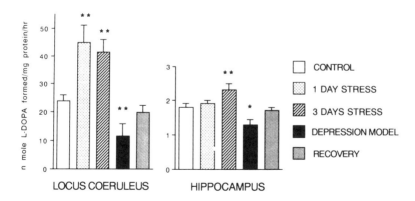

FIGURE 2. Activity of tyrosine hydroxylase in the locus coeruleus and the hippocampus of stressed rats.

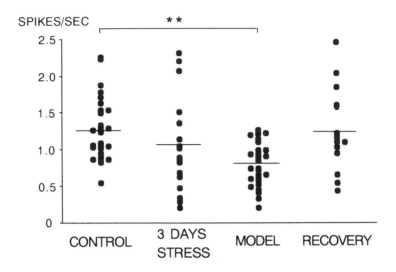

FIGURE 3.   Spontaneous firing rate of locus coeruleus neurons in stressed rats.

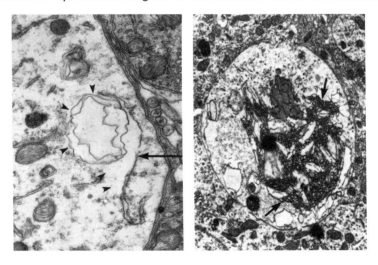

FIGURE 4.   Electron microscopic photographs of a locus coeruleus neuron of a depression model rat.   A low density area (surrounded by arrowheads) and a destruction of endoplasmic reticulum (ER) membrane (large arrow) in the cell body (x24,000) and an aggregation of microtubules or ER (small arrow) in the dendrite (x12,000) are evident.

as well as an aggregation of microtubules or ER in dendrites (Figure 4). No remarkable structural changes were observed in the spontaneous recovery rats.

β-Adrenergic receptors. ^{3}H-dihydroalprenolol (^{3}H-DHA) binding was examined in discrete brain regions. No change of B_{max} was found in the hippocampus and the cerebral cortex of rats exposed to 1 day of stress. However, with 3 days of stress, long-term stress, and even the stage of depression, B_{max} continued to be decreased in both regions (Figure 5). It was also decreased in the cerebral cortex of the spontaneous recovery rats, but not in the hippocampus.

In view of increased TH activity and increased turnover rate of NE, it seems likely that the activity of NE neurons was increased during short-term stress. When the stressor was prolonged, however, TH activity, the turnover rate of NE and the firing rate of NE neurons decreased and newly synthesized NE accumulated in the neurons. Thus, neuronal function converted from an increase to a decrease in the course of long-term stress and remained decreased in the depression model rats. Even degenerative changes were partly observed in NE neurons. Moreover, noradrenergic transmission appeared to be diminished not only by decreased presynaptic function but also by reduced numbers of β-adrenergic receptor binding sites.

Neuroendocrine Function in the HPA Axis

Adrenal weights of the 2 day stressed rats, the depression model rats and the spontaneous recovery rats were all increased when they were measured immediately after perfusion. Immunoreactivities (IRs) of CRH, ACTH, glucocorticoid receptor (GR) and TH were measured by means of immunocyto-chemistry and microdensitometry. CRH-IR was not altered in the PVN of stressed rats. However, it was markedly reduced in the ME of 2 day stressed rats and depression model rats (Figure 6). ACTH-IR in the anterior pituitary was significantly reduced in the 2 day stressed rats but not in depression model rats. GR- and TH-IRs were not altered in the PVN and LC of any stressed rats.

It is generally accepted that CRH and ACTH are hypersecreted in response to acute stress. Thus the reduction of CRH-IR in the ME in both short-term and long-term stress as well as the reduction of ACTH-IR in the pituitary during short-term stress may be an outcome

I. KITAYAMA *et al*

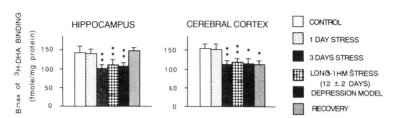

FIGURE 5. B_{max} of ^{3}H-DHA binding (property of ß-adrenergic receptor) in the hippocampus and the cerebral cortex of stressed rats.

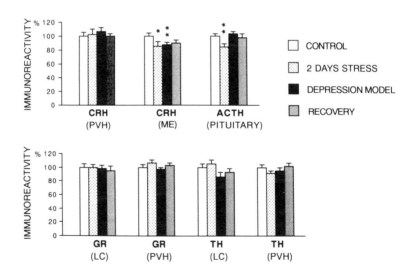

FIGURE 6.   Immunoreactivities of corticotropin releasing hormone (CRH), adrenocorticotropic hormone (ACTH), glucocorticoid receptor (GR) and tyrosine hydroxylase (TH). PVN: paraventricular nucleus, ME: median eminence, LC: locus coeruleus.

of their excessive release from axon terminals and corticotrophs, respectively. The unaltered pituitary ACTH-IR in depression model animals may imply that ACTH secretion could be adaptively normalized after receiving continuous CRH stimulation. These speculations should be confirmed by further investigation.

Kendall's correlation study disclosed that ACTH-IR in the pituitary had a trend of inverse correlation with adrenal weight. This relation seems to be reasonable when one takes into account ACTH hypersecretion and adrenocortical hypertrophy brought about under the stress conditions. ACTH-IR in the pituitary inversely correlated with GR-IR in the PVN. This is interpreted as hypersecretion of ACTH elevates serum corticosterone, which in turn increases GR-IR in cell nuclei of the PVN by a feedback mechanism. ACTH-IR in the pituitary also inversely correlated with TH-IR in the LC. In the PVN there was a positive correlation between TH-IR in NE nerve terminals and CRH-IR in the CRH-producing cells. These two correlations suggest an excitatory action of NE neurons on synthesis and release of CRH. This hypothesis is in line with recent reports hypothesizing an excitatory rather than an inhibitory role of NE neurons on CRH synthesis.

SUMMARY AND CONCLUSIONS

During short-term stress, the function of ascending NE neurons, hypothalamic CRH, pituitary ACTH, circulating glucocorticoid and their negative feedback systems are thought to be activated. When the duration of the stressor is prolonged, NE neurons may be exhausted and ACTH secretion normalized. However, hypothalamic CRH and adrenal glucocorticoids may still be hypersecreted. Why does the CRH hypersecretion occur, when stimulatory NE neurons are hypofunctioning? It is interesting to note in a few reports that CRH synthesis is regulated by such biphasic transmissions of NE neurons as α stimulatory and β inhibitory. One of the possibilities is that β-adrenergic receptor down-regulation in the depression model animals may lead to disinhibition of CRH synthesis. Another candidate may be a dysregulation of feedback control of CRH synthesis by glucocorticoids, particularly a maladaptation of GR. Our previous study revealed that GR immunoreactivity in cell nuclei was increased by chronic immobilization stress (Kitayama et al, 1989), but was not increased in depression model rats. It might be promising to study how GR responds to circulating glucocorticoids which are continuously elevated in the depression model animals.

REFERENCES

Kitayama, I, Cintra, A., Janson, A.M., Fuxe, K., Agnati, L.F., Eneroth, P., Aronsson, M., Härfstrand, A., Steinbush, H.W.M., Visser, T.J., Goldstein, M., Vale, W., and Gustafsson, J-Å. (1989).   Chronic immobilization stress: evidence for decreases of 5-hydroxytryptamine immunoreactivity and for increases of glucocorticoid receptor immunoreactivity in various brain regions of the male rat. *Journal of Neural Transmission* **77**, 93-130.

Kitayama, I., Koishizawa, M., Nomura, J., Hatotani, N., and Nagatsu, I. (1984). Changes in behavior and central catecholamines in rats after long-term severe stress. In E. Usdin, R. Kvetnansky and J. Axelrod (Eds.). "Stress: The Role of Catecholamines and Other Neurotransmitters," pp. 125-135. New York: Gordon and Breach.

Stress: Neuroendocrine and Molecular Approaches
Edited by R. Kvetnansky, R. McCarty and J. Axelrod

1992 Gordon and Breach Science
Publishers S.A., New York, USA.
Photocopying permitted by license only.

SOCIAL DOMINANCE: ROLE OF GENOTYPE AND BRAIN NOREPINEPHRINE

L. I. Serova and O. N. Kozlova

Institute of Cytology and Genetics, Siberian Branch of the
Academy of Science of the USSR, Novosibirsk, USSR

INTRODUCTION

Catecholamines (CA) are involved in the control of endocrine function during stress and in mediating reproductive (Rodriguez *et al*, 1984) and aggressive activities (Gottfries, 1989). These factors are very important for dominance relations (Benus *et al*, 1990).

Current knowledge relating to the role of endogenous CAs in regulation of social behavior is fragmentary, and attempts to elucidate brain CA influences on dominant-subordinant behaviors are of fundamental importance. An advantageous approach to address this issue is the use of genetically defined animals. Moreover, genetic differences in numbers of CA neurons in brain have been revealed (Reis, 1983).

This chapter describes the role of NE systems in the establishment of dominant behaviors in inbred male mice in micropopulations.

METHODS

Mature male mice of 6 strains (A/He, C57Bl/6j, CBA/Lac, DD, YT, and PT) weighing 22-23 grams were used. Six days before the experiment, animals were caged individually. To study hierarchic relations between mice, 6 male mice (one from each genotype) were grouped in a special clear population cage (74x44x17 cm). Social behavior testing was carried out in the morning immediately after grouping and one day later with three 20 minute observations at 20 minute intervals. The male having the maximum number of victories in social encounters was considered to be the dominant individual. Movement about the cage and aggregation with other mice were taken into account as well. The other animals showed much less social activity and their social rank was not detailed further and they were classified as subordinants.

Animals were killed by decapitation after 6 days of isolation (control) and 1, 7, and 48 hours after grouping. Levels of NE and VMA were assayed by HPLC with electrochemical detection. The HPLC system included a Pump and Bioanalytical Detector by Waters and a Shimadzu integrator. Tissue samples were homogenized in 0.1M perchloric acid and separated on a Zorbax ODS-5 μm reverse-phase column. TH activity in hypothalamus was assayed by the method of Yamauchi and Fujiawa (1987). Data were analyzed using ANOVA, Student's t test, and criterion X^2.

RESULTS AND DISCUSSION

Interactive effects of genotype may have an essential influence on social behavior in groups (Hughes, 1989). This was the main reason why genetically heterogeneous populations should be used for analysis of social dominance. Tests of social behavior showed that the

TABLE 1.   Distribution of different inbred strains on hierarchical classes.

Social rank	Number of mice						
	PT	C57Bl/6j	DD	YT	A/He	CBA/Lac	X_5^2
Dominate	72	15	15	5	2	1	202[a]
Subordinate	38	95	95	105	108	109	52[a]

[a]p<0.01

distribution of males in hierarchical classes depended on genotype and was nonrandom (Table 1). The relative number of high social rank animals was maximum is PT strain mice and did not differ from equal probability levels in DD and C57Bl/6j strains. Males of the A/He, CBA/Lac and YT mouse strains expressed dominant behavior very rarely.

To study the role of NE in control of social dominance, we employed two approaches. In one, we examined whether brain levels of NE were correlated with genetically determined dominance relations. The data point to an influence of genotype on NE content in olfactory bulb, hypothalamus, hippocampus, striatum and brain stem (Table 2). NE levels were minimal in most brain areas of C57Bl/6j

TABLE 2. NE levels (μg/g tissue) in mice of different strains.

Strain	Olfactory bulb	Hypothalamus	Hippocampus
A/He	0.35±0.03	1.10±0.19	0.76±0.13
CBA/Lac	0.42±0.09	0.94±0.12	0.49±0.07
C57Bl/6j	0.23±0.02	0.96±0.15	0.62±0.08
DD	0.35±0.06	1.53±0.24	0.73±0.17
YT	0.29±0.07	1.76±0.29	0.49±0.06
PT	0.55±0.07	2.08±0.31	0.55±0.08
ANOVA	$F(5,42)=2.26$	$F(5,41)=3.16$	$F(5,40)=3.14$
P Value	<0.05	<0.025	<0.025

Strain	Brain stem	Cortex	Striatum
A/He	0.81±0.09	0.33±0.06	0.38±0.04
CBA/Lac	0.70±0.12	0.21±0.05	0.30±0.05
C57Bl/6j	0.64±0.10	0.21±0.05	0.97±0.18
DD	0.82±0.09	0.14±0.03	1.00±0.18
YT	0.82±0.011	0.19±0.06	0.40±0.09
PT	1.10±0.05	0.16±0.03	0.95±0.10
ANOVA	$F(5,40)=2.73$	$F(5,40)=1.65$	$F(5,42)=2.3$
P Value	<0.05	>0.05	<0.05

and CBA/Lac strains and maximal in PT mice. PT mice had the highest brain NE levels and the highest percentage of dominant animals. On the contrary, very few CBA/Lac mice had dominant social ranks in micropopulations and their NE levels in brain were quite low. We suggest that genotype-dependent capacity for social dominance is related to functions of central NE neurons. This suggestion was confirmed by intra- and interstrain rank correlations between these two features (Serova et al, 1989).

The second approach was applied to study the role of NE during formation of hierarchical structure in micropopulations under social stress. In the first step, we investigated the hypothalamic NE system in mice as related to dominance. The hypothalamic NE system is involved in regulation of general neuroendocrine reaction during stress and NE neurons are activated in response to attacks (Kruk et al, 1984). PT mice showed significantly higher activity of TH when compared with CBA/Lac under control conditions and 1 hour after grouping (Table 3).

Moreover, the turnover of NE in the medial hypothalamus of these two strains was different (Table 4). VMA levels in dominant PT mice were elevated 1 hour after 6 mice were placed in a population cage. In contrast, VMA levels decreased in submissive CBA/Lac mice. In both cases, NE content was unchanged. Taking into account TH activity and turnover of NE, we conclude that activation of the hypothalamic NE system depends on the hierarchical state of male mice. Social stress in mice predisposed to dominance results in activation of NE system, whereas in subordinate mice the response is the opposite. TH increases in CBA/Lac are probably connected with

TABLE 3. Hypothalamic TH activity during establishment of hierarchical structure of populations.

Strains	TH activity nmol DOPA/minute mg protein			
	Control	1 Hour	7 Hours	48 Hours
PT	19.1 ± 2.8	30.4 ± 0.5^b	16.1 ± 2.9	14.0 ± 2.2
CBA/Lac	9.5 ± 1.3	13.0 ± 1.2^a	7.0 ± 1.1	10.5 ± 1.3
P Value	<0.05	<0.001	<0.001	

$^a p<0.05$, $^b p<0.001$ versus controls of the same strain.

TABLE 4. Turnover of NE in medial hypothalamus one hour after social stress.

Strains	NE and VMA (% control)	
	NE	VMA
PT	-20	$+100^a$
CBA/Lac	-10	-100^a

^{a}p < 0.001 versus controls of the same strain.

the dopaminergic system. NE system activation seems to link the dominant phenotype with behavioral and endocrine systems. It would be an over simplification to consider this system as the only one important in the regulation of dominant social behaviors. However, NE neurons are involved in tuning different neurotransmitter-specific neurons in the CNS (Oades, 1985).

REFERENCES

Benus, R.F., Daas, S., Roolhaas, J.M., and Ootmerssen, G.A. van. (1990). Routine information and flexibility in social and nonsocial behavior of aggressive and nonaggressive male mice. *Behaviour* 112, 176-193.

Gottfries, C.G. (1983). Neurotransmitters in the brain. *Arzneim.-Forsch./Drug Research* 39, 1025-1028.

Hughes, A.L. (1989). Interaction between strains in the social relations of inbred mice. *Behavior Genetics* 19, 685-700.

Kruk, M.P., Van der Laan, C.F., and Meelis, W. (1984). Brain-stimulation induced agonistic behavior: novel paradigm in ethopharmacological aggression research. In: K. Miczek., M. Kruk., and B. Oliver (Eds.), "Ethopharmacological Aggression Research," pp. 157-177. New York: Alan R. Liss.

Oades, R. (1985). The role of noradrenaline in tuning and dopamine in switching between signals in the CNS. *Neuroscience and Biobehavioral Reviews* 9, 261-280.

Reis, D.T. (1983). Genetic differences in numbers of chemically specified neurons in brain: A possible biological substrate for variations in onset of symptoms of cerebral aging in man. In: D. Samuel *et al.* (Eds), "Aging of the Brain," pp. 257-269. New York: Raven Press.

Rodriguez, M., Castro, R., Hernandez, G., and Mas, M. (1984). Different role of catecholaminergic and serotoninergic neurons of the medial forebrain bundle on male rat sexual behavior. *Physiology and Behavior* 39, 5-11.

Serova, L.I., Osadchuk, A.V., and Naumenko, E.B. (1989). Genetic-neurochemical analysis of social dominance in male laboratory mice. The role of brain catecholamines. *Genetics* **25**, 691-698.

Yamauchi, T., and Fujisawa, H. (1978). A simple and sensitive fluorometric assay for tyrosine hydroxylase. *Analytical Biochemistry* **89**, 143-150.

Stress: Neuroendocrine and Molecular Approaches
Edited by R. Kvetnansky, R. McCarty and J. Axelrod

1992 Gordon and Breach Science
Publishers S.A., New York, USA.

EFFECTS OF PROLONGED STRESS ON THE ACTIVITY OF MONOAMINERGIC SYSTEMS IN THE ANTERIOR HYPOTHALAMIC AREA OF EWES DURING THE EARLY AND MID PERIODS OF SEASONAL ANESTRUS

L. Chomicka and F. Przekop

Institute of Animal Physiology and Nutrition,
Polish Academy of Sciences, 05-110 Jablonna, Poland

INTRODUCTION

It has been recently found that in ewes, brain dopamine (DA) and serotonin (5-HT) modulate the release of hormones of the hypothalamo-pituitary-ovarian axis (HPOA) (Dailey *et al*, 1987) and that the activity of the HPOA is affected by repetitive stress (Przekop *et al*, 1984; 1988). These data suggest that brain DA and 5-HT systems may be involved in the neuroendocrine responses of ewes to prolonged stress. It is not yet known, however, if prolonged stress affects brain DA and 5-HT in ewes. The present study was undertaken to evaluate the effects of repetitive stress on the activity of DA and 5-HT systems in the anterior hypothalamic area (AHA) in anestrus ewes. The AHA was examined due to evidence that this brain region participates in the control of HPOA activity in ewes (Domanski *et al*, 1980). Considering the fact that in females, the effects of prolonged stress on the HPOA (Przekop *et al*, 1984) and brain 5-HT (Chomicka, 1984) depend on the phase of the reproductive cycle, and in sheep the release of luteinizing hormone varies during the course of seasonal anestrus (Jackson *et al*,

1979), ewes were examined during the early and mid-periods of anestrus.

MATERIALS AND METHODS

Polish Merino ewes 2-2.5 years old were exposed to stress during the early (beginning of March) or middle stages (end of April) of anestrus. The animals were stressed by applying electric footshocks for 3 consecutive days, from 0100-1000 hours according to a procedure previously described (Przekop *et al*, 1984). Immediately after the third exposure to footshock stress, ewes were decapitated. Control ewes were sacrificed at the same time of day. The AHA was dissected out, homogenized in an ice-cold mobile phase (1:10; w:v), and subsequently centrifuged at 14,000 g for 15 minutes at 4°C. A 10 μl aliquot of filtered supernatant was injected directly into the HPLC system (LKB, Sweden) to determine the concentrations of DA, 5-HT, DOPAC, HVA and 5-HIAA. The HPLC system was equipped with a Supelcosil LC-18 column (75 x 4.6 mm; 3 μm particles; protected by a guard column - 20 x 4.6 mm; 5 μm particles) and an electrochemical detector (LKB) with a glassy carbon working electrode set at a potential of 0.65 V versus an Ag/AgCl reference electrode. The HPLC mobile phase consisted of acetate buffer (0.1 M; pH = 3.60), EDTA disodium salt (1.5 mM/l), NaCl (0.010 M/l), heptane sulfonic acid sodium salt (3 mM/l) and 5% (v:v) methanol. Flow rate was set at 0.7 ml/minutes. Peak heights of the eluted substances were compared to those of pure standards. Statistical analyses were conducted using a Kruskal-Wallis one-way ANOVA by ranks, followed by a two-tail Wilcoxon rank sum test. Differences were considered significant if the probability of error was less than 5%.

RESULTS

DA and DOPAC levels and DOPAC/DA ratios were higher (Figure 1 - b versus a), whereas levels of 5-HT and 5-HIAA were lower (Figure 1 - b' versus a') in the AHA of control ewes during mid anestrus compared to levels found in control ewes during early anestrus. The mean concentrations of HVA (Figure 1 - b versus a) and 5-HIAA/5-HT ratios (Figure 1 - b' versus a') in the AHA were

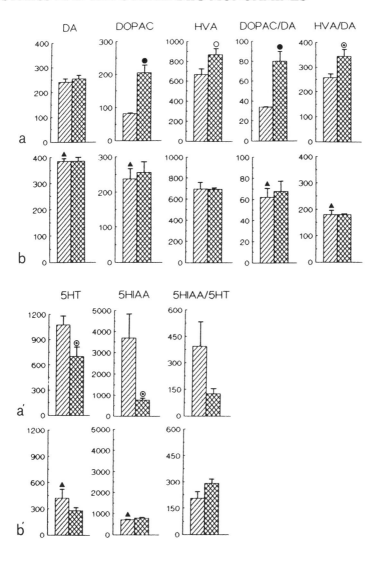

FIGURE 1. Mean levels ± SEM of DA, 5-HT and their metabolites (pg/mg wet weight tissue) and the ratios of concentrations of metabolite/amine (x 100; in %) in the AHA in control ewes (single cross-hatched bars) and in ewes exposed to footshock stress (double cross-hatched bars) during early (a, a') or mid-anestrus (b, b'). ▲ , different from control ewes during the early anestrus, P< 0.008. ,⊙ ,O , different from control ewes during identical anestrus period, P<0.008, P<0.032, P<0.05, respectively. For abbreviations see text.

similar in control sheep during the early and mid-period of anestrus.

Footshock stress was attended by elevations in DOPAC and HVA levels and by increases in DOPAC/DA and HVA/DA ratios in the AHA of ewes during early anestrus (Figure 1a), but not in ewes during the mid-anestrus (Figure 1b). Footshock stress did not markedly affect DA concentrations in the AHA of anestrus ewes (Figure 1a,b). Footshock stress led to decreases in 5-HT and 5-HIAA levels in the AHA of ewes during early anestrus (Figure 1a'), but not in ewes during the mid-anestrus (Figure 1b'). 5-HIAA/5-HT ratios were similar in control and stressed ewes (Figure 1a', b').

DISCUSSION

In ewes, the activity of 5-HT and DA systems in the AHA varied during the course of seasonal anestrus, as indicated by differences in the levels of 5-HT, DA and their metabolites in the AHA of the ewes examined during early and mid-anestrus. Based on the assumption that 5-HIAA represents primarily intraneuronal metabolism of nonreleased 5-HT and that 5-HIAA levels reflect 5-HT synthesis (Kuhn *et al*, 1986), our findings of decreased levels of 5-HIAA in the AHA of ewes during mid-anestrus may reflect a depletion of 5-HT due to a greatly decreased synthesis of the amine. This supposition is supported by the observation that 5-HT levels in the AHA were lower in ewes during mid-anestrus than those found in ewes during early anestrus. Increases in DOPAC, which occurred without elevations in HVA, point to an enhancement of intraneuronal metabolism of DA that was not related to the release of the amine (Commissiong, 1985). Therefore, increased DOPAC concentrations and DOPAC/DA ratios in the AHA of ewes during mid-anestrus imply an enhancement of the catabolism of non-released DA in the AHA of ewes during this period of anestrus. Concomitant increases in DA levels in the AHA of these ewes suggest that the enhanced catabolism of DA may have resulted from increased synthesis of the amine.

Footshock stress stimulated and suppressed the activity of DA and 5-HT systems, respectively, in the AHA of ewes during early anestrus. These effects were reflected by concomitant increases in DOPAC and HVA levels and DOPAC/DA and HVA/DA ratios (that occurred with no corresponding declines in DA levels) and decreases in 5-HT and 5-HIAA concentrations in the AHA. In contrast, no clear influence of footshock stress on DA and 5-HT systems in the AHA was found in ewes during mid-anestrus. This observation suggests at least two explanations: 1) some compensatory changes in the activity of DA and 5-HT systems may have occurred in ewes subjected to prolonged stress during mid-anestrus or 2) factors which determined differences in the activity of DA and 5-HT systems in the AHA during the course of anestrus may have affected the responses of these systems to prolonged stress.

In summary, the results of this study indicate that in seasonally anestrus ewes, the basal activity and the responsiveness of DA and 5-HT systems in the AHA to prolonged stress depend on the period of anestrus. In addition, prolonged stress stimulates and inhibits the activity of DA and 5-HT systems, respectively, in the AHA of ewes during early anestrus.

REFERENCES

Chomicka, L.K. (1984). Effects of stress on the activity of the pituitary-gonadal axis and brain serotonin. *Acta Physiologica Polonica Suppl.* **27**, 103-112.

Commissiong, J.W. (1985). Monoamine metabolites: Their relationship and lack of relationship to monoaminergic neuronal activity. *Biochemical Pharmacology* **34**, 1127-1131.

Dailey, R.A., Deaver, D.R., and Goodman, R.L. (1987). Neurotransmitter regulation of luteinizing hormone and prolactjn secretion. *Journal of Reproduction and Fertility Suppl.* **34**, 17-26.

Domanski, E., Przekop, F., and Polkowska, J. (1980). Hypothalamic centers involved in the control of gonadotrophin secretion. *Journal of Reproduction and Fertility* **58**, 493-499.

Jackson, G.L., and Davis, S.L. (1979). Comparison of luteinizing hormone and prolactin levels in cycling and anestrus ewes. *Neuroendocrinology* **28**, 256-263.

Kuhn, D.M., Wolf, W.A., and Youdim, M.B.H. (1986). Serotonin neurochemistry revised: A new look at some old axioms. *Neurochemistry International* **2**, 141-154.

Przekop, F., Wolinska-Witort, E., Mateusiak, K., Sadowski, B., and Domanski, E. (1984). The effect of prolonged stress on the estrous cycles and prolactin secretion in sheep. *Animal Reproduction Science* **7**, 333-342.

Przekop, F., Polkowska, J., and Mateusiak, K. (1988). The effect of prolonged stress on the hypothalamic luteinizing hormone-releasing hormone (LHRH) in the anestrus ewe. *Experimental and Clinical Endocrinology* **91**, 334-340.

Stress: Neuroendocrine and Molecular Approaches
Edited by R. Kvetnansky, R. McCarty and J. Axelrod

1992 Gordon and Breach Science
Publishers S.A., New York, USA.
Photocopying permitted by license only.

STRESS-INDUCED CHANGES IN NOREPINEPHRINE, DIHYDROXYPHENYLGLYCOL AND 3,4-DIHYDROXYPHENYLACETIC ACID LEVELS IN THE HYPOTHALAMIC PARAVENTRICULAR NUCLEUS OF CONSCIOUS RATS

K. Pacak[1], I. Armando[1], R. Kvetnansky[1], M. Palkovits[2]
D. S. Goldstein[1] and I.J. Kopin[1]

[1]Clinical Neuroscience Branch, National Institute of Neurological
Disorders and Stroke and [2]Laboratory of Cell Biology
National Institute of Mental Health, Bethesda, Maryland 20892 USA

INTRODUCTION

Central mechanisms determining integrated pituitary-adrenocortical and sympathoadrenal responses during stress are incompletely understood. The paraventricular nucleus (PVN) of the hypothalamus has a role in regulating pituitary-adrenocortical and sympathoadrenal outflow. The parvocellular division of the PVN, in which most cell bodies contain corticotropin-releasing hormone (CRF), receives noradrenergic innervation mainly from the brainstem. CRF in the PVN is thought to be under noradrenergic regulation, suggesting a role of norepinephrine in this brain region in stress responses (Plotsky, 1987; Hillhouse *et al*, 1975). NE synapses have been described in association with CRF immunoreactive neuronal elements (Liposits *et al*, 1986). IMMO of rats has been shown to be a useful model of stress

93

(Kvetnansky *et al*, 1970). We examined whether IMMO affects extracellular fluid (ECF) levels of endogenous catechols in the PVN. *In vivo* microdialysis was used to measure simultaneously PVN ECF concentrations of norepinephrine (NE), its neuronal metabolite dihydroxyphenylglycol (DHPG), and the dopamine metabolite dihydroxyphenylacetic acid (DOPAC) during acute IMMO in conscious, freely-moving rats.

MATERIALS AND METHODS

Animal Preparation

Male Sprague-Dawley rats (body weight 300±15 g, n = 6) from Taconic Farms (Germantown, NY) were housed at room temperature (20° C) and with food and water *ad libitum* in a room with a 12 hour light/dark cycle.

Animals were anesthetized with pentobarbital (50 mg/kg i.p.) and placed in a stereotaxic frame (David Kopf Instruments, Tujunga, CA), with incisor bar 3.2 mm below the interaural line. The skull was exposed and a small hole drilled over the right PVN. The microdialysis probe (BAS/Carnegie Medicine, West Lafayette, IN, 1 mm, cutoff value: 20,000 Dalton) was placed with the following coordinates with respect to the bregma: Post: 1.8, Lat: 0.4, Vert: -8.7 according to the atlas of Paxinos *et al* (1986). The implanted microdialysis probe was anchored to the skull with 3 skull screws and acrylic dental cement. After surgery, animals recovered in cylindrical Plexiglas cages for 20-24 hours prior to the acute experiment.

Microdialysis Procedure

Microdialysis probes were connected to a microinfusion pump (CMA 100, BAS/Carnegie Medicine, West Lafayette, IN) and perfused at a rate 1.0 ml/minute with artificial cerebrospinal fluid (189 mM NaCl, 3.9 mM KCl and 3.37 mM CaCl$_2$, pH 6.3). Collection vials contained 5 ml of 0.2 N acetic acid. After each collection period, the dialysate was frozen rapidly. Average *in vitro* recoveries determined for all probes at room temperature before use were: NE 13%, DHPG 13%, and DOPAC 13%.

Experimental Protocol

To establish basal levels of NE, DHPG, and DOPAC, the dialysis experiment began with collection of 3 samples during 30-minute intervals. Then animals were immobilized for 120 minutes.

IMMO consisted of taping the rat's limbs to a special metal frame using adhesive tape (Kvetnansky *et al*, 1970). Samples were collected during and for 2 hours after 8 30-minute intervals of IMMO. At the end of the experiment, dialysis probes were perfused with a solution of toluidine blue for 5 min, and brains were removed and stored in 10% formalin solution for later histological verification of probe position.

Assays

NE, DHPG, and DOPAC concentrations in the microdialysate samples were measured using reversed-phase liquid chromatography with electrochemical detection after partial purification by adsorption on alumina (Eisenhofer *et al*, 1986).

Statistical analyses

Results are presented as means ± S.E.M. Effects of IMMO were analyzed by one-factor analyses of variance for repeated measures, with Dunnett's post-hoc test. A p value less than 0.05 defined statistical significance.

RESULTS

Basal levels of NE averaged 177±13 pg/ml, DHPG 1436±137 pg/ml, and DOPAC 4228±648 pg/ml. IMMO produced rapid, marked, proportionately similar increases in ECF NE and DHPG levels and a more delayed increase in DOPAC levels (Figure 1). Compared to baseline levels, the maximum mean increases during stress were 246% for NE, 192% for DHPG and 198% for DOPAC. Increased levels of NE, DHPG, and DOPAC persisted throughout the duration of IMMO. After termination of IMMO, the values slowly declined; by 2 hours after IMMO, levels were only slightly above baseline.

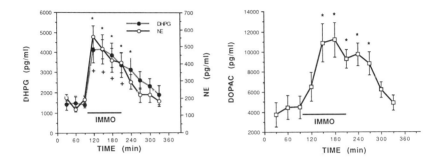

FIGURE 1. Norepinephrine (NE), dihydroxyphenylglycol (DHPG, left panel) and dihydroxyphenylacetic acid (DOPAC, right panel) levels in microdialysate samples from acutely immobilized rats. (+) P<0.05 for NE and (*) P<0.05 for DHPG and DOPAC versus unstressed basal values.

DISCUSSION

Using *in vivo* microdialysis during acute IMMO stress in conscious rats, we found increased extracellular fluid concentrations of NE and its main intraneuronal metabolite, DHPG. These results indicate increased release and reuptake of NE in the PVN during IMMO stress.

By 2 hours after IMMO, levels of both NE and DHPG returned to near baseline values. These results are consistent with a previous report by Kvetnansky *et al* (1977) describing decreased tissue levels of NE in the PVN and other hypothalamic nuclei in rats exposed to a single 20-minute period of IMMO, with a return to control values after release from restraint. The more rapid increase in NE than in DHPG is consistent with DHPG as a product of NE metabolism. The decline in NE release during the interval of IMMO may reflect partial depletion of easily releasable NE stores, with a new steady state developing in which lowered NE stores are maintained by enhanced NE synthesis.

DHPG levels reflect two intraneuronal processes: metabolism of NE after its release and neuronal reuptake (Goldstein *et al*, 1991) and metabolism of NE leaking from storage vesicles into the axonal cytoplasm (Eisenhofer et al, 1988). Simultaneous measurements of NE and DHPG therefore provide information about different aspects of noradrenergic function. Our data suggest increased NE release and turnover in the PVN during IMMO stress.

The PVN also receives dopaminergic fibers from A11 cells in the incerto-hypothalamic dopaminergic system (Bjorklund et al, 1975). Although dopamine was rarely detected in microdialysate fluid, there were high DOPAC concentrations. The increments in DOPAC levels during IMMO stress were delayed compared with those of NE and DHPG. In brain regions innervated only by noradrenergic nerves, extracellular DOPAC may reflect a balance between cytoplasmic metabolism by monoamine oxidase and transport of dopamine into NE storage vesicles within the NE terminals (Abercrombie et al, 1989). In brain areas with dopaminergic innervation, DOPAC is derived mainly from an intraneuronal pool of newly-synthetized dopamine (Zotterstrom et al, 1988). In a region containing both NE and dopamine neurons, DOPAC levels during IMMO may reflect noradrenergic or dopaminergic activation. Progressively increasing DOPAC levels during IMMO stress may reflect enhanced tyrosine hydroxylation and catecholamine biosynthesis in noradrenergic or dopaminergic neurons.

CRF plays a prominent role in the regulation ACTH responses to a variety of stressors (Plotsky, 1987). Jezova et al (1989) showed that increased plasma ACTH concentrations in response to IMMO were markedly attenuated in rats with lesions of the medial-basal hypothalamus. Effects of NE on CRF release must be interpreted with caution, since release of ACTH is mediated by arginine-vasopressin as well as CRF, but evidence for noradrenergic synapses on neurons containing immunoreactive CRF suggests that effects of NE on release of CRF may mediate the pituitary-adrenocortical response to IMMO stress.

In summary, the present study demonstrates that PVN noradrenergic activity is increased during acute IMMO stress in conscious rats. This leads to the hypothesis that increased noradrenergic activity may contribute to CRF release during stress.

REFERENCES

Abercrombie, E.D., and Zigmond, M. J. (1989). Partial injury to central noradrenergic neurons: reduction of tissue norepinephrine content is greater than reduction of extracellular norepinephrine measured by microdialysis. *Journal of Neuroscience* **9**, 4062-4067.

Bjorklund, A., Lindvall, O., and Nobin, A. (1975). Evidence of an incertohypothalamic dopamine neurone system in the rat. *Brain Research* **89**, 29-42.

Eisenhofer, G., Goldstein, D.S., Stull, R., Keiser, H.R., Sunderland, T., Murphy, D.L., and Kopin, I.J. (1986). Simultaneous liquid-chromatographic determination of 3,4-dihydroxyphenylglycol, catecholamines, and 3,4-dihydroxyphenylalanine in plasma, and their responses to inhibition of monoamine oxidase. *Clinical Chemistry* **32**, 2030-2033.

Eisenhofer, G., Goldstein, D.S., Ropchak, T.G., Nguyen, H. Q., Keiser, H.R., and Kopin, I.J. (1988). Source and physiological significance of plasma 3,4-dihydroxyphenylglycol and 3-methoxy-4-hydroxyphenylglycol. *Journal of the Autonomic Nervous System* **24**, 1-14.

Goldstein, D.S., Cannon, R.O., Quyyumi, A., Chang, P., Duncan, M., Brush, J.E., and Eisenhofer, G. (1991). Regional extraction of circulating norepinephrine, DOPA, and dihydroxyphenylglycol in humans. *Journal of the Autonomic Nervous System* **34**, 17-36.

Hillhouse, E.W., Burden, J., and Jones, M. T. (1975). The effect of various putative neurotransmitters on the release of corticotropin releasing hormone from the hypothalamus of the rat *in vitro*. I. The effect of acetylcholine and norepinephrine. *Neuroendocrinology* **17**, 1-11.

Jezova, D., Kvetnansky, R., Tilders, F.J.H., and Makara, G. (1989). Interaction of circulating catecholamines, CRF and AVP in the control of ACTH release during stress. In: G.R. Van Loon, R. Kvetnansky, R. McCarty and J. Axelrod (Eds.), "Stress: Neurochemical and Humoral Mechanisms," pp. 409-424. New York: Gordon and Breach.

Kvetnansky, R., and Mikulaj, L. (1970). Adrenal and urinary catecholamines in rats during adaptation to repeated immobilization stress. *Endocrinology* **87**, 738-743.

Kvetnansky, R., Palkovits, M., Mitro, A., Torda, T., and Mikulaj, L. (1977). Catecholamines in individual hypothalamic nuclei of acutely and repeatedly stressed rats. *Neuroendocrinology* **23**, 257-267.

Liposits, Z. S., Phelix, C., and Paull, W. K. (1986). Electron microscopic analysis of tyrosine hydroxylase, dopamine-ß-hydroxylase and phenylethanolamine-N-methyltransferase immunoreactive innervation of the hypothalamic paraventricular nucleus in the rat. *Histochemistry* **84**, 105-120.

Paxinos, G., and Watson, C. (1986). "The Rat Brain in Stereotaxic Coordinates". Orlando: Academic Press.

Plotsky, P. M. (1987). Facilitation of immunoreactive corticotropin-releasing factor secretion into the hypophysial-portal circulation after activation of catecholaminergic pathways or central norepinephrine injection. *Endocrinology* **121**, 924-930.

Zetterstrom, T., Sharp, T., Collin, A.K., and Ungerstedt, U. (1988). *In vivo* measurement of extracellular dopamine and DOPAC in rat striatum after various dopamine-releasing drugs; implications for the origin of extracellular DOPAC. *European Journal of Pharmacology* **148**, 327-334.

Stress: Neuroendocrine and Molecular Approaches
Edited by R. Kvetnansky, R. McCarty and J. Axelrod

1992 Gordon and Breach Science
Publishers S.A., New York, USA.
Photocopying permitted by license only.

LEARNING STRESS AND THE BRAIN CHOLINERGIC SYSTEM

H. Saito, N. Nishiyama and M. Segawa

Department of Chemical Pharmacology, Faculty of Pharmaceutical
Sciences, The University of Tokyo, Tokyo, Japan

INTRODUCTION

Numerous studies have demonstrated central cholinergic systems may play an important role in the retention and/or acquisition process Gerson and Baldessarini, 1980; Haroutunian *et al*, 1985; Miyamoto *et al*, 1987), although relatively little attention has been directed towards the stress inevitably involved in the learning tasks (learning stress). Therefore, in the present investigation, various cholinergic activity parameters, choline acetyltransferase (CAT), acetylcholine (ACh) and muscarinic binding sites, were studied in discrete brain areas to examine a possible correlation between retention and/or acquisition performance, which reflects the degree of stress, and neuronal activity of cholinergic systems. In addition, we wished to clarify whether these parameters behave similarly during the retention process in two types of passive avoidance tasks and the acquisition process in two types of active avoidance tasks.

MATERIALS AND METHODS

Male Std-ddy mice (7 weeks old) were obtained from SLC (Hamamatsu, Japan). They were housed in groups of five in aluminum cages (22 x 36 x 11 cm) with wooden chip bedding, and had free access to food and water. After one week of acclimation, they were used for experiments on learning behaviors. The temperature and humidity of the room were controlled at 22 ± 1 °C and 55 ± 2 %, respectively.

[1-^{14}C]-Acetyl-Coenzyme A (5 mCi/mmol) was obtained from Amersham (Bucks, England), and L-[1-^{14}C]-tyrosine (53.8 mCi/mmol), DL-[1-^{14}C]-dihydroxyphenylalanine (54.8 mCi/mmol) and [^{3}H]-quinuclidinyl benzilate (43.9 mCi/mmol) from NEN (Boston, MA, USA). Ethylhomocholine was a gift from BAS (Tokyo, Japan). All other reagents were of the highest purity obtainable commercially.

Learning Tasks

Step Through (ST) test. The apparatus and procedure for this task have been described in detail elsewhere (Segawa et al, 1990c). Each mouse was placed in the light compartment of the apparatus. When the mouse entered the dark compartment, an electric shock (max. 0.2 mA, 36 V AC) was delivered automatically through the floor grid and then the mouse was allowed to escape back to the light compartment. The time taken to enter the dark compartment after the mouse had been in the light compartment was recorded. The retention test was performed by replacing the mouse in the light compartment, 24 hours after the acquisition test. Mice which did not enter the dark compartment within 5 minutes in the first retention test were given a retention test daily for another 9 days using an identical procedure. Mice which never entered the dark compartment in the 10 retention tests were classified as good learners and the others as poor learners.

Step Down (SD) test. The apparatus and procedure have been described elsewhere (Segawa et al, 1990d). Each mouse was placed on the platform and allowed to explore the apparatus for 10 minutes. For the second half of the 10 minute exploration, the number of shocks received was counted. The mouse was then returned to its home cage. The retention test was given by replacing the mouse on the platform for 3 minutes, 24 hours later. Mice which did not step down within 3 minutes in the first retention test were given daily retention tests for the next 9 days using an identical procedure. Mice were classified by criteria similar to those in the ST test.

Shuttle Box (SB) test. Acquisition of two way avoidance response was assessed with an automated shuttle box (GT-8332S; O'Hara Co. Ltd., Tokyo, Japan) as previously described (Segawa et al, 1990a). The temporal parameters were as follows: intertrial interval of 40 seconds; warning (a duration of 30 msec and a frequency of 1 Hz) duration of 20 seconds; shock (an intensity of 36 V AC) duration of 10 seconds in the latter half of warning. The indices of avoidance behavior were the avoidance rate (the number of avoidance responses/the number of trials) and the response rate (the number of movings/minute) during each session. Each session consisted of 60 trails per day and sessions were performed daily for 7 days at the same time of day. The trained mice were classified into two groups on the basis of their performance; those with average avoidance rates over 7 sessions of more than 60-70% were taken as good learners, and those with less than 40-65% as poor ones.

Lever Press (LP) test. Acquisition of LP responses was determined in an apparatus (GT-832; O'Hara Co. Ltd., Tokyo, Japan) as previously described (Segawa *et al*, 1990b). The temporal parameters, intensity of electric shock and indices of avoidance response were similar to those of the SB test. Each session consisted of 60 trials per day and sessions were held daily for 10 days at the same time of day. Mice were classified by criteria similar to those used in the SB test.

Biochemical Determinations

CAT activity. On the day following the last session, mice were decapitated after cervical dislocation, and each brain was rapidly excised from the skull. Each brain was rinsed with ice cold 0.32 M sucrose and dissected by the method of Gispen *et al*, (1972). The brain samples included the cortex, hippocampus, amygdala with overlaying cortex piriformis, septum, hypothalamus and striatum. Each tissue was homogenized in 10 volumes of 0.32 M sucrose, and CAT activity was measured according to the method of Fonnum (1975).

ACh determination. On the day of the final session, mice were killed by microwave irradiation. The brain was rapidly removed and dissected on ice into 6 regions. ACh was estimated by HPLC method with slight minor modification (Fujimori and Yamamoto, 1987).

Receptor binding assay. Following the last session, mice were killed by decapitation. The brain was rapidly removed, and dissected into 6 regions on ice. Each tissue collected from 3-4 mice was combined.

The specific binding of [³H]-quinuclidinyl benzilate (QNB) to discrete brain regions was measured (Yamamura and Snyder, 1974).

RESULTS

Figure 1 shows regional CAT activity in brains of mice given the ST, SD, SB and LP tests, respectively. In the ST test, there were no differences in CAT activity in brain regions among groups except for striatum, the CAT activity of which in good learners was increased as compared with non-trained mice (exposed to the apparatus). In the SD test, CAT activities in the hippocampus and amygdala of trained mice were increased in comparison with those of non-trained mice, and the increase of CAT activity in amygdala of good learners was significantly larger than that of poor ones. In the SB test, CAT activity was increased in the cortex and septum, and decreased in the striatum as compared with non-trained mice. CAT activity was not altered in other brain regions among groups. In the LP test, CAT activity in the cortex, hippocampus and amygdala was lower in trained mice than in non-trained mice. In addition, CAT activity in the cortex of good learners was lower than that of poor ones. In the other brain regions examined, CAT activity was not different among groups.

Data on regional ACh content in brains of mice trained in ST, SD, SB and LP tasks are shown Figure 2. In the ST test, ACh content in the amygdala and septum of the good learners was higher than in poor ones. Furthermore, there was a positive correlation between increase in ACh content of amygdala and duration (days) of retention of the avoidance response in mice given the ST test (r = .409, p < .02, data not shown). In the SD test, ACh content was increased in the hippocampus of trained and good learner mice, and in the hypothalamus of the trained, good and poor mice, but in the cortex of the poor mice it was decreased, as compared with non-trained mice. A significant increase in ACh content was found in the cortex, hippocampus and amygdala of good learners relative to poor ones. An increase in ACh content of hippocampus and amygdala was correlated with the duration of retention of the avoidance response in mice given the SD test (r = .396 and .385, p < .02, respectively, data not shown). In the SB test, ACh content in the septum and hypothalamus of good learner mice was increased as compared with the non-trained and poor learner ones. The increase in ACh content in the amygdala of good

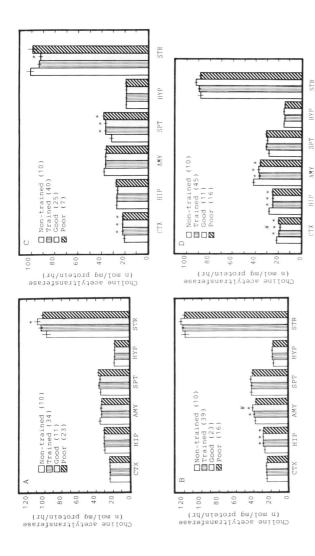

FIGURE 1.   Changes in choline acetyltransferase activity in various regions of mice after task performance in ST- (A), SD- (B), SB- (C) AND LP- (D) type avoidance behaviors. Data for non-trained (exposed to the apparatus), trained, good and poor learner mice are shown for each brain region (CTX: cortex, HIP: hippocampus, AMY: amygdala, SPT: septum, HYP: hypothalamus, and STR: striatum).   Numbers of mice used are shown in parentheses.   Values are expressed as mean±SEM and significance level were determined using ANOVA followed by Duncan's multiple range test.   Linear regression was calculated by the least-square method.   A P-value of less than .05 was regarded as being statistically significant. *P <.05 compared with the non-trained mice, and #P <.05 compared with poor learners.

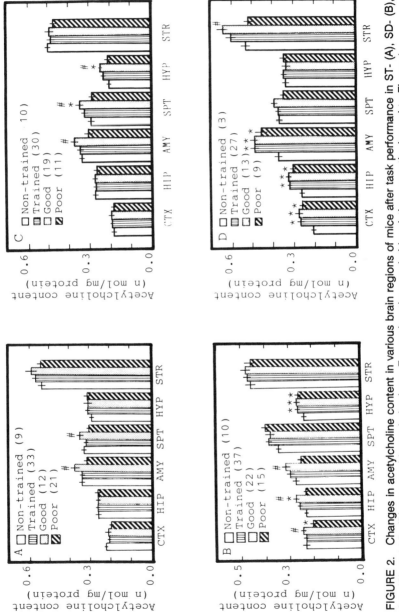

FIGURE 2. Changes in acetylcholine content in various brain regions of mice after task performance in ST- (A), SD- (B), SB- (C) and LP- (D) type avoidance behaviors. For explanations and abbreviations, see the legend to Figure 1.

learners was significantly different from that in poor ones, and there was a positive correlation between ACh content in the amygdala and average avoidance rates for 7 sessions in the SB test ($r = .576$, $p < .02$, data not shown). In the LP test, ACh content was increased in cortex, hippocampus and amygdala of trained mice and in striatum of good learners as compared with the non-trained mice. In the striatum, ACh content of good learners was significantly greater than that of poor ones, and there was a positive correlation between striatal ACh content and average avoidance rates for 10 sessions in the LP test ($r = .519$, $p < .02$, data not shown).

We examined regional QNB binding in brains of mice given ST, SD, SB and LP tests. In the ST test, the density of QNB binding sites was decreased in cortex, amygdala and striatum of good learners compared with poor learners. In the SD test, the number of QNB binding sites in hippocampus, amygdala and hypothalamus of good learners were significantly lower than those of poor learners. Numbers of QNB binding sites in the hypothalamus of trained and poor learner mice were decreased as compared with those of the non-trained. In the SB test, the maximal numbers of QNB binding sites in the amygdala of the trained and good mice were decreased in comparison with those of non-trained and poor learner mice, respectively. In the LP test, the maximal density of QNB binding sites was increased in cortex and amygdala, but decreased in striatum, of trained mice as compared with non-trained mice. In four types of avoidance tests, there was no significant difference in the dissociation constant or Hill's number of QNB binding sites among groups in any of the regions examined in this study.

DISCUSSION

The present observations indicate that brain regions in which CAT activity, ACh content and QNB binding sites were modulated during retention and/or acquisition processes depend upon the nature of the avoidance tasks. Passive and active avoidance tasks can be considered as stressful situations. Therefore, any neurochemical changes that occur during retention and/or acquisition processes may be related either to the stress induced by the experiment itself, the learning, the retrieval of memory, or the task performance, or some combination of these four. Since in SD tests as contrasted with ST, alterations in cholinergic parameters were found in certain brain areas of trained mice in the present study, it seems likely that SD-type avoidance tasks are more stressful than ST-type. LP-type active avoidance task is also thought to be more stressful than SB. On the other side, these differential alterations in cholinergic parameters are thought to be due to the modalities of the tasks used in the present studies; whereas SB-type avoidance behavior appears to be based on a type of innate behavior, i.e., escaping reaction from aversive stimuli, animals in LP-type avoidance task must learn to push the lever to avoid a forthcoming shock, which they have not yet experienced. These observations indicate that differential neuronal mechanisms may be involved in SB- and LT-type active avoidance behaviors, and also between ST- and SD-type passive avoidance behaviors. In the ST, SD and SB tasks, a significant difference between good and poor learners was found for cholinergic parameters in the amygdala, which has been implicated in learning and memory processes (Berman and Kesner, 1976; Gold *et al*, 1975). The function of these structures is probably related to emotional and motivational components involved in information processing pathways. Since the nucleus basalis magnocellularis sends axonal projections to cortex and amygdala (Kimura *et al*, 1981; Masulam *et al*, 1983ab), the data suggest that increased neuronal impulse flow related to retention and/or acquisition may have taken place only in a selective population of the cholinergic projection of the nucleus basalis magnocellularis. Moreover, there was a positive correlation between ACh content in amygdala and duration of retention of avoidance response in both passive avoidance tasks and average avoidance rates in the SB-type active avoidance task. These results support the view that the amygdala has a modulatory role in retention and/or acquisition processes (Bresnahan and Routtenberg, 1972; Liang *et al*, 1982; Thompson *et al*, 1983). In the case of the LP test, further experiments are necessary to elucidate the reason for the difference.

REFERENCES

Berman, R.F., and Kesner, R.P. (1976). Post trial hippocampal, amygdaloid, and lateral hypothalamic electrical stimulation: Effects of short- and long-term memory of an appetitive experience. *Journal of Comparative and Physiological Psychology* **90**, 260-267.
Bresnahan, E., and Routtenburg, A. (1972). Memory disruption by unilateral low level

sub-seizure stimulation of medial amygdaloid nucleus. *Physiology and Behavior* **9**, 513-525.

Fonnum, F. (1975). A rapid radiochemical method for the determination of choline acetyltransferase. *Journal of Neurochemistry* **24**, 407-409.

Fujimori, K., and Yamamoto, K. (1987). Determination of acetylcholine and choline in percholate extracts of brain tissue using liquid chromatography-electrochemistry with an immobilized-enzyme reactor. *Journal of Chromatography* **414**, 167-173.

Gerson, S.C., and Baldessarini, R.J. (1980). Motor effects of serotonin in the central nervous system. *Life Sciences* **27**, 1435-1451.

Gispen, W.H., Scotman, P., and de Kloet, E.R. (1972). Brain RNA and hypophysectomy; a topographical study. *Neuroendocrinology* **9**, 285-296.

Gold, P.E., Edwards, R.M., and McGaugh, J.L. (1975). Amnesia produced by unilateral, subseizure, electrical stimulation of amygdala in rats. *Behavioral Biology* **15**, 95-105.

Haroutunian, V., Kanof, P., and Davis, K.L. (1985). Pharmacological alleviation of cholinergic lesion induced by memory deficits in rats. *Life Sciences* **37**, 945-942.

Kimura, H., McGeer, P.L., Pneg, J.H., and McGeer, E.G. (1981). The central cholinergic system studied by choline acetyltransferase immunohistochemistry in the cat. *Journal of Comparative Neurology* **200**, 151-201.

Liang, K.C., McGaugh, J.L., Martinez Jr., J.L., Jensen, R.A., Vasquez, B.J., and Messing, R.B. (1982). Post-training amygdaloid lesions impair retention of an inhibitory avoidance response. *Behavioural Brain Research* **4**, 237-249.

Masulam, M.-M., Mufson, E.J., Wainer, B.H., and Levey, A.I. (1983a). Central cholinergic pathway in the rat: an overview based on an alternative nomenclature. *Neuroscience* **10**, 1185-1201.

Masulam, M.-M., Mufson, E.J., Levey, A.I., and Wainer, B.H. (1983b). Cholinergic innervation of cortex by the basal forebrain: cytochemistry and cortical connections of the septal area, diagonal band nuclei, nucleus basalis (substantia innominata) and hypothalamus in the rheusus monkey. *Journal of Comparative Neurology* **214**, 170-197.

Miyamoto, M., Kato, J., Narumi, S., and Nagaoka, A. (1987). Characteristics of memory impairment following lesioning of the basal forebrain and medial septal nucleus in rats. *Brain Research* **419**, 19-31.

Segawa, M., Saito, H., and Nishiyama, N. (1990a). Alterations in choline acetyltransferase and tyrosine hydroxylase activities of various brain areas after the acquisition of active avoidance tasks in mice. *Biogenic Amines* **7**, 171-180.

Segawa, M., Saito, H., and Nishiyama, N. (1990b). Alterations in acetylcholine content and muscarinic binding site of various brain areas in mice. *Biogenic Amines* **7**, 181-190.

Segawa, M., Saito, H., and Nishiyama, N. (1990c). Alterations in choline acetyltransferase and tyrosine hydroxylase activities of various brain areas after passive avoidance performance in mice. *Biogenic Amines* **7**, 191-197.

Segawa, M., Saito, H., and Nishiyama, N. (1990d). Alterations in acetylcholine content and muscarinic cholinergic binding of various brain regions after passive avoidance behaviors in mice. *Biogenic Amines* **7**, 199-207.

Thompson, R.F., Berger, T.W., and Madden IV, J. (1983). Cellular processes of learning and memory in the mammalian CNS. *Annual Review of Neuroscience* **6**, 447-491.

Yamamura, H.I., and Snyder, S.H. (1974). Muscarinic cholinergic binding in rat brain. *Proceedings of the National Academy of Sciences* (USA) **71**, 1725-1729.

Stress: Neuroendocrine and Molecular Approaches
Edited by R. Kvetnansky, R. McCarty and J. Axelrod

1992 Gordon and Breach Science
Publishers S.A., New York, USA.
Photocopying permitted by license only.

VOLTAMMETRIC DETERMINATION OF CATECHOL SUBSTANCES IN CORPUS STRIATUM OF RATS AFTER ELECTROCONVULSIVE SHOCK

K. Murgas and J. Pavlasek

Institute of Experimental Endocrinology
Institute of Normal and Pathological Physiology
Slovak Academy of Sciences
Bratislava, Czecho-Slovakia

INTRODUCTION

The fact that electrical stimulation of the brain can cause seizures has been known for over a century (Fritsch and Hitzig, 1870; Albertoni, 1882). Electrical stimulation, and specifically electroconvulsive shock (ECS), produces many neurochemical alterations which may be related to its efficacy in the treatment of various psychiatric disorders. However, the neurophysiological mechanisms underlying ECS are still unknown. The purpose of this report is to describe voltammetric measurements of neurotransmitter substances in corpus striatum in rats evoked by ECS.

MATERIALS AND METHODS

Eight SPF male Wistar rats weighing 300 g were used. The animals were anesthetized with chloral hydrate (400 mg/kg b.w.) and 30 minutes later they were mounted in a stereotaxic apparatus. A hole was drilled into the skull, the dura and the pia mater were pierced, and

109

an electrochemically prepared working microelectrode (Gonon *et al*, 1981) was stereotaxically inserted (Fifkova and Marsala, 1957) into the corpus striatum (AP - 1.0 mm, L 2.5 mm, V 4.0 mm). The working carbon microelectrode was made of pyrolytic carbon fiber (7 μm in diameter) insulated in a glass micropipette with the length of the exposed tip 200 μm. The additional and reference Ag/AgCl electrodes were in contact with the dura, and were attached to the bone with steel screws. Two other steel screws were fixed to the temporal side of the scull. They were connected through cables to an ECS device. This could operate in two modes giving either a minimal electroshock stimulus (MinECS = 50 Hz AC, intensity between 20 and 30 mA for 0.2 sec) or a maximal electroshock stimulus (MaxECS = 50 Hz, around 150 mA for 0.2 sec.) (Browning, 1987). The intensity of the AC current used for MinECS was just below the facial and forelimb clonus threshold while the current intensity used for MaxECS produced convulsions for approximately 5 minutes. Details can be found elsewhere (Swinyard, 1972; Browning, 1987).

A differential pulse voltammetry technique was used for electrochemical analyses (PA4, LP Praha) with the following parameters: linear potential sweep from -100 mV to 500 mV, scan rate 100 mV /second, pulse amplitude 50 mV, pulse duration 60 msec, pulse period 0.2 seconds. Measurements were repeated at one minute intervals.

The oxidation potential peak values obtained during a baseline period of 30 minutes were taken as 100%. Subsequently, animals received either MinECS or MaxECS and immediately (from 3 to 5 seconds) afterwards the first poststimulating voltammogram was recorded, followed by measurements at 1 minute intervals.

Changes in dopamine oxidation potential were expressed as percentages of control values. Student's t test was used to evaluate the results. Mean values and standard errors are shown.

RESULTS

The center of the ascorbic acid oxidation potential peak was detected at + 120 mV, and catechol substances, in corpus striatum mainly dopamine (Wightman *et al*, 1988), at + 320 mV polarizing voltage. At preliminary calibration, the + 320 mV peak increased upon microinjection of a 2 nM solution of dopamine standard into the vicinity of the working microelectrode within the striatum.

Immediately (from 3 to 5 sec) after MaxECS, the catechol oxidation current peak increased to 1300% compared with the baseline (100%). One minute after Max ECS application the peak was 300% above baseline level, one minute later 130% above baseline and the next minute only slightly (30%) above the baseline. (Figure 1, open circles).

Application of MinECS resulted in a 30% increase in the catechol oxidation current above baseline immediately after the application, with no effect afterwards. (Figure 1, solid circles.)

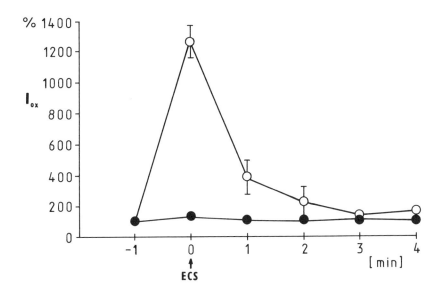

FIGURE 1. Increase of the catechol substances oxidation current I_{ox} from control level (100% marked as -1 min) immediately after electroconvulsive shock (ECS) application (MaxECS = open circles, and MinECS = solid circles, time 0 minutes) and during the next 4 minutes.

DISCUSSION

None of the techniques used previously to monitor neurotransmitter concentrations in neuronal tissue have fully reflected the dynamics of utilization. For example, the widely used punch technique (Palkovits, 1973) allows post-mortem biochemical analysis of neurotransmitter concentrations in defined nuclei or substructures of nuclei; however, the data obtained indicate total content in the respective structure at a given time, but cannot provide information on dynamic patterns (Murgas et al, 1989). The push-pull cannulation technique (Gaddum, 1961), while offering the possibility of abandoning post-mortem analysis of neurotransmitter turnover, presents a number of problems, mainly concerning the establishment of a balanced inflow/outflow of the perfusion liquid and washout of neurotransmission products and is not suitable to detect rapidly changing events (Redgrave, 1977).

The voltammetric technique is only moderately invasive as far as mechanical damage to the brain structures investigated is concerned. On the other hand, acute experiments require anesthesia, the effects of which are incompletely defined. In chronic experiments, restraint is unavoidable to prevent the animals from biting at the leads.

In our experiments we found that the increase of the oxidation current of catechol substances (in corpus striatum mainly dopamine) as a result of MaxECS application was enormous. This sort of release was previously observed and described only under special conditions, such as terminal anoxic depolarization (Murgas and Pavlasek, 1989; Gonzales-Mora et al, 1989), ischemia (Ogura et al, 1989), severe poisoning by ethanol (Murgas and Diaz, 1991), direct local application of KCl solution or detergents into brain structures (e.g., TRITON X - 100) (Murgas and Pavlasek, 1989, 1991) and in our recent study dealing with experimental head injury. Pathological brain states (injury, anoxia, ischemia, intoxication) are associated with large scale changes of the brain microenvironment in which loss of selective permeability of membranes and massive leakage of transmitter substances into extracellular space play a crucial role (Gleiter and Nutt, 1989). We suggest a possible relation between excessive release of neurotransmitters into the extraneuronal space and clinical manifestations of comatose states.

REFERENCES

Albertoni, P. (1982). Untersuchung uber die Wirkungen einiger Arztneimittel auf die Erregbarkeit des Grosshirns nebst Beitragen zur Therapie der Epilepsie. *Archives of Experimental Pathology and Pharmacology* **15**, 248-288.

Browning, R.A. (1987). The role of neurotransmitters in electroshock seizure models. In: P. C. Jobe and H. E. Laird II (Eds.), "Neurotransmitters and Epilepsy", pp. 277-320. The Humana Press.

Fifkova, E., and Marsala, J. (1960). "Stereotaxic Atlas for the Cat, Rabbit, and Rat Brain." Prague: Statni Zdravotnicke Nakladatelstvi.

Fritsch, G., and Hitzig, E. (1870). Uber die electrische Erregbarkeit des Grosshirns. Archives fur Anatomic und Physiologie der *Wissenschaftlichen Medizine* **37**, 300-332.

Gaddum, J.H. (1961). Push-pull cannulae. Journal of Physiology (London) 155, 1P.

Gleiter, C.H., and Nutt, D.J. (1989). Chronic electroconvulsive shock and neurotransmitter receptors: An update. *Life Sciences* **44**, 985-1006.

Gonon, F., Fombarlet, C.M., Buda, M., and Pujol, F. J. (1981). Electrochemical treatment of pyrolytic carbon fibre electrodes. *Analytical Chemistry* **53**, 1386-1389.

Gonzales-Mora, J.L., Maidment, N.T., Guadalupe, T., and Mas, M. (1989). Postmortem dopamined dynamics assessed by voltammetry and microdialysis. *Brain Research Bulletin* **23**, 323-327.

Murgas, K., Kostal, L., and Jurani, M. (1989). Methodical aspects of voltammetry (In Slovak). *Ceskoslovenska Fyziologie* **36**, 309-328.

Murgas, K., and Pavlasek, J. (1990). Early postmortal changes in the rat brain: Increase of catecholamine content in extraneuronal space as determined by voltammetry. *Cellular and Molecular Neurobiology* **9**, 406.

Murgas, K., and Diaz, N. (1991). Effect of acute alcohol treatment on dopamine concentration in corpus striatum of rats: A voltammetric study. (In Slovak). *Proceedings of the VI Congress of Czechoslovakian Neurochemical Society*, p. 59.

Murgas, K., Orlicky, J., and Pavlasek, J. (1991). Monitoring of potassium-stimulated catecholamine changes in striatal synaptosomal preparations and in corpus (In press). striatum of rats: A voltammetric study. *General Physiology and Biophysics* **10**, 423-434.

Ogura, K., Shibuya, M., Sizuki, Y., Kanamori, M., and Ikegaki, I. (1989). Changes in striatal dopamine metabolism measured by *in vivo* voltammetry during transient brain ischemia in rats. *Stroke* **20**, 783-787.

Palkovits, M. (1973). Isolated removal of hypothalamic and other brain nuclei of rat brain. *Brain Research* **59**, 449-450.

Redgrave, P. (1977). A modified push-pull system for the localized perfusion of brain tissue. *Pharmacology, Biochemistry and Behavior* **6**, 471-474.

Swinyard, E.A. (1972). Electrically induced convulsions. In: D.P. Purpura, J.K. Penry, D.M. Woodbury, D.B. Tower, and R.D. Walter (Eds.), "Experimental Models of Epilepsy". New York: Raven Press.

Wightman, R.M., May, L.J., and Michael, A.C. (1988). Detection of dopamine in the brain. *Analytical Chemistry* **60**, 769A.

Stress: Neuroendocrine and Molecular Approaches
Edited by R. Kvetnansky, R. McCarty and J. Axelrod

1992 Gordon and Breach Science
Publishers S.A., New York, USA.
Photocopying permitted by license only.

CENTRAL PEPTIDERGIC MECHANISMS OF CARDIOVASCULAR CONTROL: C- AND N-TERMINAL FRAGMENTS OF SUBSTANCE P

R. Richter[1], F. Qadri[1], P. Oehme[2] and T. Unger[1]

[1]Department of Pharmacology, University of Heidelberg
[2]Institute of Drug Research, Berlin, Germany

INTRODUCTION

Evidence has accumulated that the neuropeptide substance P [SP(1-11)] is co-localized with catecholamines in brain and that it mediates cardiovascular, behavioral and stress responses (for a review, see Pernow, 1983; Unger *et al*, 1985). Although a functional interaction between central catecholamines and SP(1-11) has been suggested, its nature remains unclear. Moreover, several enzymes have been isolated from central nervous tissue which cleave SP(1-11) into C- and N- terminal fragments (Blumberg *et al*, 1980). From cardiovascular findings, we know that icv injections of SP(1-11) cause strong pressor and heart rate responses (Unger *et al*, 1981; 1985) while C- and N-terminal fragments of SP(1-11) may exert opposite effects (Hall *et al*, 1987).

The present study was designed to elucidate central mechanisms of action of SP(1-11) on catecholamine release from the anterior hypothalamus (AH) in conscious Wistar rats using microdialysis and HPLC-ED. Furthermore, we compared the cardiovascular responses to C- and N- terminal fragments of SP with those of the full SP(1-11) sequence in conscious rats.

MATERIALS AND METHODS

Experiments were performed in conscious, unrestrained male Wistar rats (300 - 340 g, Thomae, FRG). Two days before the experiments, animals were anesthetized with chloral hydrate (400 mg/kg i.p.; Sigma) and chronically instrumented with femoral artery catheters (PE10 in PE50, Portex) as described previously (Unger *et al*, 1981). On the next day, animals were anesthetized and mounted in a Kopf stereotaxic apparatus; the skull was exposed and a cannula (PE20) was inserted into the left lateral brain ventricle (A-0.6 mm; L-1.3 mm; H-5.0 mm). The dialysis probe was implanted into the AH (A-1.5 mm; L+0.7 mm; H-9.0 mm). All probes were perfused with an artificial cerebrospinal fluid (CSF) (140 mM NaCl; 3.35 mM KCl; 1.15 mM $MgCl_2$; 1.26 mM $CaCl_2$; 1.2 mM Na_2PO_4; 0.3 mM NaH_2PO_4; pH 7.4) at a flow rate of 2 μl/minute. The perfusate was injected directly into the HPLC-ED system to measure the concentration of catecholamines (Badoer *et al*, 1989).

Perfusate collected over the first 30 minutes was discarded. After collection of 4 samples over consecutive 15 minute intervals to establish basal levels (prevehicle), animals received an icv injection of vehicle (CSF) and 4 more samples were collected over 15 minute intervals (vehicle). Thereafter, 4 more samples were obtained (predrug). Rats then received an icv injection of 500 pmol SP(1-11) and the procedure of sampling was identical to that for vehicle injections. Parallel to brain microdialysis, cardiovascular responses (MAP, HR) to icv injections of vehicle and SP(1-11) were recorded.

All peptides were a gift from Dr. M. Bienert and Dr. M. Beyermann (Institute of Drug Research, Berlin, Germany). Since there were no significant differences between samples within each group, data were pooled for statistical analysis. The data were analyzed using one-way analysis of variance (ANOVA) with Dunnett's multiple comparison test.

RESULTS

The selective inhibitory actions of icv injected SP(1-11) on catecholamine release from the AH are shown in Figure 1. To exclude artifacts, we compared only the values for icv injections of SP(1-11) versus vehicle. Before each injection basal levels of the

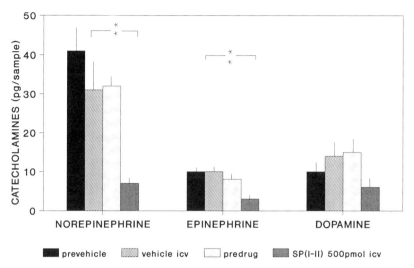

FIGURE 1. Inhibitory effect of subtance P (500 pmol icv) on catecholamine release from rat anterior hypothalamus *in vivo*. Data were pooled from 4 consecutive 15 minute samples per rat. Data are means ± S.E.M. (n=6-8); ** p< 0.01.

released catecholamines were measured (prevehicle and predrug).

Basal (prevehicle) levels of catecholamines were as follows -- norepinephrine (NE): 41 ± 5.4 pg/sample (n=32), epinephrine (EPI): 10 ± 1.2 pg/sample (n=28) and dopamine (DA): 10 ± 2.5 pg/sample (n=28). SP(1-11) resulted in an inhibition of NE and EPI release of immediate onset and of about 1 hour duration (data not shown). Both NE and EPI release were significantly reduced by 77% and 70%, respectively (p< 0.01). There was no significant reduction of DA levels in this brain region (Figure 1).

Figures 2 and 3 show a dose-dependent and long-lasting increase in MAP (20 ± 2 mmHg) and HR (161 ± 12 bpm) after icv administration of SP(1-11). Both of the C- terminal fragments SP(4-11) and SP(5-11) were as potent as SP(1-11) (all p< 0.05). In contrast, the N- terminal tetrapeptides were not active (SP(1-4)NH$_2$) or showed little pressor activity (SP(1-4)COOH) at the highest dose when compared with SP(1-11). However, both tetrapeptides elicited a modest tachycardia in the dose range tested (p< 0.05).

MAP and HR changes induced by the neuropeptides in general occurred after a latency of 20 - 30 seconds. Notably, the HR response usually lasted for 10 -25 minutes and featured 2 -3 stages of activation,

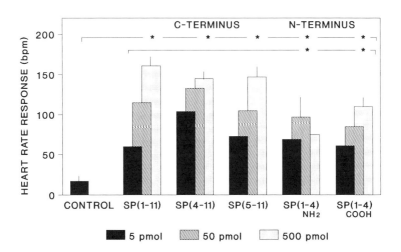

FIGURE 2.   Blood pressure responses after icv injection of C- and N- terminal fragments of substance P (5-500 pmol). Data are means ± S.E.M. (n = 7-14); * p< 0.05.

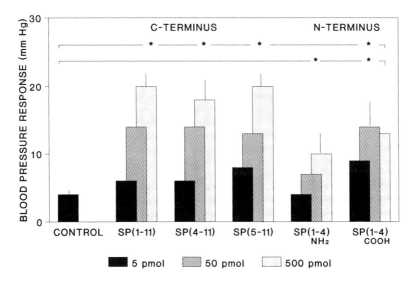

FIGURE 3.   Heart rate responses after i.c.v. injection of C- and N-terminal fragments of substance P (5-500 pmol). Data are means ± S.E.M. (n = 7-14); * p< 0.05.

suggesting consecutive interactions with different brain areas. To determine the effects of different peptide fragments on the duration of tachycardia, we calculated the recovery of HR to basal levels following icv injection of 500 pmol of each peptide (Table 1). Of all peptides tested, SP(5-11) yielded the longest recovery time, followed by SP(4-11) ($p < 0.05$), while the N-terminal tetrapeptides and full-length SP(1-11) were much less potent.

DISCUSSION

In the present study, *in vivo* brain microdialysis was used to monitor extracellular levels of NE, EPI and DA in the rat AH in response to icv injected SP(1-11). Basal catecholamine levels that were obtained agree well with those presented in previous reports for rat hypothalamus (Routledge *et al*, 1987; Kapoor *et al*, 1987). After icv administration of SP(1-11), a selective and rapid decline of hypothalamic NE and EPI levels occurred. The time course of this inhibition also indicates that peptidergic mechanisms regulating extracellular NE and EPI levels are quite efficient and long-lasting. By contrast, DA levels in the AH remained relatively constant following SP(1-11) administration. This heterogeneity may result from differential profiles of catecholaminergic transmission in the forebrain. Moreover, there is evidence that NEergic and EPIergic neurons in brainstem and AH exert inhibitory effects on blood pressure (Howes, 1984). We suggest that the decrease in NEergic and EPIergic activity

TABLE 1. Recovery of heart rate (HR) after icv injection of SP(1-11) and of C- and N- terminal fragments.

Treatment	N	Recovery of HR (in minutes)
CSF (Control)	6	3.5 ± 1.0
SP(1-11)	10	12.7 ± 0.8*
SP(4-11)	8	23.6 ± 2.8*#
SP(5-11)	14	27.3 ± 4.0*#
SP(1-4)NH$_2$	8	17.3 ± 4.1*
SP(1-4)COOH	7	17.0 ± 4.3*

Data are mean ± S.E.M. *$p < 0.05$ versus corresponding value for vehicle; #$p < 0.05$ versus corresponding value for SP(1-11) by Dunnett's test.

in AH due to SP(1-11) may impair the inhibitory influence of this region on cardiovascular responses to stressful stimuli.

In view of the considerable evidence that the neuropeptide SP(1-11) as well as SP-fragments participate in cardiovascular control, the present study compared the effects of icv injected C- and N-terminal fragments of SP(1-11) on MAP and HR responses in conscious rats. The C- terminal fragments, SP(4-11) and SP(5-11), both caused dose-dependent increases in MAP and HR, similar to that induced by SP(1-11). Thus, the C- terminal fragments may act on the same receptor as the parent molecule. Cridland and co-workers (1988) suggested an activation of SP(1-11) by enzymatic cleavage to SP(1-7) at the receptor site in the spinal cord. The significant prolonged recovery of HR to both C- terminal fragments seems to support this hypothesis. SP(5-11) has been detected in the spinal cord and hypothalamus and was more potent than SP(1-11) in inducing behavioral responses when administered intrathecally (Sakurada, 1988). The N- terminal tetrapeptides investigated [SP(1-4)NH$_2$ and SP(1-4)COOH] both lacked significant effects. These results are in conflict with previous findings (Hall *et al*, (1987) of a brief depressor effect after injection of SP(1-4) into the 4th cerebral ventricle in rats. This discrepancy may reflect an altered circulatory control induced by anesthesia.

In conclusion, the neuropeptide SP(1-11) selectively reduces NE and EPI levels in rat AH monitored by *in vivo* microdialysis. This inhibitory effect might be important for central processing of stressful stimuli. Furthermore, the central cardiovascular effects of SP(1-11) are linked to the C-terminus of the neuropeptide.

REFERENCES

Badoer, E., Würth, H., Qadri, F., Itoi, K., and Unger, Th. (1988). Central noradrenergic pathways are not involved in the pressor response to intracerebroventricular substance P. *European Journal of Pharmacology* **154**, 105-108.

Badoer, E., Würth, H., Türck, D., Qadri, F., Itoi, K., Dominiak, P., and Unger, T. (1989). The K+-induced increase in noradrenaline and dopamine release are accompanied by reductions in the release of their intraneuronal metabolites from the rat anterior hypothalamus. *Naunyn-Schmiedeberg's Archives of Pharmacology* **339**, 54-59.

Blumberg, S., Teichberg, V.I., Charli, J.L., Hersh, L.B., and McKelvy, J.F. (1980). Cleavage of substance P to an N- terminal tetrapeptide and a C- terminal heptapeptide by a post-proline cleaving enzyme from bovine brain. *Brain Research* **192**, 477-486.

Cridland, R.A., and Henry, J.L. (1988). N- and C- terminal fragments of substance P: Spinal effects in the rat tail flick test. *Brain Research Bulletin* **20**, 429-432.

Hall, M.E., Miley, F.B., and Stewart, J.M. (1987). Modulation of blood pressure by substance P: opposite effects of N- and C-terminal fragments on anesthetized rats. *Life Sciences* **40**, 1909-1914.

Howes, L.G. (1984). Central catecholaminergic neurones and spontaneously hypertensive rats. *Journal of Autonomic Pharmacology* **4**, 207-217.

Kapoor, V., and Chalmers, J.P. (1987). A simple, sensitive method for the determination of extracellular catecholamines in the rat hypothalamus using *in vivo* dialysis. *Journal of Neuroscience Methods* **19**, 173-182.

Pernow, B. (1983). Substance P. *Pharmacological Review* **35**, 86- .

Routledge, C., and Marsden, C.A. (1987). Adrenaline in the CNS: *in vivo* evidence for a functional pathway innervating the hypothalamus. *Neuropharmacology* **26**, 823-830.

Sakurada, T., Kuwahara, H., Takahashi, K., Sakurada, S., Kisara, K., and Terenius, L. (1988). Substance P(1-7) antagonizes substance P-induced aversive behaviour in mice. *Neuroscience Letters* **95**, 281-285.

Unger, Th., Rascher, W., Schuster, C., Pavlovitch, R., Schömig, A., Dietz, R., and Ganten, D. (1981). Central blood pressure effects of substance P and angiotensin II: role of the sympathetic nervous system and vasopressin. *European Journal of Pharmacology* **71**, 33-42.

Unger, Th., Becker, H., Petty, M., Demmert, G., Schneider, B., Ganten, D., and Lang, R.E. (1985). Differential effects of central Angiotensin II and Substance P on sympathetic nerve activity in conscious rats. *Circulation Research* **56**, 563-575.

Stress: Neuroendocrine and Molecular Approaches
Edited by R. Kvetnansky, R. McCarty and J. Axelrod

1992 Gordon and Breach Science
Publishers S.A., New York, USA.
Photocopying permitted by license only.

DECREASE IN THYROLIBERIN IN RAT SPINAL CORD AFTER IMMOBILIZATION AND EFFECTS OF PARAVENTRICULAR NUCLEUS (PVN) LESIONS

M. Dobrakovova, K. Povazanova, N. Michajlovskij
and V. Strbak

Institute of Experimental Endocrinology, Slovak Academy of Sciences,
Bratislava, Czecho-Slovakia

INTRODUCTION

Thyroliberin (TRH) was originally characterized as a hypothalamic tripeptide hormone which regulates pituitary thyroid stimulating hormone (TSH) secretion and was subsequently shown to produce neuromodulatory effects (Brown, 1981). Neurons in the paraventricular nucleus (PVN) project to the intermediolateral cell column (IML) of spinal cord (Swanson and Sawchenko, 1980). Many TRH-containing neurons also project to IML which may influence sympathetic preganglionic neurons (Hirsch and Helke, 1988). TRH immunoreactivity has also been described in the ventral horn of spinal cord in close association with motoneurons (Hokfelt *et al*, 1975) and in the dorsal horn (Harkness and Brownfield, 1986) where a role for the peptide in sensory processing has been suggested.

The aim of the present study was to examine the role of the PVN in regulation of TRH content in spinal cord and to see if immobilization stress affected its content. Surprisingly rapid changes in TRH content in thoracic and lumbar spinal cord suggest a role for this peptide in neuroendocrine stress responses. A portion of TRH content in the thoracic and lumbar spinal cord was found to be PVN-dependent.

MATERIAL AND METHODS

SPF male Wistar rats (VELAZ, Prague) weighing 250-300 g were used. The rats were housed in wire-mesh cages, (6 animals per cage) with a 12 hour light-12 hour dark photoperiod. Animals were provided a standard laboratory diet and tap water *ad libitum*.

PVN lesions were performed with a rotating knife as described previously (Makara *et al*, 1981). The knife was lowered into the brain 1.8 mm behind the bregma in the midline through the sagittal sinus, with the blade pointing caudally until the tip touched the base of the skull. Then the lesion was made by turning the knife 360^0 to the left and 360^0 to the right. For sham operations, the knife was lowered 5 mm below the surface of the brain and was not rotated. The animals were left to recover for 10 days until the experiment was begun.

The stress procedure involved forced immobilization (IMO) of rats by taping all four limbs to metal mounts attached to a board (Kvetnansky and Mikulaj, 1970). Rats were decapitated before and 1, 30 or 150 minutes after the beginning of IMO. Decapitation occurred while rats were immobilized.

Cervical (C 2-7), thoracic (TH 7-13) and lumbar (L 1-6) portions of SC were cleaned within 2 minutes after decapitation, weighed and homogenized in 20 volumes of 0.25% acetic acid. Homogenates were boiled for 15 minutes and then lyophilized. They were dissolved in assay buffer on the day of the RIA.

TRH was quantified by RIA using our own highly specific antibody which does not cross-react with TRH free acid, His-Pro-di-keto-piperazine, LHRH, GHRH and various amino acids.

After a logarithmic transformation, data were analyzed by two-way analysis of variance for repeated measures followed by paired comparisons (Dunn, 1976). Results are expressed as the geometrical means with SEMs.

RESULTS

TRH concentrations were one-third lower in the cervical than in thoracic and lumbar portions of SC (Figure 1). Lesions of PVN resulted in a decrease of TRH in the lower thoracic and lumbar SC, but not in the cervical part.

One minute after the beginning of IMO, TRH concentrations were diminished by one-third in thoracic and lumbar parts of SC (Figure 2). The levels were restored by the 30th minute of IMO and remained relatively constant out to 150 minutes. In PVN-lesioned animals, an increase of lumbar TRH concentration was found at the 30th minute of IMO as compared to baseline values (Figure 2).

DISCUSSION

The decrease of TRH concentrations in the lower thoracic and lumbar SC after PVN lesions suggests that a part of the neuropeptide content here may either originate in PVN or depend on PVN regulation. We have measured only total TRH content in several parts of SC. Therefore, we can only speculate on which of the TRH-containing neurons were affected by both PVN lesions and IMO. Our results suggest that neurons from PVN projecting to IML (Swanson and Sawchenko, 1980) contain TRH. It was found that TRH and other

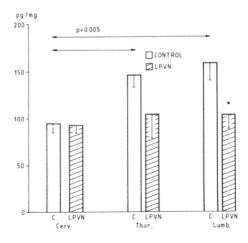

FIGURE 1.   TRH concentrations in cervical (C 2-7), lower thoracic (Th 7-12) and lumbar (L 1-6) parts of spinal cord 10 days after lesions of PVN (LPVN).

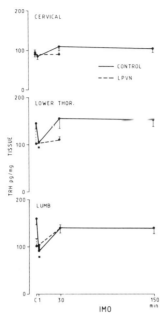

FIGURE 2.   Effects of IMO on TRH concentrations in spinal cord. The decrease in
controls at the 1st minute of IMO was significant in thoracic and lumbar parts,
where also a return to basal values by 30 minutes was noted. In PVN-lesioned
animals, the value for cervical cord at 1 minute was not measured.

neurotransmitter-specific neurons appose populations of preganglionic
sympathetic neurons in IML (projecting either to adrenal medulla or
cervical sympathetic trunk--Appel *et al*, 1987). It may be that there is
a link between immediate release of TRH in SC and release of
catecholamines during IMO (Kvetnansky *et al*, 1978). On the other
hand, the presence of TRH in ventral (Hokfelt *et al*, 1975) and dorsal
horns (Harkness and Brownfield, 1986) of SC suggest a possible role
of TRH-containing neurons in motor activity or sensory processing,
respectively, at the beginning of IMO. The definition of the neuron
population engaged in early TRH changes during IMO requires more
detailed study.

ACKNOWLEDGEMENTS

This project has been partially supported by SAS Research grant

431/1991 and by the International Atomic Energy Agency in Vienna, Austria (Research Contract # 6010/RB).

REFERENCES

Appel, N.M., Wessendorf, M.W., and Elde, R.P. (1987). Thyrotropin-releasing hormone in spinal cord: coexistence with serotonin and with substance P in fibers and terminals apposing identified preganglionic sympathetic neurons. *Brain Research* **415**, 137-143.

Brown, M.R. (1981). Thyrotropin releasing factor: a putative CNS regulator of the autonomic nervous system. *Life Sciences* **28**, 1789-1795.

Dunn, O.J. (1976). Multiple comparisons among means. *Journal of the American Statistical Association* **56**, 52-64.

Harkness, D.H., and Brownfield, M.S. (1986). A thyrotropin-releasing hormone-containing system in the rat dorsal horn separate from serotonin. *Brain Research* **384**, 323-333.

Hirsch, M.D., and Helke C.J. (1988). Bulbospinal thyrotropin-releasing hormone projections to the intermediolateral cell column: a double fluorescence immunohistochemical-retrograde tracing study in the rat. *Neuroscience* **2**, 625-637.

Hokfelt, T., Fuxe, K., Johansson, O., Jeffcoate, S., and White, N. (1975). Thyrotropin releasing hormone (TRH)-containing nerve terminals in certain brain stem nuclei and in the spinal cord. *Neuroscience Letters* **1**, 133-139.

Kvetnansky, R., Sun, C.L., Lake, C.R., Thoa, N.B., Torda, T., and Kopin I.J. (1978). Effect of handling and forced immobilization on rat plasma levels of epinephrine, norepinephrine and dopamine-beta-hydroxylase. *Endocrinology* **103**, 1868-1874.

Makara, G.B., Stark, E., Karteszi, M., Palkovits, M., and Rappay, G. (1981). Effects of paraventricular lesions on stimulated ACTH release and CRF in stalk-median eminence of the rat. *American Journal of Physiology* **200**, E441-E445.

Swanson L.W., and Sawchenko P.E. (1980). Paraventricular nucleus: A site for the integration of neuroendocrine and autonomic mechanisms. *Neuroendocrinology* **31**, 410-417.

Stress: Neuroendocrine and Molecular Approaches
Edited by R. Kvetnansky, R. McCarty and J. Axelrod

1992 Gordon and Breach Science
Publishers S.A., New York, USA.
Photocopying permitted by license only.

ROLE OF DOPAMINE AND SEROTONIN IN EMETIC RESPONSES OF SUNCUS MURINUS TO MOTION STRESS

N. Matsuki, Y. Torii, F. Okada and H. Saito

Department of Chemical Pharmacology,
Faculty of Pharmaceutical Sciences,
University of Tokyo, Tokyo, Japan.

INTRODUCTION

Emesis is generally considered as a defensive reflex to expel toxic food or substances ingested accidentally. However, there are a number of emetogenic stimuli which apparently do not fulfill the defensive purpose. Emesis caused by cancer chemotherapeutic drugs as a side effect produces serious discomfort in patients. Motion sickness also induces emesis, and space motion sickness has been an obstruction of space flight. However, research on emesis and motion sickness has been restricted partly because only a few experimental animals are able to vomit. Rodents (e.g., rats, mice, guinea pigs) and lagomorphs (e.g., rabbits) are the most widely used experimental animals and their biological characteristics have been documented well, but they never vomit.

We have shown that *Suncus murinus*, the musk shrew can vomit in response to various emetogenic drugs and motion sickness (Ueno *et al*, 1987, 1988). *Suncus murinus* is a species of insectivore, and its small size makes it ideal for handling and experimentation. *Suncus* is currently the most sensitive to motion sickness among experimental mammalian animals studied. In the present study, the effects of dopaminergic and serotonergic drugs on emetic responses were investigated.

SUNCUS MURINUS

Suncus murinus belongs to the family Soricidae of the order Insectivora and is distributed widely in tropical and subtropical areas. The insectivores are the most primitive and the earliest eutherians and regarded as the direct ancestors of the Primates (Colbert, 1958). *Suncus* is currently available as an experimental animal in Japan, and breeding is not very difficult. The body weight of adult animals is 50 to 80 g for males and 30 to 50 g for females. The gestation period is 30 days, and the average number of offspring is 3 to 4 per litter. They are sexually mature at about 2 months of age. Their life span has not been systematically studied but is probably longer than 2 years.

Recent studies from our laboratory indicate that *Suncus* is a unique experimental animal model for various types of research, such as studies on nerve growth factor (Ueyama *et al*, 1981), fatty liver degeneration (Yasuhara *et al*, in press), ß-adrenoceptor-mediated responses (Abe *et al*, 1988; Nagata *et al*, 1990), emesis (Ueno *et al*, 1987; Matsuki *et al*, 1988; Torii *et al*, 1991) and motion sickness (Ueno *et al*, 1988; Kaji *et al*, 1990, in press). Table 1 shows various stimuli which induce emesis in *Suncus*. Unlike rats, *Suncus* possesses steroid 17α-hydroxylase and synthesizes cortisol (Lin *et al*, 1986), and tyrosine hydroxylase levels in the adrenal are high (Maruoka *et al*, 1988, 1989), suggesting that the animal may be a good model for stress studies as well.

Experimental conditions are similar to those reported from our laboratory (Ueno *et al*, 1987, 1988; Kaji *et al*, 1990; Torii *et al*, 1991). Healthy adult *Suncus* weighing 40-80 g were used. The motion stimulus employed was reciprocal shaking with an amplitude of 40 mm and frequency of 1.0 Hz for 5 minutes. Each animal was transferred to an individual cage and after 5 to 10 minutes of acclimation, the emetogenic stimulus was started. Number of vomiting episodes and latency to the first emesis were recorded. Prophylactic drugs were administered 30 minutes prior to the emetogenic stimuli.

Effects of Dopaminergic Agonists and Antagonists

Apomorphine is a widely used emetogenic drug in animal research. However, species dependent variations in sensitivities to apomorphine are well known. Dogs are the most sensitive but monkeys are not and cats are less sensitive. Sensitivities of ferrets are controversial

TABLE 1. Various stimuli which induce emesis in Suncus murinus.

Stimuli	$ED_{50 \ (mg/kg)}$
Emetogenic Drugs	
Copper Sulfate (p.o.)	21.4
Emetine (s.c.)	47.6
Lobeline (s.c.)	2.8
Nicotine (s.c.)	7.9
Pilocarpine (s.c.)	ND
Veratrine (s.c.)	0.4
Cancer Chemotherapeutics	
Bleomycin (i.v.)	ND
Cisplatin (i.v.)	8.4
Cisplatin (i.p.)	10.0
Cyclophosphamide (i.v.,s.c.)	ND
5-Fluorouracil (i.v.)	ND
Methotrexate (i.v.)	ND
Mitomycin C (i.v.)	ND
Dopaminergic Agonists	
Bromocriptine (s.c.)	12.3
SKF 38393 (s.c.)	29.9
Serotonergic Agonists	
Serotonin (i.p.)	2.7
Serotonin (s.c.)	4.7
2-Methyl-Serotonin (i.p.)	0.97
Other Stimuli	
Motion	ND
X-irradiation	429 (cGy)

ED_{50} values were determined using up-and-down method. ND: not determined.

(Andrews *et al*, 1990). Ueno and co-workers (1987) reported that apomorphine does not cause emesis in *Suncus* but does increase spontaneous movements of animals.

As shown in Table 2, SKF 38393, a specific D_1 receptor agonist,

TABLE 2.   Emetic effects of dopaminergic agonists in *Suncus*.

ED$_{50}$ Values

Drug	Route	ED$_{50}$ (mg/kg)
Apomorphine	s.c.	No Vomiting
L-DOPA	s.c.	> 200
Bromocriptine	s.c.	12.3
SKF 38393	s.c.	29.9

Synergism

Drug	Dose (mg/kg s.c.)	Number of *Suncus* Vomited / Tested
SKF 38393	1.0	0 / 5
Bromocriptine	5.0	2 / 5
SKF 38393 + Bromocriptine	1.0 5.0	5 / 5

or bromocriptine, a specific D$_2$ receptor agonist, caused emesis. A combination of SKF 38393 and bromocriptine augmented the emetic response synergistically. Therefore, failure of apomorphine to induce emesis is not due to the lack of dopaminergic receptor(s)-mediated emetic responses. Additional pharmacological effects of apomorphine may mask or prevent the emesis. Both SCH 23390, a specific D$_1$ receptor antagonist, and YM01951-2, a specific D$_2$ receptor antagonist, prevented emesis induced by various drugs, motion and X-irradiation (Table 3). These emetogenic stimuli either act on the central nervous system (nicotine, veratrine, motion) or peripheral tissues (copper sulfate, cisplatin, serotonin, X-irradiation). Therefore, dopaminergic system(s) may be involved in the final common pathway for the vomiting reflex.

Effects of Serotonergic Agonists and Antagonists

Serotonergic 5-HT$_3$ receptor antagonists, such as ICS 205-930,

TABLE 3. Effects of dopaminergic and serotonergic drugs and vagotomy on emetic responses in *Suncus*.

Emetic Stimuli	Dopaminergic		Serotonergic		block by
	D_1 Antagonist SCH 23390	D_2 Antagonist YM01951-2	5-HT$_{1A}$ Agonist 8-OH-DPAT	5-HT$_3$ Antagonist ICS 205-930	Surgical Vagotomy
Motion (1Hz, 40mm)	< 1.0	< 2.0	0.037	-	NO
Nicotine (mg/kg, s.c.)	N.T.	N.T.	0.047	-	NO
Veratrine (mg/kg, s.c.)	< 1.0	< 4.0	0.121	-	N.T.
Copper Sulfate (40 mg/kg, p.o.)	1.3	1.0	0.283	-	YES
Cisplatin (20 mg/kg, i.p.)	2.8	1.1	0.857	0.025	YES
Serotonin (10 mg/kg, i.p.)	N.T.	N.T.	N.T.	0.008	YES
X-irradiation (800 cGy)	N.T.	N.T.	N.T.	0.020	YES

Values indicate ID$_{50}$ (mg/kg). -: no antiemetic effect. N.T. : not tested.

zacopride, BRL43694, and GR38032F, prevented emesis induced by cisplatin (Torii *et al*, 1991), serotonin (Torii *et al*, submitted) and X-irradiation (Table 3). However, the antagonists had no effect on veratrine-, nicotine-, copper sulfate- and motion-induced emesis. Cisplatin-, serotonin- and X-irradiation-induced emesis was completely prevented by surgical vagotomy. Our preliminary experiments showed an increase in serum serotonin levels after administration of cisplatin. These results suggest that a $5-HT_3$ receptor-mediated mechanism is involved only in the emesis caused by peripheral serotonin.

Recently serotonergic $5-HT_{1A}$ agonists have been introduced as anxiolytic drugs. One of the most potent $5-HT_{1A}$ agonists, 8-hydroxy-2-(di-n-propylamino)tetraline (8-OH-DPAT), strongly inhibited both central and peripheral emetic stimuli, but peripheral stimuli seemed to be less sensitive. Motion sickness- induced emesis was also prevented strongly, and 8-OH-DPAT is so far the most potent anti-motion sickness drug in *Suncus*. These results suggest that a $5-HT_{1A}$-receptor-mediated mechanism(s), probably located in the central nervous system, is involved in the emetic response common to various stimuli.

Adaptation to Motion Stress

Adaptation to motion stress was observed when animals were shaken repetitively (Ueno *et al*, 1988; Kaji *et al*, 1990). The adaptation seems to occur in sensory systems rather than motor systems. Table 4 indicates serum ACTH and cortisol levels just after a single motion stimulus or 2-weeks of daily motion stress. The motion stress itself did not affect serum content of ACTH or cortisol, but ACTH levels of the adapted animals were significantly lower. *Suncus* can be a good model for the study of adaptation.

TABLE 4. Serum ACTH and cortisol levels after single or repetitive motion stress in the *Suncus*.

	ACTH (pg/ml)	cortisol(ng/ml)
Control	231.1 ± 21.0	42 ± 5
Single Motion Stress[a]	238.1 ± 17.9	32 ± 3
Repetitive Motion Stress[b]	161.5 ± 15.6*	23 ± 5

Number of animals: 3 to 8. a: 2 Hz, 40 mm, 2 min. b: Severe motion stress (2 Hz, 40 mm, 15 min.) daily for 2 weeks. *: significantly different from control ($p < 0.05$).

REFERENCES

Abe, K., Wang, C-H., Tanaka, H., Saito, H., and Matsuki, N. (1988). Characteristics of cardiac ß-adrenoceptors in *Suncus murinus*. *Chemical and Pharmaceutical Bulletin* (Tokyo) **36**, 4081-4087.

Colbert, E.H. (1958). "Evolution of the Vertebrates", pp. 249-261. New York: John Wiley and Sons.

Kaji, T., Saito, H., Ueno, S., and Matsuki, N. (1990). Comparison of various motion stimuli on motion sickness and acquisition of adaptation in *Suncus murinus*. *Experimental Animals* **39**, 75-79.

Kaji, T., Saito, H., Ueno, S., Yasuhara, T., Nakajima, T., and Matsuki, N. (1991). Role of histamine in motion sickness of *Suncus murinus*. *Aviation, Space and Environmental Medicine*, **62**, 1053-1057.

Lin, S-C, Shiga, H., Kato, Y., Saito, H., and Kamei, S. (1986). Serum constituents of *Suncus murinus*. *Experimental Animals* **35**, 77-85.

Maruoka, Y., Saito, H., Tanaka, H., Nakajima, T., and Yazawa, K. (1988). Determination of catecholamines and their metabolites in urine of *Suncus murinus*. *Biogenic Amines* **5**, 483-488.

Maruoka, Y., Nishiyama, N., Saito, H., Nakajima, T., and Yazawa, K. (1989). Determination of catecholamines and their metabolites in brain, heart and adrenal of *Suncus murinus*. *Biogenic Amines* **6**, 135-143.

Matsuki, N., Ueno, S., Kaji, T., Ishihara, A., Wang, C-H., and Saito, H. (1988). Emesis induced by cancer chemotherapeutic agents in the *Suncus murinus*: A new experimental model. *Japanese Journal of Pharmacology* **48**, 303-306.

Nagata, K., Abe, K., Wang, C-H., Saito, H., and Matsuki, N. (1990). Deficiency of ß-adrenoceptor mediated relaxation in *Suncus* trachea. *Japanese Journal of Pharmacology* **52**, 115-121.

Torii, Y., Saito, H., and Matsuki, N. (1991). Selective blockade of cytotoxic drug-induced emesis by 5-HT3 receptor antagonists in *Suncus* murinus. *Japanese Journal of Pharmacology* **55**, 107-113.

Torii, Y., Saito, H., and Matsuki, N. (1991). Serotonin is emetogenic in *Suncus murinus*. *Naunyn-Schmiedeberg's Archives of Pharmacology*, in press.

Ueno, S., Matsuki, N., and Saito, H. (1987). *Suncus murinus*: a new experimental model in emesis research. *Life Sciences* **41**, 513-518.

Ueno, S., Matsuki, N., and Saito, H. (1988). *Suncus murinus* as a new experimental animal model for motion sickness. *Life Sciences* **43**, 413-420.

Ueyama, T., Saito, H., and Yohro, T. (1981). *Suncus murinus* submandibular gland and prostate are new sources of nerve growth factor. *Biomedical Research* **2**, 438-441.

Yasuhara, M., Ohama, T., Matsuki, N., Saito, H., Shiga, J., Inoue, K., Kurokawa, K., and Teramoto, T. (1991). Induction of fatty liver by fasting in *Suncus*. Journal of Lipid Research, **32**, 887-891.

PART TWO

PERIPHERAL MONOAMINE AND NEUROPEPTIDE RESPONSES TO STRESS

Stress: Neuroendocrine and Molecular Approaches
Edited by R. Kvetnansky, R. McCarty and J. Axelrod

1992 Gordon and Breach Science
Publishers S.A., New York, USA.
Photocopying permitted by license only.

STRESS-INDUCED CHANGES IN PLASMA DOPA LEVELS DEPEND ON SYMPATHETIC ACTIVITY AND TYROSINE HYDROXYLATION

R. Kvetnansky[1,2], K. Fukuhara[1], V. K. Weise[1], I. Armando[1], I. J. Kopin[1] and D. S. Goldstein[1]

[1]Clinical Neuroscience Branch, National Institute of Neurological Disorders and Stroke, National Institutes of Health Bethesda, Maryland, 20892 USA
[2]Institute of Experimental Endocrinology, Slovak Academy of Sciences 833 06 Bratislava, Czecho-Slovakia

INTRODUCTION

DOPA is the precursor of all the endogenous catecholamines and the immediate product of the enzymatic rate-limiting step in catecholamine biosynthesis, hydroxylation of tyrosine. Dopa concentration in plasma is higher than that of the sympathetic neurotransmitter norepinephrine (Goldstein *et al*, 1984; Grossman *et al*, 1990). Since DOPA is converted rapidly to dopamine by L-aromatic-amino-acid decarboxylase in catecholamine-synthesizing cells, traditional views about catecholamine biosynthesis failed to consider a relationship between plasma DOPA levels and the turnover of endogenous catecholamines.

The source and meaning of plasma DOPA in terms of sympathoneural function have been a controversial issue. Whereas Goldstein *et al* (1987) and Eisenhofer *et al* (1988b) found that several manipulations that alter the rate of tyrosine hydroxylation produced directionally similar changes in plasma DOPA levels, Eldrup *et al* (1989) reported that sympathectomy did not affect tissue

139

concentrations of DOPA and that manipulations producing large changes in plasma norepinephrine (NE) levels did not affect plasma DOPA levels. Devalon *et al* (1989) found an increase in plasma DOPA levels during exercise, a situation known to activate the sympathoadrenal system.

The present study evaluated effects of some stressors; i.e., immobilization and handling, which produce profound sympathoadrenal activation (Kvetnansky *et al*, 1978), on plasma DOPA and catecholamine levels in conscious rats. Groups of rats were pretreated with chlorisondamine to block ganglionic neurotransmission or with α-methyl-para-tyrosine to inhibit tyrosine hydroxylation. If plasma DOPA levels increase acutely during stress, and if pretreatment with chlorisondamine attenuates stress-evoked increases in plasma DOPA levels, then changes in plasma DOPA levels depend on post-ganglionic sympathoneural activity. If pretreatment with α-methyl-para-tyrosine decreases DOPA levels to baseline values and attenuates responses of plasma DOPA during stress, DOPA in plasma emanates from catecholamine-synthesizing cells and increases in DOPA levels during sympathoadrenal activation reflect increases in the rate of tyrosine hydroxylation. Goldstein *et al* (1987) reported that treatment with α-methyl-para-tyrosine decreased plasma DOPA levels in conscious dogs. The effect of this treatment on plasma DOPA responses during stress was not studied.

Decapitation almost instantaneously stimulates especially adrenomedullary secretion, resulting in extremely high plasma levels of epinephrine (EPI) (Kvetnansky *et al*, 1978). If plasma DOPA levels are derived partly from adrenomedullary secretion, then decapitation can be expected to increase plasma DOPA levels. On the other hand, if plasma DOPA levels are derived partly from sympathetic nerve endings, then those levels will be increased in adrenalectomized animals, both before and during immobilization, since adrenalectomy augments sympathoneural responses to stress (Kvetnansky *et al*, 1979a; Udelsman *et al*, 1987).

The aim of this study was to establish 1) whether stress affects plasma DOPA levels, and if so, whether there is a correlation with changes in plasma catecholamine levels; 2) whether the sympatho-neural system is a major source of DOPA in plasma; and 3) whether changes in plasma DOPA levels indicate altered catecholamine synthesis at the step of tyrosine hydroxylation.

METHODS

Male, murine pathogen-free, Sprague-Dawley rats (280-340 g) were obtained from Taconic Farm (Germantown, NY). The experiments began at least 7 days after arrival of the animals. The animals were housed 5-6 per cage under light-controlled conditions (light on from 0700-1900 hours) with a room temperature of 23° ± 2° C. Food and water were available *ad libitum*. All experiments were done between 0800-1200 hours.

Adrenalectomized or sham-operated rats were also obtained from the same source. Animals were tested at least 16 days after surgery. No mineralocorticoid was administered; 0.9% saline solution was provided *ad libitum* to adrenalectomized animals.

Blood samples from awake rats were collected into heparinized tubes via a chronically indwelling polyethylene catheter that had been inserted into the tail artery 20-24 hours prior to the experiment. This allowed us to obtain plasma catecholamine levels when the animals were unstressed by manipulations related to blood sampling. Details of the cannulation procedure were described previously (Chiueh and Kopin, 1978; Kvetnansky *et al*, 1978). After surgery, each rat was housed in an individual plastic cage, with the catheter extending out of the cage.

Blood samples (0.5-0.8 ml) were collected via the catheter at the indicated times, and the same volume of heparinized saline (100 IU/ml) was administered intra-arterially after each blood sample was obtained. The blood was centrifuged (3000 x g) at 4° C for 15 minutes and the plasma stored at -70° C until assayed. In some animals, blood was not obtained via an indwelling cannula but by decapitation.

Immobilization stress as conducted under this protocol was approved by the NINDS Animal Care and Use Committee. Immobilization was done by taping all four limbs of the rat to metal mounts attached to a board, as described previously (Kvetnansky and Mikulaj, 1970). Two types of immobilization experiments were performed. In some animals, blood was collected before and then after the 5th, 20th, 60th and 120th minutes of immobilization. In others, blood was collected before and then after the 1st, 3rd and 5th minutes of immobilization. In the latter group, after the 5 minute immobilization period, the animals were released from the boards and returned to their home cages. Twenty minutes later, a final blood sample was obtained. In animals without indwelling cannulas, decapitation was used to obtain blood after 2 hours of immobilization.

After collection of the control blood sample via the catheter, animals were picked up, handled gently for 1 minute and then returned to their home cages. Blood samples were collected immediately at the end of handling and 5 and 15 minutes after return of rats to their home cages.

Chlorisondamine (Ecolid™, Ciba Pharmaceutical, Summit, NJ) dissolved in saline or saline alone (n = 9 per group) was administered intraperitoneally at a dose of 10 mg/kg. Animals were cannulated as described above. After a baseline blood sample was obtained, chlorisondamine was injected. One hour later, another blood sample was obtained, and then the animals were immobilized for 2 hours. Blood was sampled at 5, 20, 60, and 120 minutes after initiation of immobilization.

D,L-α-methyl-para-tyrosine methyl ester hydrochloride (Research Biochemicals, Inc., Natick, MA) or saline was administered in two doses (100 or 200 mg/kg) intraperitoneally in 0.1 ml of saline per 100 g body weight. In one set of experiments, α-methyl-para-tyrosine or saline was administered to animals with indwelling arterial cannulas, and blood was sampled at baseline and 1, 2, and 3 hours after drug administration.

In a second set of experiments, α-methyl-para-tyrosine or saline was administered to animals (n = 7-8 per group) with indwelling arterial cannulas (100 mg/kg). One hour later, after collection of an arterial blood sample, the animals were immobilized as described above. Blood was sampled again at 5, 20, 60, and 120 minutes of immobilization.

In a third set of experiments, blood was collected after decapitation 3 hours after drug administration. There were 4 treatment groups, with 7-8 rats per group: saline, with decapitation 3 hours later; α-methyl-para-tyrosine (200 mg/kg), with decapitation 3 hours later; saline, followed 1 hour later by immobilization for 2 hours; and α-methyl-para-tyrosine (200 mg/kg) followed 1 hour later by immobilization for 2 hours.

Adrenalectomized animals or sham-operated animals (n = 7 per group) were tested 20-24 hours after insertion of an arterial cannula as described above. After a baseline blood sample was obtained, the animals were subjected to immobilization, with blood sampled at 5, 20, 60, and 120 minutes.

Plasma concentrations of catechols--norepinephrine (NE),

epinephrine (EPI), dopamine (DA), 3,4-dihydroxyphenylalanine (DOPA), 3,4-dihydroxyphenylglycol (DHPG), and 3,4-dihydroxyphenylacetic acid (DOPAC)--were assayed using liquid chromatography with electrochemical detection after batch alumina extraction, as described by Eisenhofer *et al* (1986), with minor modifications

Results are presented as mean values ± SEM. Data obtained from blood samples from the same cannulated rat were analyzed by one- or two-factor repeated measures ANOVA, as appropriate, followed by Fisher's PLSD test. The significance of differences at the specific time points was evaluated by one-factor ANOVA with Fisher's PLSD post-hoc test. Data obtained from blood samples after decapitation of animals were analyzed by one-factor ANOVA followed by Fisher's PLSD post-hoc test. Statistical significance was considered at p values less than 0.05.

RESULTS

Plasma DOPA levels were significantly increased at one minute of immobilization and were further increased at the 5th minute (Figure 1). In contrast, plasma NE and EPI levels reached maximum values at one minute of IMO. By 20 minutes after release from restraint, plasma levels of DOPA, NE, and EPI had decreased to values slightly above those of baseline. (Figure 1).

Gentle handling of rats for one minute significantly increased plasma levels of DOPA, NE and EPI (Figure 2) which rapidly returned to baseline values within 5 minutes after handling.

Administration of α-methyl-para-tyrosine at both the 100 mg/kg and 200 mg/kg doses significantly decreased plasma DOPA levels by about 50% (Figure 3). The decreases in DOPA levels persisted for at least three hours. Similarly, when blood was obtained by decapitation, plasma levels of DOPA were significantly lower in rats treated with α-methyl-para-tyrosine than in rats treated with saline (Figure 5).

Pretreatment with α-methyl-para-tyrosine markedly affected the responses of plasma DOPA levels during immobilization (Figure 4). Instead of rapid increases in DOPA levels, peaking at 5 minutes of immobilization, DOPA levels in α-methyl-para-tyrosine-treated animals were unchanged at 5 minutes of immobilization, compared with the pre-immobilization baseline values 1 hour after administration of the drug. Thereafter, plasma DOPA levels gradually increased during the

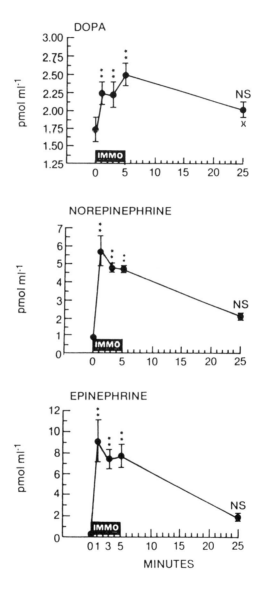

FIGURE 1. Effects of brief immobilization (1-5 minutes) and post-immobilization interval (20 minutes) on plasma DOPA and catecholamine levels. Mean values ± SEM for 7-8 rats. (**)$p<0.01$; NS -- not significantly different from pre-immobilization baseline. (x) $p<0.05$ compared to 5 minutes of immobilization.

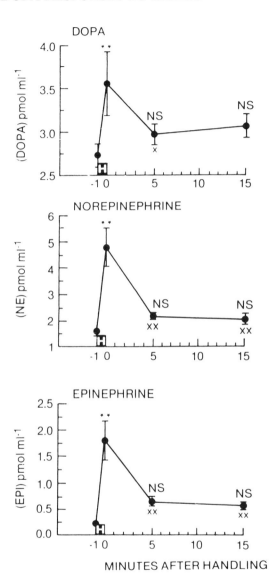

FIGURE 2.   Effects of one minute of handling and post-handling intervals on plasma DOPA and catecholamine levels.   Mean values ±SEM for 6-7 animals. (**)p<0.01 compared to pre-handling baseline NS -- not significantly different from baseline.   (x)p<0.05 and (xx)p<0.01 compared to one minute of handling.

R. KVETNANSKY *et al*

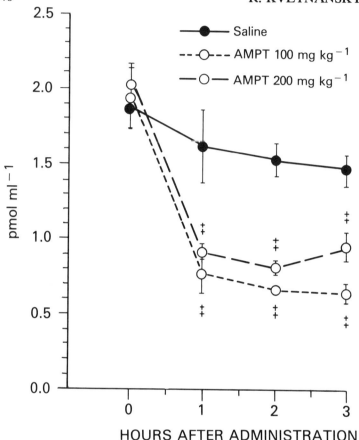

FIGURE 3. Effects of α-methyl-para-tyrosine (AMPT) on plasma DOPA levels in rats, with blood obtained via an indwelling arterial cannula. Mean values ± SEM for 5 rats per group. (+ +) p<0.01 compared to saline-treated group.

2 hour immobilization period but did not return to levels obtained before α-methyl-para-tyrosine treatment.

In decapitated rats, a two hour IMO produced only slight elevations in plasma DOPA levels which significantly dropped after AMPT administration (Figure 5). Animals pre-treated with α-methyl-para-tyrosine also had markedly attenuated responses of plasma levels of DA (Figure 6) and the DA metabolites DOPAC and HVA (Figure 7).

In marked contrast to the attenuation or abolition of DOPA, DA, and DOPAC responses during immobilization in animals

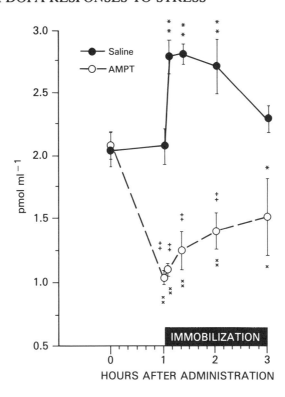

FIGURE 4. Effects of immobilization on plasma levels of DOPA in rats pre-treated with saline or with α-methyl-para-tyrosine (AMPT; 100 mg/kg), with blood obtained via an indwelling arterial cannula. Mean values ± SEM for 7-8 rats per group. (*) $p < 0.05$ and (**) $p < 0.01$ compared to pre-immobilization baseline 1 hour after administration of AMPT or saline (T = 1). (+ +) $p < 0.01$ compared to saline-treated group. (x) $p < 0.05$ and (xx) $p < 0.01$ compared to pre-immobilization baseline before drug treatment (T = 0).

pretreated with α-methyl-para-tyrosine, responses of plasma NE (Figure 8) and MHPG (Figure 9) levels were basically unchanged and responses of plasma EPI (Figure 8) were augmented. Although acute increases in plasma DHPG levels during immobilization were largely preserved in rats pretreated with α-methyl-para-tyrosine, as immobilization continued, the levels declined to values only slightly higher than those obtained just before the immobilization (Figure 9).

Treatment with chlorisondamine significantly decreased plasma DOPA and NE levels (Figure 10) and virtually abolished responses of

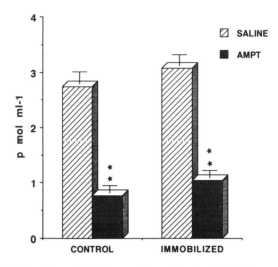

FIGURE 5.   Effects of immobilization on plasma levels of DOPA in rats pre-treated with saline or with α-methyl-para-tyrosine (AMPT; 200 mg/kg), with blood obtained by decapitation.   Mean values ± SEM for 7-8 rats per group.   (**) p<0.01 compared to saline treatment.

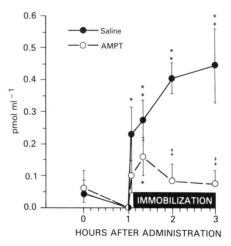

FIGURE 6.   Effects of immobilization on plasma levels of dopamine in rats pre-treated with saline or with α-methyl-para-tyrosine (AMPT), with blood obtained via an indwelling arterial cannula.   Mean values ± SEM for 7-8 rats per group. (*)p<0.05 and (**) p<0.01 compared to pre-immobilization baseline 1 hour after administration of AMPT or saline (T=1).   (++)p<0.01 compared to saline-treated group.

plasma DOPA and NE during immobilization (Figure 10). In adrenalectomized animals, baseline plasma levels of DOPA and NE were increased compared with levels in sham-operated animals (Figure 11). During the 2 hour period of immobilization, plasma DOPA levels progressively increased in adrenalectomized animals, whereas in sham-operated animals plasma DOPA levels tended to decrease. Plasma NE levels were significantly higher in adrenalectomized than in sham-operated animals at all time points during the immobilization period (Figure 11).

DISCUSSION

The present results demonstrate that during acute immobilization stress in conscious rats, large, rapid increases in plasma levels of catecholamines and their metabolites are attended by rapid increases in plasma levels of the precursor, DOPA. Since chlorisondamine treatment decreased baseline DOPA levels, plasma DOPA concentrations depend at least in part on post-ganglionic sympathoneural outflow, and since the increases in plasma DOPA levels were abolished in rats pretreated with chlorisondamine, increments in plasma DOPA levels during immobilization depend entirely on increases in post-ganglionic sympathetic activity.

These results generally agree with those of Szemeredi et al (1991) in pithed rats, where increments in plasma DOPA levels during electrical stimulation of the spinal cord were virtually abolished in animals pretreated with chlorisondamine, and with those of Garty et al (1989) in anesthetized rats, where reflexive increases in plasma DOPA levels during nitroprusside-induced hypotension were attenuated in ganglion-blocked animals.

Administration of α-methyl-para-tyrosine to inhibit tyrosine hydroxylation decreased plasma DOPA levels by over 50%, consistent with the findings of Goldstein et al (1987) in conscious dogs and indicating that DOPA in plasma is derived substantially from catecholamine-synthesizing cells. The virtual abolition of rapid increases in DOPA levels during immobilization in rats treated with α-methyl-para-tyrosine suggests that increases in plasma DOPA levels during exposure to this stressor depend on tyrosine hydroxylation.

Although tissue content of tyrosine hydroxylase increases only slowly during stress (Kvetnansky et al, 1970, 1971), tyrosine

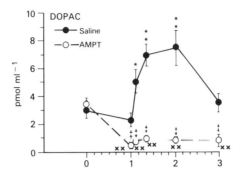

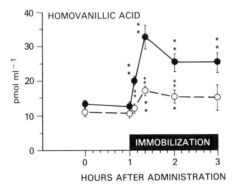

FIGURE 7.   Effects of immobilization on plasma levels of dihydroxyphenylacetic acid (DOPAC) and homovanillic acid in rats pre-treated with saline or with α-methyl-para- tyrosine (AMPT), with blood obtained via an indwelling arterial cannula. Mean values ± SEM for 7-8 rats per group. (*) $p < 0.05$ and (**) $p < 0.01$ compared to pre-immobilization baseline 1 hour after administration of AMPT or saline (T = 1). (+ +) $p < 0.01$ compared to saline-treated group. (xx) $p < 0.01$ compared to pre-immobilization baseline before drug treatment (T = 0).

hydroxylase activity increases rapidly during sympathetic stimulation (Weiner *et al*, 1978) and during the influence of some stressors (Fluharty *et al*, 1985), apparently due to phosphorylation-induced increases in affinity of the enzyme for its pterin co-factor (Fluharty *et al*, 1985). If there were an available neuronal (or extraneuronal) store of DOPA that could be released during immobilization, then pretreatment with α-methyl-para- tyrosine would not produce such a marked attenuation of plasma DOPA responses. The results therefore suggest that the rapid release of DOPA into the circulation during

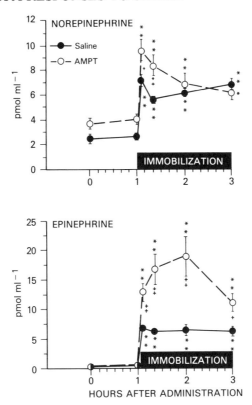

FIGURE 8.   Effects of immobilization on plasma levels of norepinephrine and epinephrine in rats pre-treated with saline or with α-methyl-para-tyrosine (AMPT), with blood obtained via an indwelling arterial cannula. Mean values ± SEM for 7-8 rats per group. (*) $p < 0.05$;(**) $p < 0.01$ compared to pre-immobilization baseline 1 hour after administration of AMPT or saline (T = 1). (+) $p < 0.05$; (+ +) $p < 0.01$ compared to saline-treated group.

immobilization is from transiently increased, axoplasmic, newly-synthesized DOPA.

In animals pretreated with α-methyl-para-tyrosine, plasma DOPA levels gradually increased as immobilization continued. Since α-methyl-para-tyrosine produced sustained decreases in plasma DOPA levels in animals that did not undergo immobilization, the gradual increases in DOPA levels cannot be explained simply by decreases over time in the extent of inhibition of tyrosine hydroxylase. Moreover, in animals pretreated with α-methyl-para-tyrosine, plasma

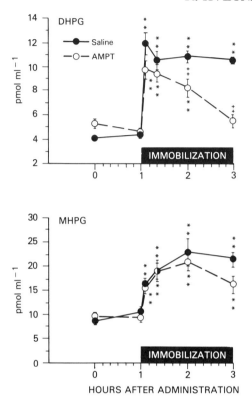

FIGURE 9. Effects of immobilization on plasma levels of dihydroxyphenylglycol (DHPG) and methoxyhydroxyphenylglycol (MHPG) in rats pre-treated with saline or with α-methyl-para-tyrosine (AMPT), with blood obtained via an indwelling arterial cannula. Mean values ± SEM for 7-8 rats per group. (**) $p < 0.01$ compared to pre-immobilization baseline 1 hour after administration of AMPT or saline (T = 1). (+ +) $p < 0.01$ compared to saline-treated group.

levels of DA, DOPAC, and HVA remained markedly decreased throughout the immobilization period, indicating persistent suppression of catecholamine biosynthesis. Since the slow increases in plasma DOPA concentrations during prolonged immobilization in α-methyl-para-tyrosine-treated animals seem to reflect neither a return of tyrosine hydroxylase activity towards baseline levels nor an elimination of the competitive inhibition during sympathoadrenal activation, there may be a decrease in DOPA clearance or an increased contribution to plasma DOPA from an additional source.

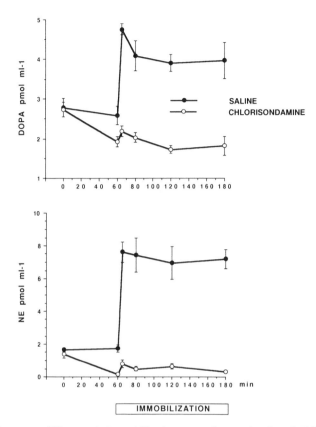

FIGURE 10. Effects of immobilization on plasma levels of DOPA and norepinephrine (NE) in rats pre-treated with saline or with chlorisondamine (10 mg/kg), with blood obtained via an indwelling arterial cannula. Mean values ± SEM for 9 rats per group.

Pretreatment with α-methyl-para-tyrosine markedly attenuated responses of plasma DA and its metabolites without attenuating responses of plasma levels of NE and its metabolites during immobilization stress. Thus, relatively little of the DA released during sympathoadrenal stimulation appears to be derived from a stored pool of DA in dopaminergic or noradrenergic nerves and increases in DA release are presumably not due to decreases in intra-neuronal metabolism of DA; instead, it appears that immobilization-induced increments in DA release are related to enhanced synthesis of DA from DOPA.

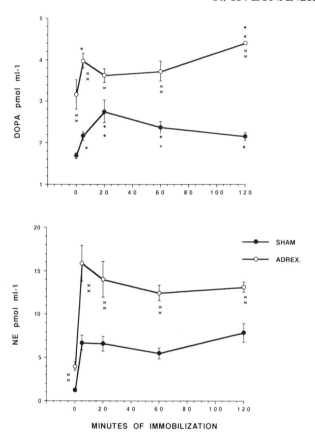

FIGURE 11. Effects of immobilization on plasma levels of DOPA (upper panel) and norepinephrine (NE, lower panel) in adrenalectomized (ADREX) or in sham-operated rats, with blood obtained via an indwelling arterial cannula. Mean values ± SEM for 7 rats per group. (*) $p < 0.05$ and (**) $p < 0.01$ compared to baseline value (T = 0). (x) $p < 0.05$ and (xx) $p < 0.01$ compared to sham-operated animals.

Decapitation, which is well-known to produce dramatic increases in plasma catecholamine concentrations (Kvetnansky *et al*, 1978), produced no change in plasma DOPA levels, indicating that even profound sympathetic-adrenomedullary stimulation does not immediately release DOPA into the bloodstream. The more rapid increases in plasma NE and EPI than in DOPA levels during decapitation and immobilization indicate that DOPA is not co-released by exocytosis of vesicles in sympathetic nerve terminals or the adrenal medulla. It has been suggested that DOPA is released by diffusion from the axoplasm (Devalon *et al*, 1989).

Elevated DOPA and NE levels in adrenalectomized animals, both at baseline and during immobilization, confirm the impression that sympathoneural outflow during stress is enhanced in the presence of low circulating concentrations of adrenal corticosteroids (Kvetnansky *et al*, 1979a; Udelsman *et al*, 1987) and are also consistent with the view that plasma DOPA is derived from sympathetic nerve terminals. Adrenal demedullation does not enhance responses of plasma NE (Kvetnansky *et al*, 1979b) or DOPA levels (unpublished observations) during immobilization stress.

Whereas in the study of Goldstein *et al* (1987), plasma levels of DHPG and DOPAC were both decreased by about 90% in dogs treated with α-methyl-para-tyrosine, in the present study with rats, α-methyl-para-tyrosine treatment differentially decreased plasma DOPAC levels. This difference can be explained by the delay of the first blood sample by 24 hours in the dog study. Since noradrenergic terminals store much more NE than DA, blockade of tyrosine hydroxylase may be expected to decrease rapidly the content of DA in sympathetic nerve terminals, and so plasma DOPAC levels would also fall rapidly. In contrast, since under resting conditions the majority of DHPG in plasma is from leakage of NE from vesicles into the axoplasm (Eisenhofer *et al*, 1988a; Goldstein *et al*, 1988), DHPG levels may be maintained partially by continued NE leakage. Eventually, inhibition of catecholamine biosynthesis results in profoundly decreased levels of both DHPG and DOPAC.

In conclusion, immobilization stress or handling rapidly increased plasma DOPA levels in conscious rats. The increase was abolished by chlorisondamine, indicating dependence of the DOPA responses on post-ganglionic sympathoneural activity. The findings with α-methyl-para-tyrosine suggest that plasma DOPA levels under baseline conditions are derived at least partly from catecholamine-synthesizing cells and that during immobilization, rapid increases in plasma DOPA levels reflect rapid increases in tyrosine hydroxylation. Plasma DOPA levels increase gradually in α-methyl-para-tyrosine-treated animals as immobilization continues; the mechanism for this increase is obscure. The results generally support the view that acute changes in plasma DOPA levels during stress reflect changes in the rate of catecholamine biosynthesis in sympathetic nerve terminals.

REFERENCES

Chiueh, C.C., and Kopin I.J. (1978). Hyperresponsiveness of spontaneously hypertensive rats to indirect measurement of blood pressure. *American Journal of Physiology* **234**, H690-H695.

Devalon, M.L., Miller, T.D., Squires, R.W., Rogers, P.J., Bove, A.A., and Tyce G.M. (1989). Dopa in plasma increases during acute exercise and after exercise training. *Journal of Laboratory and Clinical Medicine* **114**, 321-327.

Eisenhofer, G., Goldstein, D.S., Ropchak, T.G., Nguyen, H.Q., Keiser, H.R., and Kopin, I.J. (1988a). Source and physiological significance of plasma 3,4-dihydroxyphenylglycol and 3-methoxy-4-hydroxyphenylglycol. *Journal of the Autonomic Nervous System* **24**, 1-14.

Eisenhofer, G., Ropchak, T., Nguyen, H., Keiser, H.R., Kopin, I.J., and Goldstein, D.S. (1988b). Source and physiological significance of plasma 3,4-dihydroxyphenylalanine in the rat. *Journal of Neurochemistry* **51**, 1204-1213.

Eldrup, E., Christensen, N.J., Andreasen, J., and Hilsted, J. (1989). Plasma dihydroxyphenylalanine (DOPA) is independent of sympathetic activity in humans. *European Journal of Clinical Investigation* **19**, 514-517.

Fluharty, S.J., Snyder, G.L., Zigmond, M.J., and Stricker, E.M. (1985). Tyrosine hydroxylase activity and catecholamine biosynthesis in the adrenal medulla of rats during stress. *Journal of Pharmacology and Experimental Therapeutics* **233**, 32-38.

Garty, M., Deka-Starosta, A., Chang, P.C., Eisenhofer, G., Zukowska-Grojec, Z., Stull, R., Kopin, I.J., and Goldstein, D.S. (1989). Plasma levels of catechols during reflexive changes in sympathetic nerve activity. *Neurochemical Research* **14**, 523-531.

Goldstein, D.S., Eisenhofer, G., Stull, R., Folio, C.J., Keiser, H.R., and Kopin, I.J. (1988). Plasma dihydroxyphenylglycol and the intraneuronal disposition of norepinephrine in humans. *Journal of Clinical Investigation* **81**, 213-220.

Goldstein, D.S., Stull, R.W., Zimlichman, R., Levinson, P.D., Smith, H., and Keiser, H.R. (1984). Simultaneous measurement of DOPA, DOPAC, and catecholamines in plasma by liquid chromatography with electrochemical detection. *Clinical Chemistry* **30**, 815-816.

Goldstein, D.S., Udelsman, R., Eisenhofer, G., Keiser, H.R., and Kopin, I.J. (1987). Neuronal source of plasma dihydroxyphenylalanine. *Journal of Clinical Endocrinology and Metabolism* **64**, 856-861.

Grossman, E., Hoffman, A., Chang, P.C., Keiser, H.R., and Goldstein, D.S. (1990). Increased spillover of dopa into arterial blood during dietary salt loading. *Clinical Science* **78**, 423-429.

Kvetnansky, R., and Mikulaj, L. (1970). Adrenal and urinary catecholamines in rats during adaptation to repeated immobilization stress. *Endocrinology* **87**, 738-743.

Kvetnansky, R., Sun, C., Lake, C.R., Thoa, N.B., Torda, T., and Kopin, I.J. (1978). Effect of handling and forced immobilization on rat plasma levels of epinephrine, norepinephrine and dopamine-ß-hydroxylase. *Endocrinology* **103**, 1868-1874.

Kvetnansky, R., Weise, V.K., Gewirtz, G.P., and Kopin, I.J. (1971). Synthesis of adrenal catecholamines in rats during and after immobilization stress. *Endocrinology* **89**, 46-49.

Kvetnansky, R., Weise, V.K., and Kopin,I.J. (1970). Elevation of adrenal tyrosine hydroxylase and phenylethanolamine-N-methyltransferase by repeated immobilization of rats. *Endocrinology* **87**, 744-749.

Kvetnansky, R., Weise, V.K., and Kopin, I.J. (1979a). The origins of plasma epinephrine, norepinephrine and dopamine levels in stressed rats. In: E. Usdin, I.J.Kopin, and J. Barchas (Eds.), "Catecholamines: Basic and Clinical Frontiers", Volume I, pp. 684-686. New York: Pergamon Press.

Kvetnansky, R., Weise, V.K., Thoa, N.B., and Kopin, I.J. (1979b). Effects of chronic guanethidine treatment and adrenal medullectomy on plasma levels of catecholamines and corticosterone in forcibly immobilized rats. *Journal of Pharmacology and Experimental Therapeutics* **209**, 287-291.

Szemeredi, K., Pacak, K., Kopin, I.J., and Goldstein, D.S. (1991). Sympathoneural and skeletal muscle contributions to plasma dopa responses in pithed rats. *Journal of the Autonomic Nervous System*, in press.

Udelsman, R., Goldstein, D.S., Loriaux, D.L., and Chrousos, G.P. (1987). Catecholamine- glucocorticoid interactions during surgical stress. *Journal of Surgical Research* **43**, 539-545.

Weiner, N., Lee, F.L., Dreyer, E., and Barnes, E. (1978). The activation of tyrosine hydroxylase in noradrenergic neurons during acute nerve stimulation. *Life Sciences* **22**, 1197-1216.

Stress: Neuroendocrine and Molecular Approaches
Edited by R. Kvetnansky, R. McCarty and J. Axelrod

1992 Gordon and Breach Science
Publishers S.A., New York, USA.
Photocopying permitted by license only.

SOURCES OF DOPA AND SULFATED CATECHOLAMINES IN PLASMA

G. M. Tyce[1], B. Banwart[6], S. L. Chinnow[1], S. L. Chritton[4],
M. K. Dousa[2], L. W. Hunter[3], E. W. Kristensen[5]
and D. K. Rorie[3]

Departments of [1]Physiology, [2]Laboratory Medicine,
[3]Anesthesiology, [4]Graduate School of Medicine
Mayo Clinic and Foundation, Rochester, Minnesota, USA
[5]Abbott Labs, Abbott Park, Illinois, USA
[6]University of Iowa, Iowa City, Iowa, USA

INTRODUCTION

Dopa is present in plasma of man and several experimental animals (Dousa and Tyce, 1988). The concentrations of dopa in plasma exceed the concentrations of free catecholamines. It has been suggested (Goldstein *et al*, 1987) that concentrations of dopa in plasma reflect the activity of tyrosine hydroxylase, the enzyme that is rate-limiting in catecholamine biosynthesis. Tyrosine hydroxylase is restricted to noradrenergic neurons in the peripheral nervous system, to chromaffin cells in the adrenal medulla, and to catecholaminergic tracts in the brain. Tyrosine hydroxylase activity is increased during nerve stimulation (Morgenroth, Boadle-Biber and Roth, 1974), so dopa concentrations in plasma may be expected to reflect indirectly sympathetic nerve stimulation. However, there is some disagreement that neurons are the sole source of plasma dopa (Anton, 1991; Kuchel *et al*, 1990). Catecholamine sulfates, especially dopamine (DA) sulfate, are also very abundant constituents of plasma (Dousa and Tyce, 1988), and the sources of these catecholamine sulfates are not known. Diet

159

has been considered to be a source of dopa and of conjugated catecholamine, and results of experiments on the effects of diet on the concentrations of these compounds in plasma are presented herein. This communication is also a review of recent experiments in which the production of dopa and of sulfated catecholamine was studied in a number of perfused or superfused isolated tissue preparations.

METHODS

Effects of Feeding on Plasma Dopa and Conjugated Catecholamines

Dogs of either sex weighing 22 to 24 kg were fasted for 24 hours. They were fed a 0.4 kg can of dog food, and blood samples (3-4 ml) were drawn from the jugular vein before and at hourly intervals after feeding.

Perfusion of Isolated Dog Adrenal Glands

The method for isolation and perfusion of dog adrenal glands has been described (Chritton *et al*, 1991). Briefly, adrenals were removed from dogs and were retrogradely perfused with warmed, oxygenated Krebs-Ringer solution. After a 60 minute stabilization period the glands were perfused for 2 minutes with Krebs-Ringer to which carbachol (3 x 10^{-5}M) had been added. After this time the glands were perfused with plain Krebs-Ringer for a further 40 to 60 minutes. Superfusate was collected continuously in intervals into tubes containing sodium metabisulfate (10 μl of a 5% solution per ml of perfusate collected) as an antioxidant.

Superfused Minces of Dog Sympathetic Ganglia

Lumbar ganglia were removed from dogs, minced, and superfused as described by Kristensen *et al* (1990). Superfusate was collected in 10 minute intervals. After a 60 minute stabilization period tissues were exposed to a stimulus for 10 minutes. The stimuli employed were 100 μM amphetamine or 80 mM K^+. In some experiments Ca^{++}-free Krebs-Ringer was used. In these studies $CaCl_2$ was replaced with NaCl to maintain isotonicity, and 1 μM EGTA was added (Kristensen *et al*, 1990).

Superfused Segments of Arteries and Veins

Segments of pulmonary and mesenteric arteries and of portal and saphenous veins were removed from mongrel dogs. Helical strips of these vessels were prepared and superfused with Krebs-Ringer solution aerated with 95% O_2/5% CO_2 (Rorie and Tyce, 1987). After a 60 minute stabilization period, superfusate was collected continuously in 10 minute intervals for 7 intervals. Transmural stimulation at 2 Hz and at 10 Hz was applied during the second and the fifth intervals, respectively.

Determination of Dopa and Free and Sulfated Catecholamines in Plasma and in Perfusates

Dopa was isolated from fluids by a variety of methods and was then quantified by HPLC with electrochemical detection. Dopa, and in some experiments its major metabolite, 3-0-methyldopa, were isolated from plasma of man and dogs on columns of Dowex 50x4 (200 to 400 mesh) as described by Dousa and Tyce (1989). In the feeding experiments and in the perfusions of dog adrenals or of dog sympathetic ganglia, dopa was isolated on alumina columns at pH 8.6 and eluted from the columns with dilute perchloric acid (Banwart et al, 1989; Kristensen et al, 1990; Chritton et al, 1991). In the experiments in which isolated segments of arteries or veins were superfused, dopa was isolated from perfusates on Sep-Pak C-18 cartridges as described by Hunter et al (1988). Elution was with a mixture of 0.15% trifluoracetic acid, 0.2mM disodium EDTA, and 10% acetonitrile. This eluate was lyophilized and dissolved in a small volume of mobile phase before analysis by HPLC.

Free catecholamine in perfusates and in plasma were separated as described by Dousa and Tyce (1989) by adsorption on alumina and cation exchange chromatography. Catecholamines in perfusates of veins and arteries were separated on Sep-Pak C-18 cartridges (Hunter et al, 1989).

Effluents from the alumina columns contained the conjugated catecholamines. These were hydrolyzed to free catecholamines either by acid boiling (Dousa and Tyce, 1989) or by incubation with sulfatase (Devalon et al, 1989). Catecholamines freed from sulfate conjugation were isolated on alumina and cation-exchange columns as described for the free catecholamines. The interference of free catecholamines in the sulfated catecholamine fraction was determined during each assay by adding standard amounts of catecholamines to aliquots of plasma and measuring its "bleed through" into the alumina effluent. The bleed-through was usually 1% but could vary from 1% to 3%. Data have been corrected by subtraction of this percentage of free catecholamine from the sulfated catecholamine value.

Amino acids and catecholamines in their respective fractions were quantified by HPLC. Analytes were separated on a reversed-phase column (Ultrasphere IP 5μm, 4-6 mm i.d. x 25 cm length, Beckman, Inc., Berkeley, California, U.S.A.) using a variety of mobile phases (Hunter *et al*, 1988; Dousa and Tyce, 1988; Kristensen *et al*, 1990).

Data are corrected for appropriate recoveries through the extraction procedures. Means (±S.D.) are presented.

RESULTS

Catecholamines and Related Amino Acids in Plasma

High concentrations of dopa were present in plasma of man and dog (Table 1). 3-0-Methyldopa, the major metabolite of dopa (Tyce,

TABLE 1.   Dopa, 3-0-methyldopa and catecholamines in plasma.

Compound	Man	Dog
DOPA	1.7 ± 0.3	10.4 ± 3.1
3-0-Methyldopa	17.9 ± 0.9	256.2 ± 31.9
Norepinephrine		
free	0.3 ± 0.04	0.4 ± 0.1
sulfated	1.2 ± 0.2	0.1 ± 0.02
Epinephrine		
free	0.03 ± 0.02	0.2 ± 0.04
sulfated	n.d.	n.d.
Dopamine		
free	n.d.	n.d.
sulfated	5.6 ± 1.1	4.5 ± 1.6

Data are means (±S.D.) ng/ml plasma. (Data from Dousa and Tyce, 1988, with permission.)

1971; Sharpless *et al*, 1972) was also present in these plasma samples, and the concentrations of 3-0-methyldopa greatly exceeded the dopa concentrations. Conjugated dopa and conjugated 3-0-methyldopa were not detected in plasma.

Free and conjugated catecholamines in plasma are also shown in Table 1. The concentrations of conjugated catecholamine exceeded those of the free catecholamines. The amounts of dopa in plasma were greater than those of free norepinephrine (NE) and epinephrine (EPI) in both man and dog and less than those of sulfated DA in man but not in dog.

Effects of Feeding on Amino Acids and Catecholamines in Plasma

The concentrations of dopa and 3-0-methyldopa in plasma of fasted dogs did not change in the 5 hours after feeding (Figure 1). The mean (±S.D.) concentration of sulfated DA in plasma before feeding was 2.5±2.1 ng/ml and the concentration of free NE was 0.34±0.11 ng/ml. After feeding, sulfated DA increased in the first 3 hours to 3.6±1.5 ng/ml, and free NE decreased to 0.26±0.11 ng/ml. These small changes were significant ($p < 0.05$). Other analytes in plasma were unaffected by feeding.

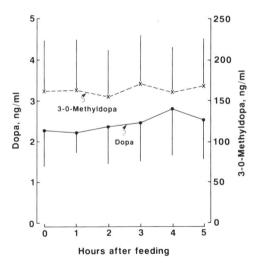

FIGURE 1. Concentrations of dopa and 3-0-methyldopa in plasma of dogs at different times after eating 0.5 kg of canned dog food. Means (±S.D.) are shown. (Data from Banwart *et al*, 1989, with permission.)

Analysis of the canned dog food showed little free or conjugated DA (0.8±0.1 and 0.4±0.2 ng/g, respectively), but after strong acid boiling for 12 hours, considerable amounts of free dopa were measured in the food (1132 and 4412 ng/g in two analyses, Banwart *et al*, 1989).

Dopa and Catecholamines in Superfused Lumbar Sympathetic Ganglia of the Dog

The concentrations of dopa in nonsuperfused ganglia (21.0±8.1 ng/g) exceeded those of DA (12.5±6.9 ng/g) and EPI (11.1±9.6 ng/g) but were less than those of NE (121.7±47.2 ng/g) (Kristensen *et al*, 1990).

Dopa was released from ganglia by 80 mM K^+, and this release was Ca^{++}-dependent (Figure 2). DA was also released, but the amounts released were less than those of dopa. Release of other analytes (dihydroxyphenylacetic acid [DOPAC], dihydroxyphenylglycol [DOPEG], NE and EPI) exceeded the release of dopa. Amphetamine released DA, NE, EPI and DOPEG but did not release dopa or DOPAC. No sulfated catecholamines were detected in superfusates of dog sympathetic ganglia.

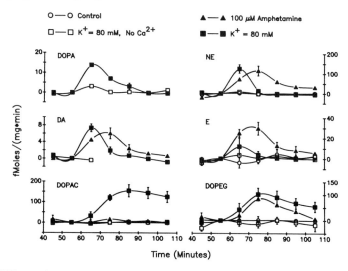

FIGURE 2. Average net evoked overflow profiles in response to the specified stimuli for dopa, DA, DOPAC, NE, EPI, and DOPEG from minced dog lumbar sympathetic ganglia. The stimulus was exposed to the tissue for 10 minutes starting at t=60 minutes. (o) control; (□) 80 mM K^+, no Ca^{++}; (▲) 100 µM amphetamine; (■)80 mM K^+. (Data from Kristensen *et al*, 1990, with permission.)

Releases of Dopa and Sulfated Catecholamines from Dog Adrenal Gland

Dopa was present in perfusates of dog adrenal gland which were collected under basal conditions 60 minutes after setting up the preparation. The basal efflux of dopa (18 pmol/minute) was less than those of the catecholamine (4500 pmol/minute for EPI, 680 pmol/minute for NE, 57 pmol/minute for DA). Carbachol (3×10^{-5}M) induced a two-to-three-fold increase in dopa efflux, and this increase was Ca^{++}-dependent. The increases in dopa were less than the increases in the catecholamines.

Small amounts of sulfated catecholamines were detected in perfusates of dog adrenal glands under basal conditions (Table 2). Acetylcholine (3×10^{-5}M) predictably evoked the releases of free NE, EPI, and DA. Although the amounts of sulfated catecholamines in perfusate appeared to increase during evoked release, only the increase in sulfated EPI achieved significance ($p < 0.05$).

Overflow of Dopa from Canine Blood Vessels

Small amounts of dopa overflowed under basal conditions from isolated superfused segments of dog pulmonary artery, saphenous vein, portal vein, and mesenteric artery. The amounts of dopa in superfusions were 20% to 40% of the amounts of NE and 3% to 7% of the amounts of DOPEG.

During electrical stimulation of these vessels, the amounts of NE and of dopa in superfusate were increased, and these increases were frequency-dependent (Figure 3). NE overflow peaked during the stimulation, whereas dopa overflow peaked during the interval immediately following the stimulation (Figure 3). This difference could be attributable to NE but not dopa being removed from the junctional cleft by neuronal uptake. However, the time courses of dopa and NE effluxes were still different when neuronal and extraneuronal uptakes were blocked by desipramine (1 μM) and corticosterone (40 μM), respectively. (In these studies α_2 adrenoreceptors were also blocked by yohimbine, 0.1 μM) (Figure 3).

The basal release of dopa from portal vein appeared to be Ca^{++}-independent; however, the evoked release of dopa was Ca^{++}-dependent. The evoked release of dopa from portal vein was greatly reduced in the presence of α-methyl-p-tyrosine (20 μM), an inhibitor

TABLE 2.   Free and sulfated catecholamine in perfusates ofisolated dog adrenal Gland.

| | Release (ng/min) | |
	Basal	Evoked*
Norepinephrine		
Free	79 ± 27	315 ± 106**
Sulfated	3 ± 2.5	5 ± 3.3
Epinephrine		
Free	656 ± 268	2,249 ± 655**
Sulfated	21 ± 17	44 ± 34**
Dopamine		
Free	5 ± 2	12 ± 3**
Sulfated	0.5 ± 0.4	0.3 ± 0.1

*Release by 3×10^{-5}M ACh. ** Significant differences from basal ($p < 0.05$).

of tyrosine hydroxylase, and was increased 10-fold in the presence of 3-hydroxybenzylhydrazine (NSD-1015, 20 μM), an inhibitor of aromatic L-amino acid decarboxylase. The basal and evoked release of NE was not significantly affected by the presence of either α-methyl-p-tyrosine or NSD-1015.

Under basal conditions, small amounts of conjugated NE were detected in superfusates of dog pulmonary artery (Rorie and Tyce, 1987). The amounts of conjugated NE increased during nerve stimulation, but this increase was not Ca^{++}-dependent (Rorie and Tyce, 1987).

CONCLUSIONS

Dopa in plasma does not appear to be derived from diet, at least in dogs. Canned dog food, however, contained high concentrations of dopa. Dopa was released in small but clearly measurable amounts from veins and arteries, from sympathetic ganglia, and from chromaffin cells of adrenal glands during nerve or chromaffin cell stimulation. These releases were Ca^{++}-dependent.

Release of dopa from veins and arteries was frequency-dependent during nerve stimulation, but release of dopa was slower than release of NE. Because dopa release was greatly reduced

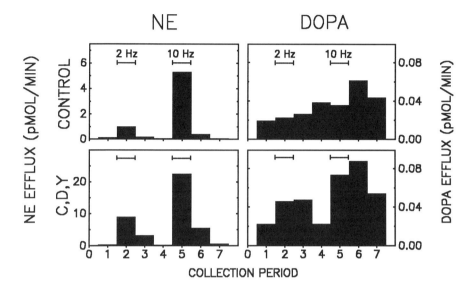

FIGURE 3.   Concentrations of norepinephrine (NE) and dopa in superfusates of canine portal vein.   The bars denote intervals during which electrical stimulation was applied.   The frequency of stimulation was 2 Hz or 10 Hz.   Each collection period was of 10 minutes duration.   Experiments were done on control segments (upper bar graphs) and on segments to which corticosterone (C, 40 μM), desipramine (D, 1 μM), and yohimbine (Y, 0.1 μM) were added to the superfusate (lower bar graphs).

when tyrosine hydroxylase was inhibited and increased when aromatic-L-amino acid decarboxylase was inhibited, it is reasonable to deduce that release of dopa reflects tyrosine hydroxylase activity. The calcium dependance of dopa release probably reflects the activation of tyrosine hydroxylase by a Ca^{++}-dependent mechanism (Vulliet *et al*, 1984).

Goshima *et al* (1988) have demonstrated release of dopa from slices of rat brain. Release of dopa in their experiments followed almost exactly the same time course as release of DA, and they deduced that release of dopa was from storage vesicles. However, it is more generally accepted that dopa is both formed and metabolized in the cytoplasm. Our experiments indicated that in blood vessels, dopa and NE were released from different compartments in cells because dopa release was almost completely inhibited in the presence of α-methyl-p-tyrosine whereas NE release was not affected to any significant extent.

It is puzzling that dopa should leave neurons, since it is metabolized at an appreciably more rapid rate than it is formed. It is also not clear by which mechanism dopa does leave cells.

The amounts of 3-0-methyldopa in plasma of man and dogs exceeded the amounts of dopa. 3-0-Methyldopa is a major metabolite of dopa (Tyce, 1971), and 0-methylation of dopa appears to be the only metabolic pathway by which this compound is synthesized. If 3-0-methyldopa + dopa originate from dopa leaving neurons, then the "loss" of dopa from the catecholamine biosynthetic pathway in neurons is very considerable.

Small but detectable amounts of sulfated NE were detected in saphenous vein, and sulfated NE, EPI, and DA were detected in small amounts in perfusates of adrenal glands. Conjugated catecholamines were not detected in sympathetic ganglia nor in superfusates of these tissues. Significant amounts of sulfated DA appeared to originate in diet in dogs. The source of the sulfated DA appeared to be dopa in the diet, rather than free or sulfated catecholamines. It appears unlikely, however, that all of the sulfated DA present in plasma was derived from precursors in the diet, since concentrations of sulfated DA were high in plasma of dogs which had been fasted for 24 hours.

ACKNOWLEDGMENTS

This work was supported by U.S. Public Health Service Grants NS17858, HL23217, and GM41797.

REFERENCES

Anton, A.H. (1991). Is plasma DOPA a valid indicator of sympathetic activity? *Journal of Laboratory and Clinical Medicine* 117, 263-264.

Banwart, B., Miller, T.D., Jones, J.D., and Tyce, G.M. (1989). Plasma dopa and feeding. *Proceedings of the Society for Experimental Biology and Medicine* 191, 357-361.

Chritton, S.L., Dousa, M.K., Yaksh, T.L., and Tyce, G.M. (1991). Nicotinic- and muscarinic-evoked release of canine adrenal catecholamine and peptides. *American Journal of Physiology* 260, R589-R599.

Devalon, M.L., Miller, T.D., Squires, R.W., Rogers, P.J., Bove. A.A., and Tyce, G.M. (1989). Dopa in plasma increases during acute exercise and after exercise training. *Journal of Laboratory and Clinical Medicine* **114**, 321-327.

Dousa, M.K., and Tyce, G.M. (1988). Free and conjugated plasma catecholamine, DOPA and 3-0-methyldopa in humans and in various animals species. *Proceedings of the Society for Experimental Biology and Medicine* **188**, 427-434.

Goldstein, D.S., Udelsman, R., Eisenhofer, G., Stull, R., Keiser, H.R., and Kopin, I.J. (1987). Neuronal source of plasma dihydroxyphenylalanine. *Journal of Clinical Endocrinology and Metabolism* **64**, 856-861.

Goshima, Y., Kubo, T., and Misu, Y. (1988). Transmitter-like release of endogenous 3,4-dihydroxyphenylalanine from rat striatal slices. *Journal of Neurochemistry* **50**, 1725-1730.

Hunter, L.W., Rorie, D.K., Yaksh, T.L., and Tyce, G.M. (1988). Concurrent separation of catecholamine, dihydroxyphenylglycol, vasoactive intestinal peptide, and neuropeptide Y in superfusate and tissue extract. *Analytical Biochemistry* **173**, 340-352.

Kristensen, E.W., Chinnow, S.L., Montreuil, R.S., and Tyce, G.M. (1990). Precursors and metabolites of norepinephrine in sympathetic ganglia of the dog. *Journal of Neurochemistry* **54**, 1782-1790.

Kuchel, O., Buu, N.T., and Edwards, D.J. (1990). Alternative catecholamine pathways after tyrosine hydroxylase inhibition in malignant pheochromocytoma. *Journal of Laboratory and Clinical Medicine* **115**, 449-453.

Morgenroth, V.H. III, Boadle-Biber, M., and Roth, R.H. (1974). Tyrosine hydroxylase: activation by nerve stimulation. *Proceedings of the National Academy of Sciences* (USA) **71**, 4283-4287.

Rorie, D.K., and Tyce, G.M. (1987). Free and conjugated catecholamine in dog pulmonary artery: presence and pharmacological action. *American Journal of Physiology* **253**, H66-H74.

Sharpless, N.S., Muenter, M.D., Tyce, G.M., and Owen, C.A., Jr. (1972). 3-Methoxy-4-hydroxyphenylalanine (3-0-methyldopa) in plasma during oral L-dopa therapy of patients with Parkinson's disease. *Clinica Chimica Acta* **37**, 359-369.

Tyce, G.M. (1971). Metabolism of 3-4-dihydroxyphenylalanine by isolated perfused rat liver. *Biochemical Pharmacology* **20**, 3447-3462.

Vulliet, P.R., Woodgett, J.R., and Cohen, P. (1984). Phosphorylation of tyrosine hydroxylase by calmodulin-dependent multiprotein kinase. *Journal of Biological Chemistry* **259**, 13680-13683.

Stress: Neuroendocrine and Molecular Approaches
Edited by R. Kvetnansky, R. McCarty and J. Axelrod

1992 Gordon and Breach Science
Publishers S.A., New York, USA.

ANTAGONISTIC EFFECTS OF ANGIOTENSIN II AND ANF ON CATECHOLAMINE RELEASE IN THE ADRENAL MEDULLA

W. Wuttke, M. Dietrich and H. Jarry

Division of Clinical and Experimental Endocrinology
Department of Obstetrics and Gynecology
University of Göttingen, Göttingen, Germany

INTRODUCTION

The chromaffin cells of the adrenal medulla in all species studied so far produce epinephrine (EPI) and some norepinephrine (NE) as well as a variety of neuropeptides (Lundberg and Hökfelt, 1983; Livett, 1984). Since the adrenal medulla is believed to be a specialized post-synaptic sympathetic structure, it is not surprising to find the same neuropeptides in chromaffin cells that have been detected within the post-synaptic sympathetic nervous system. All post-synaptic sympathetic neurons are innervated by cholinergic neurons which also co-localize a variety of neuropeptides. Stress-induced adrenomedullary catecholamine release involves therefore also the activation of presynaptic cholinergic neurons. We and others have previously shown that acetylcholine or nicotine not only stimulate catecholamine release from chromaffin cells but also co-localized peptides (Vaupel *et al*, 1988; Jarry *et al*, 1989). Relatively little is known about the functions of adrenomedullary peptides. Most experiments have been performed *in vitro* utilizing bovine chromaffin cells kept under primary culture conditions. Since the concentrations of the released peptides under

171

such conditions are low and furthermore diluted within the culture medium, direct interactions between chromaffin cells in a paracrine manner do not occur. Addition of various peptides under investigation to the culture medium has yielded information that some of them have inhibitory, others no and others stimulatory effects on catecholamine release.

One of the peptides found in chromaffin cells is angiotensin II (AII; Bianchi *et al*, 1986). It has been shown that the adrenal medulla contains its own renin-angiotensin system (Ganten *et al*, 1983) and the presence of angiotensin converting enzyme was also demonstrated, indicating local production of this peptide (Chai *et al*, 1986). Furthermore, the presence of AII receptors on chromaffin cells argues for a local action at these cells (Healy *et al*, 1985). Following systemic injections, AII increased epinephrine concentrations in the blood of cats and dogs and *in vitro* it was also shown that it has stimulatory effects on catecholamine release from chromaffin cells (Feldberg and Lewis, 1964; Peach *et al*, 1966).

In the adrenal cortex, AII has a stimulatory effect on aldosterone release, an effect which is antagonized by atrial natriuretic factor (ANF; Aguilera, 1987). Similar antagonistic effects of AII and ANF were demonstrated to occur within the hypothalamus where neurons also produce these two peptides (Jacobowitz *et al*, 1985; Skofitsch *et al*, 1985; Morel *et al*, 1988). ANF is also produced by chromaffin cells and its receptors are primarily located on EPI producing cells (Bianchi *et al*, 1985; Inagaki *et al*, 1986). About 15% of chromaffin cells contain the mRNA for ANF (Morel *et al*, 1988).

Recently, we developed a microdialysis system which allows for the study of secretion rates of catecholamines and peptides in freely moving rats (Jarry *et al*, 1985). Utilizing this method (see below) we demonstrated that stress-induced EPI release can largely be inhibited by met-enkephalin, an effect which can be antagonized by naloxone, indicating that specific opiate receptors are involved in the mediation of this stress modulatory function (Jarry *et al*, 1989). In the present study, we examined whether AII applied through the microdialysis system has effects on catecholamine release in freely moving rats. Furthermore, we studied the local effects of ANF alone and in combination with AII.

MATERIALS AND METHOD

Female, 3-6 month old Sprague-Dawley rats were implanted with a microdialysis system which has been described in detail elsewhere. In brief, a 5 mm length of dialysis tubing (outer diameter 200 μm) was extended with silastic and teflon tubings. Following laparotomy, one adrenal gland was exposed and a mandril was inserted from pole to pole of the adrenal gland. Through this mandril the microdialysis tubing surface was inserted such that the free microdialysis tubing surface was located within the adrenal medulla. The mandril was then removed and the extending tubings were exteriorized by insertion through the skin at the nape of the neck. The afferent end of this system was connected to a peristaltic pump which pumped Ringer's solution through the system. At the efferent end fractions were collected at 15 minute intervals by means of a fraction collector. At the same time, animals were also implanted with an indwelling jugular vein catheter. During the following 24 hours animals were allowed to recover. Prior to the beginning of the actual dialysis and blood withdrawal time the animals were adjusted to the presence of the experimenter for at least 2 hours during which only dialysis fractions were collected. To determine spontaneous catecholamine release rates, 4-6 fractions were collected. The Ringer's solution used for dialysis was then changed to Ringer's solution containing either AII ($2x10^{-6}$M, which results in an intraadrenal concentration of $2x10^{-9}$M), the AII receptor blocker saralasin ($2x10^{-6}$M = intraadrenal concentration of $2x10^{-9}$M) or a combination of these two peptides. In a fourth experiment, ANF ($2x10^{-6}$M = intraadrenal concentration of $2x10^{-9}$M) was added to the Ringer solution and in the final experiment ANF was applied intraadrenally from the beginning of the experiment and during the second period ANF plus AII were applied. The estimation of intraadrenal concentrations of the peptide is based on 0.1% transfer rates of the peptides through the microdialysis tubing. At the end of each fraction period a blood sample was also withdrawn through the jugular vein catheter. EPI and NE concentrations in the adrenal dialysates were measured by an HPLC-EC method (Jarry *et al*, 1989). Serum prolactin and ACTH concentrations were measured by conventional radioimmunassays. In the following figures typical examples of the response of individual animals to each treatment are shown.

RESULTS AND DISCUSSION

Figure 1 shows NE (upper panel) and EPI (lower panel) concentrations as measured in the adrenal dialysates prior to, during

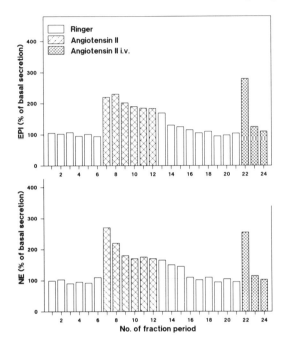

FIGURE 1. Intraadrenal application of AII had a significant stimulatory effect on the release rates of both catecholamines. In this experiment AII was also given systemically (right part of the figure). This systemic injection of AII also stimulated NE and EPI release.

and following intraadrenal application of AII. The intraadrenal release rates of the two catecholamines were stable and showed little fluctuation. When AII was administered into the adrenal medulla by means of the dialysis system, both catecholamines were significantly stimulated in a phasic/tonic manner. Following discontinuation of AII application, the release rates of both catecholamines dropped down to those observed prior to application of AII within 30 minutes. Measurement of serum prolactin and ACTH (Figure 2) in this animal indicated that the peptide did not exert systemic effects since both hormones remained unchanged during the intraadrenal application time of AII whereas they increase dramatically following systemic injection. These experiments clearly demonstrate that intraadrenal application of AII at physiological concentrations has a stimulatory

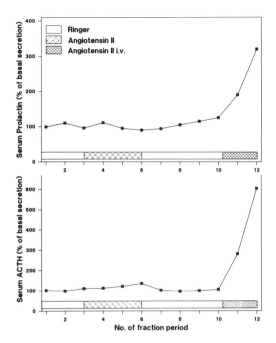

FIGURE 2. Serum prolactin and ACTH levels are not influenced by intraadrenal application of AII while systemic injection of the peptide stimulated release of both hormones significantly. These results indicate that the topical application of AII through the microdialysis system has no systemic effects.

effect on both EPI and NE release. This is a local, i.e., intraadrenal, effect since neither serum prolactin nor ACTH concentrations varied during the experimental period.

Intraadrenal application of saralasin, a specific AII receptor blocker, did not have an effect on NE or EPI release. In the following experiments saralasin was administered intraadrenally from the beginning of the experiment and this was followed by a period when saralasin and AII were given together. The pretreatment of the adrenal medulla with saralasin completely abolished the stimulatory effects of AII on both catecholamines (Figure 3), indicating that AII (as shown in Figure 1) acted via a specific, receptor-mediated mechanism.

When ANF was administered intraadrenally, the spontaneous release rates of both NE and EPI were significantly inhibited (Figure 4) and pretreatment of the adrenal medulla with ANF completely

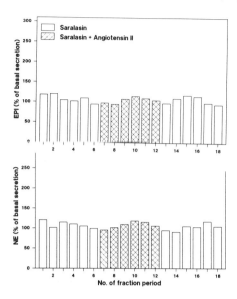

FIGURE 3.   Intraadrenal application of saralasin blocked AII-induced catecholamine release, indicating that AII acts via specific receptor-mediated mechanisms.

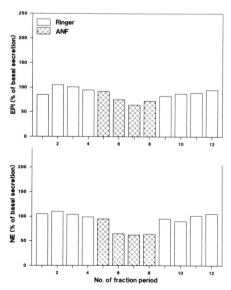

FIGURE 4.   Intraadrenal application of ANF through the microdialysis system significantly inhibited NE and EPI release.

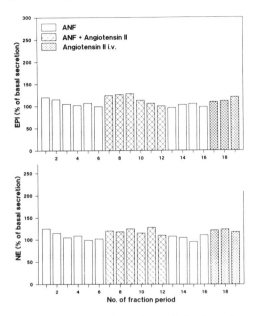

FIGURE 5.   The ANF-induced reduction of catecholamine release rates was not reversed by additional treatment with AII.

abolished the AII-mediated stimulation of both catecholamines (Figure 5).

In these experiments, we demonstrated for the first time that local application of AII into the adrenal medulla in awake, unstressed and freely moving animals had a stimulatory effect on the release of both adrenal catecholamines. This appears to be a specific receptor-mediated effect as saralasin, a known AII receptor blocker, antagonized the stimulatory effect of AII on catecholamines. Since the local production of AII within chromaffin cells is firmly established (Bianchi *et al*, 1986) and since specific AII receptors have been demonstrated on these cells (Healy *et al*, 1985), one might speculate that AII has an amplifying function on stress-induced catecholamine release. From *in vitro* experiments it is known that a nicotinic stimulus causes not only the release of catecholamines but also of AII, which undoubtedly acts in the neighborhood of its release site, thereby amplifying the stress-induced signal. On the other hand, ANF which is also produced in chromaffin cells, has an inhibitory effect on basal catecholamine release as well as an AII-stimulated catecholamine release. This peptide is probably also co-released with catecholamines in response to stress but only about 15% of all chromaffin cells produce ANF (Morel *et al*, 1988). It is therefore possible that subgroups of chromaffin cells exist which exhibit stress-induced ANF release and thereby inhibit the secretion of catecholamines in an auto- and paracrine fashion. Other cells which might not be in the neighborhood of ANF-producing cells would be unaffected. Such a mechanism could explain how different stressors can target different adrenomedullary cell types. In earlier experiments we demonstrated, for example, that electric footshock stress but not insulin-induced hypoglycemia, had differential effects on adreno-medullary release of peptides (Vaupel *et al*, 1988). Electric footshock stress resulted in massive release of substance P whereas insulin-induced hypoglycemic stress had no effect on adrenomedullary substance P release.

The presence of a variety of stimulatory and inhibitory peptides in adrenal chromaffin cells is undoubtedly of physiologic significance and may guarantee the proper response to different environmental and internal stressful situations.

REFERENCES

Aguilera, G. (1987). Differential effects of atrial natriuretic factor on angiotensin II- and adrenocorticotropin-stimulated aldosterone secretion. *Endocrinology* 120, 299-304.

Bianchi, C., Gutkowska, J., Thibault, G., Garcia, R., Genest, J., and Cantin, M. (1985). Radioautographic localization of ^{125}I-atrial natriuretic factor (ANF) in rat tissues. *Histochemistry* 82, 441-452.

Bianchi, C., Gutkowska, J., Charbonneau, C., Ballak, M., Anand-Srivastava, M.B., DeLean, A., Genest, J., and Cantin, M. (1986). Internalization and lysosomal association of (^{125}I) angiotensin II in norepinephrine-containing cells of the rat adrenal medulla. *Endocrinology* 119, 1873-1875.

Chai, S.Y., Allen, A.M., Adam, W.R., and Mendelsohn, F.A. (1986). Local actions of angiotensin II: quantitative *in vitro* autoradiographic localization of angiotensin II receptor binding and angiotensin converting enzyme in target tissues. *Journal of Cardiovascular Pharmacology* 8 (Suppl 10), 35-39.

Feldberg, W., and Lewis, G.P. (1964). The action of peptides on the adrenal medulla. Release of adrenaline by bradykinin and angiotensin. *Journal of Physiology* (London) 171, 98-108.

Ganten, D., Hermann, K., Unger, T., and Lang, R.E. (1983). The tissue renin-angiotensin systems: focus on brain angiotensin, adrenal gland and arterial wall. *Clinical and Experimental Hypertension* (A) 5, 1099-1118.

Healy, D.P., Maciejewski, A.R., and Printz, M.P. (1985). Autoradiographic localization of (^{125}I)-angiotensin II binding sites in the rat adrenal gland. *Endocrinology* 116, 1221-1223.

Inagaki, S., Kubota, Y., Kito, S., Kangawa, K., and Matsuo, H. (1986). Immunoreactive atrial natriuretic polypeptides in the adrenal medulla and sympathetic ganglia. *Regulatory Peptides* 15, 249-260.

Jacobowitz, D.M., Skofitsch, G., Keiser, H.R., Eskay, R.L, and Zamir, N. (1985). Evidence for the existence of atrial natriuretic factor-containing neurons in the rat brain. *Neuroendocrinology* 40, 92-94.

Jarry, H., Düker, E.M., and Wuttke, W. (1985). Adrenal release of catecholamines and met-enkephalin before and after stress as measured by a novel *in vivo* dialysis method in the rat. *Neuroscience Letters* 60, 273-278.

Jarry, H., Dietrich, M., Barthel, A., Giesler, A., Wuttke, W. (1989). *In vivo* demonstration of a paracrine, inhibitory action of Met-Enkephalin on adrenomedullary catecholamine release in the rat. *Endocrinology* 125, 624-629.

Livett, B. (1984). Adrenal medullary chromaffin cells *in vitro*. *Physiological Reviews* 64, 1103-1161.

Lundberg, J., and Hökfelt, T. (1983). Coexistence of peptides and classical neurotransmitters. *Trends in Neuroscience* 62, 325-333.

Morel, G., Chabot, J.G., Garcia-Caballero, T., Gossard, F., Dihl, F., Belles-Isles, M., and Heisler, S. (1988). Synthesis, internalization, and localization of atrial natriuretic peptide in rat adrenal medulla. *Endocrinology* 123, 149-158.

Peach, M.J., Cline, W.H., and Watts, D.T. (1966). Release of adrenal catecholamines by angiotensin II. *Circulation Research* 19, 571-575.

Skofitsch, G., Jacobowitz, D.M., Eskay, R.L., and Zamir, N. (1985). Distribution of atrial natriuretic factor-like immunoreactive neurons in the rat brain. *Neuroscience* 16, 917-948.

Vaupel, R., Jarry, J., Schlömer, H.T., and Wuttke, W. (1988). Differential response of substance P-containing subtypes of adrenomedullary cell to different stressors. *Endocrinology* 123, 2140-2145.

Stress: Neuroendocrine and Molecular Approaches
Edited by R. Kvetnansky, R. McCarty and J. Axelrod

1992 Gordon and Breach Science
Publishers S.A., New York, USA.
Photocopying permitted by license only.

EFFECT OF ACUTE AND REPEATED IMMOBILIZATION ON TISSUE CONTENT OF DOPA

I. Armando, V. K. Weise, K. Pacak, R. Kvetnansky and
D. S. Goldstein

Clinical Neuroscience Branch, National Institute of Neurological
Diseases and Stroke, National Institutes of Health, Bethesda, MD

INTRODUCTION

3,4-Dihydroxyphenylalanine (DOPA) is the immediate product of hydroxylation of tyrosine by tyrosine hydroxylase (TH), the rate-limiting enzymatic step in catecholamine biosynthesis. DOPA is a normal constituent of plasma in humans and experimental animals (Goldstein *et al*, 1984; Dousa *et al*, 1988). Plasma levels of DOPA exceed by far those of the free catecholamines. Several lines of evidence have indicated that a substantial proportion of DOPA in plasma is from sympathetically-innervated tissues: a) Arteriovenous increments in plasma DOPA levels in intact limbs are not observed in sympathectomized limbs (Goldstein *et al*, 1987); b) there are large decreases in plasma DOPA levels in dogs after inhibition of TH activity with α-methyl-p-tyrosine (Goldstein *et al*, 1987); and c) plasma DOPA levels increase during short-term exercise, a known stimulus of sympathetic nervous system activity (Devalon *et al*, 1989). It has also been suggested that changes in plasma DOPA levels may reflect changes in TH activity (Goldstein *et al*, 1987; Eisenhofer *et al*, 1988).

To assess TH activity, the rate of increase of tissue DOPA concentration has been used after inhibition of L-aromatic amino acid

181

decarboxylase (Carlsson *et al*, 1972).

Several stressors increase catecholamine synthesis. In particular, immobilization, which has been used as a model of stress, leads to increases in adrenal TH activity (Kvetnansky *et al*, 1980).

To examine the relationship between TH and DOPA levels in sympathetically- innervated tissues, we assessed the effects of acute and repeated immobilization stress (IMMO) on tissue contents of DOPA and TH in laboratory rats.

MATERIALS AND METHODS

Animals

Male Sprague-Dawley rats (250-350 g) were allocated randomly to the experimental groups (n=6-7 per group) and subjected to either 1 (IMMO 1) or 7 (IMMO 7) daily periods of 120 minutes of immobilization.

The animals were sacrificed at the end of the restraint period, and kidney, spleen, lung, heart, adrenals and femoral and abdominal muscles were removed, weighed and stored at -70°C. Naive, unrestrained rats and rats subjected to 6 immobilization periods and killed 24 hours later (adapted controls) served as controls for the acutely and repeatedly stressed groups, respectively.

Assays

The tissue content of DOPA, norepinephrine (NE) and dopamine (DA) was measured using high pressure, reversed-phase, ion-pairing, liquid chromatography with electrochemical detection. The tissues were thawed, homogenized in 0.3N perchloric acid, and centrifuged at 6000 g for 20 minutes. The catechols in 25-200 μl aliquots of the supernate were partially purified by batch alumina extraction, separated using high pressure liquid chromatography, and quantified amperometrically by the current produced upon exposure of the column effluent to oxidizing and then reducing potentials in series (Eisenhofer *et al*, 1986). Recovery through the alumina extraction step averaged 70-80% for catecholamines and 45-55% for DOPA. Catechol concentrations in each sample were corrected for recovery of an internal standard, dihydroxybenzylamine. Levels of DOPA were further corrected for differences in recovery of the internal standard and of DOPA in a mixture of external standards. The limit of detection was about 10 pg/volume assayed for each catechol.

The activity of TH in the tissues was assayed using a CO_2-trapping technique at a saturating substrate concentration (Waymire et al, 1971). Tissues were weighed and homogenized (spleen and adrenals) or minced (heart), and then centrifuged. Aliquots of the supernates were incubated with a mixture containing L-tyrosine (final concentration 100 μM), 0.15 μCi L-carboxyl-^{14}C-tyrosine (Amersham, S.A. 55 nCi/nmol), cofactors, catalase, and hog kidney decarboxylase at a final pH of 6.1. Before the incubation, a rubber stopper fitted with a plastic well containing filter paper soaked in liquid scintillation cocktail was placed in each incubation tube. At the end of the incubation, the paper was transferred to a scintillation vial and the radioactivity was measured by liquid scintillation spectrometry. Results are expressed as nmol (or pmol) of product formed during the incubation. Values reported were corrected for tissue inhibition.

Data Analysis

Data are presented as means ± SEM. One-way analyses of variance (ANOVA) were used to compare tissue levels of catechols and tissue activities of TH in the four groups. A p value less than 0.05 defined statistical significance.

RESULTS

Acute immobilization (IMMO 1) resulted in decreases of 30-50% in DOPA levels in all tissues except the spleen ($p < 0.02$). Repeated immobilization (IMMO 7) did not result in further reductions in DOPA levels; however, in both kidney and heart DOPA content tended to decrease further after repeated stress. Tissues from adapted controls (6 IMMO + 24 hours) also had significant reductions in DOPA content, except for the lungs.

Repeated immobilization was attended by significant ($p < 0.01$) increases in DOPA levels in the spleen, whereas levels in acutely immobilized rats or in adapted controls did not differ from levels in naive animals. DOPA in adrenals from immobilized animals were below the detection limit of the assay. Naive controls and adapted controls showed similar values (119 ± 82 and 120 ± 81 pg/mg, respectively) (Figure 1).

Figure 2 shows the TH activities of adrenal, spleen and heart tissues from naive rats and animals subjected to acute and repeated immobilization. TH activity was unchanged after acute immobilization. Repeated immobilization increased TH activity significantly in all 3 tissues.

DISCUSSION

Stress-induced decreases in tissue DOPA levels in the heart and adrenal glands of repeatedly immobilized rats contrasted with significant increases in TH activity. The results indicate that tissue DOPA does not parallel TH activity and actually changes in the opposite direction in repeatedly stressed animals. Stress-induced decreases in tissue DOPA content were also observed in skeletal muscle, kidney, and lung.

DOPA levels varied among the tissues and no relationship was observed between DOPA and NE content. These findings are consistent with the view that DOPA can be stored in and released

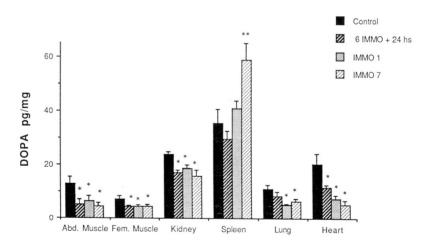

FIGURE 1. Effects of acute and repeated stress on DOPA in tissues. Values are means and vertical bars denote 1 SEM * p<0.02; ** p<0.01 compared to controls.

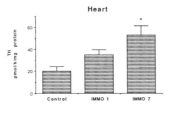

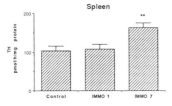

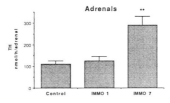

FIGURE 2. Effects of acute (IMMO 1) and repeated (IMMO 7) immobilization on tissue TH activity. Values are means and vertical bars denote 1 SEM. * $p < 0.05$; ** $p < 0.01$ compared to controls.

from non-neuronal cells. In skeletal muscle, endogenous DOPA is located in muscle cells (Eldrup *et al*, 1989). Increased TH activity and DOPA synthesis in sympathetic nerve endings could be offset by increased DOPA release into the bloodstream, increased NE synthesis to replete releasable NE stores, and increased mobilization of stored DOPA from parenchymal cells. In peripheral organs such as the pancreas, liver, and kidney, DOPA may be taken up from the bloodstream and stored in parenchymal cells (Anden *et al*, 1989). DOPA in the pancreas is present mainly in exocrine cells (Alm *et al*, 1969).

If there were large pools of non-neuronal DOPA in these tissues, the present results would indicate a mobilization of those pools in stressed rats. The existence of such a pool would also explain the observed lack of relationship between DOPA content and TH activity in some tissues. Mobilized DOPA could be released into the circulation or re-utilized for NE synthesis.

REFERENCES

Alm, P., Ehinger, B., and Falck, B. (1969). Histochemical studies on the metabolism of L-DOPA and some related substances in the exocrine pancreas. *Acta Physiologica Scandinavica* 76, 106-120.

Anden, N.E., Grabowska-Anden, M., and Schwieler, J. (1989). Transfer of DOPA from the sympatho-adrenal system to the pancreas, liver and kidney via the blood circulation. *Acta Physiologica Scandinavica* 136, 75-79.

Carlsson, A., Kehr, W., Linquist, M., Magnusson, T., and Atack, C.V. (1972). Regulation of monoamine metabolism in the central nervous system. *Pharmacological Reviews* 24, 371-384.

Devalon, M., Miller, T.D., Squires, R.W., Rogers, P.J., Bove, A.A., and Tyce, G. (1989). DOPA in plasma increases during acute exercise and after exercise training. *Journal of Laboratory and Clinical Medicine* 114, 321-327.

Dousa, M., and Tyce, G. (1988). Free and conjugated plasma catecholamines, DOPA and 3-O-methyldopa in humans and in various animal species. *Proceedings of the Society for Experimental Biology and Medicine* 188, 427-434.

Eisenhofer, G., Goldstein, D.S., Stull, R., Keiser, H.R., Sunderland, T., Murphy, D.L., and Kopin, I.J. (1986). Simultaneous liquid chromatographic determination of 3,4-dihydroxyphenylglycol, catecholamines, and 3,4-dihydroxyphenylalanine in plasma and their responses to inhibition of monoamine oxidase. *Clinical Chemistry* 32, 2030-2033.

Eisenhofer, G., Ropchak, T., Nguyen, H., Keiser, H.R., Kopin I.J., and Goldstein, D.S. (1988). Source and physiological significance of plasma 3,4-dihydroxyphenylalanine in the rat. *Journal of Neurochemistry* 51, 1204-1213.

Eldrup, E., Richter, E.A., and Christensen, N.J. (1989). DOPA, norepinephrine, and dopamine in rat tissues: No effect of sympathectomy on muscle DOPA. *American Journal of Physiology* 256, E284-E287.

Goldstein, D.S., Stull, R.W., Zimlichman, R., Levinson, P.D., Smith, H., and Keiser, H.R. (1984). Simultaneous measurement of DOPA, DOPAC, and catecholamines in plasma by liquid chromatography with electrochemical detection. *Clinical Chemistry* 30, 815-816.

Goldstein, D.S., Udelsman, R., Eisenhofer, G., Keiser, H.R., and Kopin I.J. (1987). Neuronal source of plasma dihydroxyphenylalanine. *Journal of Clinical Endocrinology and Metabolism* 64, 856-861.

Kvetnansky, R. (1980). Plasma catecholamines and stress. In: E. Usdin, R. Kvetnansky and I.J. Kopin (Eds.), "Catecholamines and Stress", pp. 7-17. New York: Elsevier-North Holland.

Waymire, J., Bjur, R., and Weiner, N. (1971). Assay of tyrosine hydroxylase by coupled decarboxylation of DOPA formed from l-^{14}C-L-tyrosine. *Analytical Biochemistry* 43, 588-600.

Stress: Neuroendocrine and Molecular Approaches
Edited by R. Kvetnansky, R. McCarty and J. Axelrod

1992 Gordon and Breach Science
Publishers S.A., New York, USA.

PLASMA LEVELS OF CATECHOLAMINES, ALDOSTERONE, ATRIAL NATRIURETIC PEPTIDE AND RENIN ACTIVITY DURING IMMOBILIZATION STRESS IN RATS

L. Macho[1], R. Kvetnansky[1], M. Fickova[1], D. Jezova[1]
B. Lichardus[1] and R. M. Carey[2]

[1]Institute of Experimental Endocrinology, Slovak Academy of Sciences,
Bratislava, Czecho-Slovakia, [2]Department of Internal Medicine, School
of Medicine, University of Virginia, Charlottesville, VA 22906 USA

INTRODUCTION

Exposure of experimental animals to hypokinesia, an immobilization stressor with forced restriction of movement (Macho *et al*, 1977, Langer *et al*, 1981) in special adjustable cages has been used as a model for simulation of some effects of microgravity (Kovalenko, 1977). It has been demonstrated that short term hypokinesia is a stressful situation in rats with attendant increases in plasma levels of corticosterone (CS), epinephrine (EPI), and norepinephrine (NE) (Macho *et al*, 1989a). The increase in circulating catecholamines correlates well with elevations of nonesterified fatty acid concentrations in plasma and with lipolytic activity in adipose tissue (Macho *et al*, 1984). However, in rats exposed to hypokinesia for longer times (7 to 75 days), plasma corticosterone levels were similar to control animals and plasma norepinephrine levels were permanently elevated during this prolonged period of immobilization. These results showed that long-term hypokinesia in rats is a model of a stressful situation with marked activation of the sympathetic nervous system but with activity of the adrenocortical system within the normal range (Macho *et al*, 1989a).

187

It is known that catecholamines are involved in regulation of hormones affecting water-electrolyte metabolism in experimental animals (for a review, see Carey *et al*, 1989). The secretion of renin is stimulated by activation of noradrenergic neurons in the kidney or by increased circulating catecholamines. During stress induced by hypoxemia, hypercapnic acidosis, anesthesia, or heat, the renin-angiotensin-aldosterone system is activated. The aim of the present experiments was to study the relationship between plasma catecholamines and hormones involved in the regulation of water-electrolyte metabolism in rats exposed to immobilization stress (hypokinesia).

METHODS

Adult male Wistar rats, SPF colony, weighing approximately 300 g were used. In Experiment I, rats were immobilized in adjustable hypokinetic cages (Macho *et al*, 1986) for 1, 7 or 60 days (8 animals per group). After this period, animals were sacrificed by decapitation (this procedure was very quick; within 10-15 seconds the animals were released from hypokinetic cages and decapitated). A control group included rats with free movement in standard laboratory cages. All rats were supplied with food and water *ad libitum*. Blood samples were collected and plasma frozen. Levels of corticosterone (Beitins *et al*, 1970), plasma renin activity (PRA) (Sealey *et al*, 1977), aldosterone (ALDO) (Buhler *et al*, 1974), and atrial natriuretic peptide (ANP) (Chevalier *et al*, 1988) were determined by radioimmunoassay.

In Experiment 2, groups of rats (8 animals per group) were exposed to hypokinesia for 7 or 60 days. One day before the end of hypokinesia, rats were anesthetized with pentobarbital (5 mg/100g body weight i.p.) and a cannula was inserted into the tail artery and the animals were returned to hypokinetic cages. A group of intact rats was anesthetized and after cannulation the animals were immobilized in hypokinetic cages for the next 24 hours. In the control group, intact rats were anesthetized, cannulated and placed into individual laboratory cages with complete freedom of movement. In all groups, blood was collected via the tail artery cannula 24 hours after cannulation. The duration of hypokinesia was 1, 7 or 60 days. Levels of NE and EPI (Peuler and Johnson, 1977), ALDO, PRA, ANP,

ACTH (Jezova-Repcekova *et al*, 1978; Jezova *et al*, 1987) and CS in plasma were determined.

RESULTS

A significant increase in plasma EPI and NE levels was found in rats exposed to hypokinesia for 1 or 7 days (Figures 1 and 2). After 60 days of hypokinesia, plasma NE levels remained elevated, while levels of EPI were not significantly different from those of controls.

A marked increase in plasma ALDO values was observed in rats exposed to hypokinesia for 1 day (Figure 3) in both experiments. Plasma ALDO levels decreased with increasing duration of hypokinesia and after 60 days ALDO levels in immobilized and control rats were not significantly different.

Values of PRA (Figure 4) were augmented in animals exposed to hypokinesia for 1 day, but after 7 or 60 days of immobilization, values of PRA returned to those observed in controls.

A decrease of plasma ANP levels (Figure 5) was found in rats exposed to hypokinesia for 7 days and in animals sacrificed by decapitation also after 1 day of immobilization.

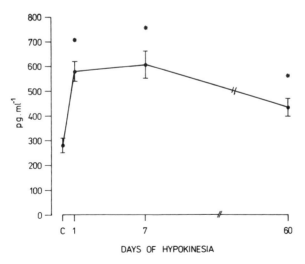

FIGURE 1. Plasma norepinephrine concentrations in rats exposed to hypokinesia for 1, 7 or 60 days. *$p < 0.05$ compared to controls.

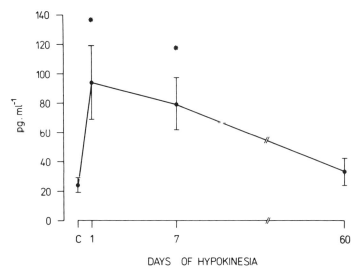

FIGURE 2.   Plasma epinephrine concentrations in rats exposed to hypokinesia for 1, 7 or 60 days. *p<0.05 compared to controls.

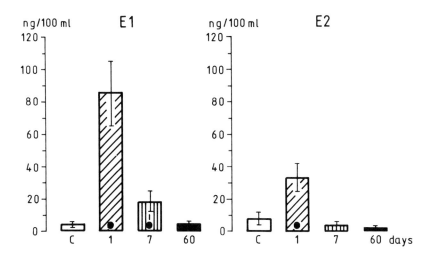

FIGURE 3.   Plasma aldosterone (ALD) concentrations in rats exposed to hypokinesia for 1, 7 or 60 days. ●p<0.05 compared to control group. E1 = Experiment 1: blood samples after decapitation; E2 = Experiment 2: blood taken from tail artery cannula.

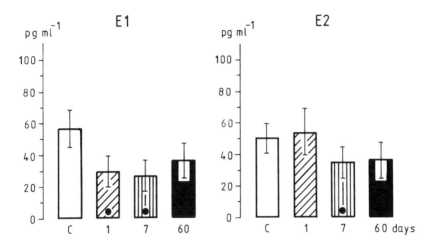

FIGURE 4. Plasma renin activity (PRA) in rats exposed to hypokinesia for 1, 7 or 60 days. ●p<0.05 compared to controls, + p<0.05 (1 versus 60 days) in Experiment 2.

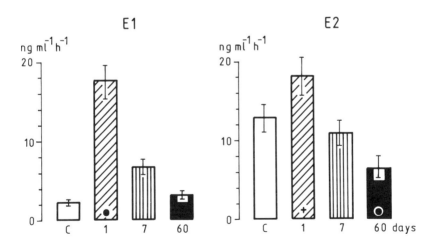

FIGURE 5. Levels of atrial natriuretic peptide (ANP) in plasma of rats exposed to hypokinesia for 1, 7 or 60 days. ●p<0.05 (versus controls).

Plasma CS levels showed a pronounced increase after one day of immobilization. However, after 7 or 60 days of immobilization, CS levels in plasma were similar to controls (Figure 6). No significant changes in plasma ACTH were observed in rats after 1, 7 or 60 days of immobilization.

DISCUSSION

The results of the present experiments dealing with changed plasma catecholamines and CS levels in rats exposed to hypokinesia are in agreement with our previous observations (Macho *et al*, 1989b) and indicate that long-term hypokinesia in rats represents a model of stress with marked and continuous activation of the sympathetic nervous system but with adrenocortical function increased only during initial exposure to hypokinesia.

These results also suggest that short-term hypokinesia is a stressful stimulus which induces increases in PRA. The increase in PRA in acutely stressed animals is in agreement with observations of Jindra and co-workers (1984) and Ganong (1991) who demonstrated an increase in PRA after stressful stimuli such as head-up tilt, immobilization for 10 minutes, low sodium diet and administration of p-chloroamphetamine. Further, it was noted that these increases in PRA were mediated by catecholamines and were inhibited by ß-adrenergic blockade with propranolol (Ganong, 1992). The increase in PRA in acutely stressed animals could be related to the increase in activity of the sympathetic nervous system. However, during long-term hypokinesia in our experiments, values of PRA decreased in spite of high circulating NE levels. Further studies are necessary to explain the mechanism relating to this decrease in PRA during long-term hypokinesia. Systemic infusion of NE in conscious dogs increased PRA and blood pressure, but both returned to control values after three days of continuous administration of NE (Katholi *et al*, 1977). In these experiments (Katholi *et al*, 1977), intrarenal NE administration was required to sustain an increase in PRA. This suggests that prolonged immobilization stress may not stimulate the renal sympathetic nervous system in continuous fashion.

In rats exposed to short-term hypokinesia plasma ALDO levels were markedly elevated. This increase in plasma ALDO probably resulted at least in part from increased angiotensin generation

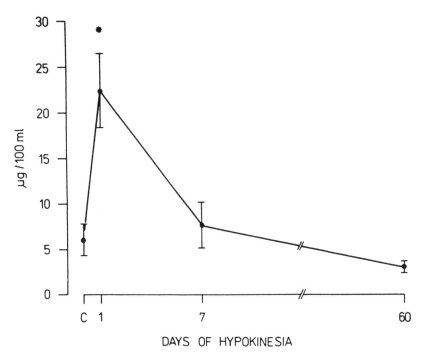

FIGURE 6. Plasma corticosterone concentration in rats exposed to hypokinesia for 1, 7 or 60 days. *p<0.05 (versus controls).

secondary to an increase in renin release from the juxtaglomerular cells of the kidney. In addition, plasma ALDO levels may have been affected by stimulation of the adrenal cortex by ACTH, because the largest increase of plasma CS levels was found after short-term immobilization (1 day) and corticosterone levels were unchanged after 7 or 60 days of hypokinesia. Stimulation of the adrenal cortex appeared to persist in spite of our findings of similar levels of ACTH in experimental and control animals after 1 day of hypokinesia. It is likely that the increase in ACTH preceded the time of plasma CS elevations and therefore it was not detected after 1 day of hypokinesia. There were differences in the increase of plasma ALDO levels in hypokinetic rats and in PRA values of control and experimental animals in our two experiments. These are probably due to different conditions of blood sampling (anesthesia and operation 1 day before experiments or decapitation of animals).

In conclusion, the results of the present experiments indicate that long-term immobilization (hypokinesia) in rats represents a model of stress with marked activation of the sympathetic nervous system. Adrenocortical function and the activity of the renin-angiotensin-aldosterone system are increased only during short-term exposure to hypokinesia.

REFERENCES

Beitins, J.Z., Shaw, M.H., Kowalski, A., and Migeon, C.J. (1970). Comparison of competitive protein binding radioassay of cortisol to double isotope dilution and Porter Silber methods. *Steroids* **15**, 765-776.

Buhler, F.R., Sealey J.E., and Laragh, J.H. (1974). Radioimmunoassay of plasma aldosterone. In: J.H. Laragh (Ed.), "Hypertension Manual," pp. 635-669. New York: Dun-Donnelley.

Carey, R.M., Rose, Jr., C.E., and Peach, M.J. (1989). Role of the renin-angiotensin-aldosterone system in stress. In: G.R. Van Loon, R. Kvetnansky, R. McCarty and J. Axelrod (Eds.), "Stress: Neurochemical and Humoral Mechanisms," pp. 833-844. New York: Gordon and Breach.

Chevalier, R.L., Gomez, R.A., Carey, R.M., Peach, M.J., and Linden, J.M. (1988). Renal effects of atrial natriuretic peptide infusion in young and adult rats. *Pediatric Research* **24**, 333-337.

Ganong, W.F. (1992). Renin as a stress hormone. In: R. Kvetnansky, R. McCarty and J. Axelrod (Eds.), "Stress: Neuroendocrine and Molecular Approaches. New York: Gordon and Breach.

Jezova-Repcekova, D., Foldes, O., Vigas, M., Brozmanova, H., and Jurcovicova, J. (1978). Preliminary results of a radioimmunoassay for ACTH. *Radiochemical and Radioanalytical Letters* **34**, 147-154.

Jezova, D., Kvetnansky, R., Kovacs, K., Oprsalova, Z., Vigas, M., and Makara, G.B. (1987). Insulin-induced hypoglycemia activates the release of adrenocorticotropin predominantly via central and propranolol insensitive mechanisms. *Endocrinology* **120**, 409-415.

Jindra, A., Kvetnansky, R., and Ripka, O. (1984). Effect of the sympathetic-adrenomedullary system on renin activation under stress. In: E. Usdin, R. Kvetnansky, and J. Axelrod (Eds.), "Stress: The Role of Catecholamines and Other Neurotransmitters," pp. 415-422. New York: Gordon and Breach.

Katholi, R.E., Carey, R.M., Ayers, C.R., and Vaughan, Jr., E.D. (1977). Production of sustained hypertension by chronic intrarenal norepinephrine infusion in conscious dogs. *Circulation Research* **35** (Suppl. I) I-118- I-126.

Kovalenko, I.E.A. (1977). The main methods of simulation of weightlessness effects on the body. *Kosmicheskaya Biologia Aviakosmicheskaya Medicina* **11**, 3-9. (in Russian).

Langer, P., Foldes, O., Macho, L., and Kvetnansky, R. (1981). Changes of iodothyronine levels in plasma after acute and long term hypokinesia (unforced restriction) in rats. *Endocrinologia Experimentalis* **15**, 139-144.

Macho, L., Murgas, K., Kadlecova, V., Strbak, V., and Fickova, M. (1977). The influence of hypokinesia on endocrine system, mobility and lipid metabolism in rats. Thesis 10. *Symposium on Space Biology and Medicine*, pp. 43-44. Suchumi, Academy of Sciences of USSR. (in Russian).

Macho, L., Kvetnansky, R., and Fickova, M. (1984). The effect of hypokinesia on lipid metabolism in adipose tissue. *Acta Astronautica* **11**, 735-738.

Macho, L., Fickova, M., Langer, P., and Kvetnansky, R. (1986). Changes in plasma levels of fatty acids during hypokinesia. *Bratislavske Lekarske Listy* **85**, 3-13. (in Slovak).

Macho, L., Kvetnansky, R., and Fickova, M. (1989a). Activation of the sympathoadrenal system in rats during hypokinesia. *Experimental and Clinical Endocrinology* **94**, 127-132.

Macho, L., Kvetnansky, R., Fickova, M., and Durisova, I. (1989b). Hypokinesia as a model of weightlessness: Effect on plasma catecholamines and their stimulation of cAMP production in rats. In: G.R. Van Loon, R. Kvetnansky, R. McCarty and J. Axelrod (Eds.), "Stress: Neurochemical and Humoral Mechanisms," pp. 1039-1052. New York: Gordon and Breach.

Peuler, J.D., and Johnson, G.A. (1977). Simultaneous single isotope radioenzymatic assay of plasma norepinephrine, epinephrine and dopamine. *Life Sciences* **21**, 625-636.

Sealey, J.E., and Laragh, J.H. (1977). How to do a plasma renin assay. *Cardiovascular Medicine* **2**, 1079-1092.

Stress: Neuroendocrine and Molecular Approaches
Edited by R. Kvetnansky, R. McCarty and J. Axelrod

1992 Gordon and Breach Science
Publishers S.A., New York, USA.
Photocopying permitted by license only.

NEURONAL, ADRENOMEDULLARY AND PLATELET-DERIVED NEUROPEPTIDE Y RESPONSES TO STRESS IN RATS

Z. Zukowska-Grojec[1], G. H. Shen[1], A. Deka-Starosta[2]
A. K. Myers[1], R. Kvetnansky[2] and R. McCarty[3]

[1]Department of Physiology and Biophysics, Georgetown University
Medical Center, Washington, DC USA
[2]National Institute of Neurological Disorders and Stroke
Bethesda, Maryland USA
[3]Department of Psychology, University of Virginia
Charlottesville, Virginia USA

INTRODUCTION

Neuropeptide Y (NPY) is a 36 amino acid tyrosine-rich peptide abundantly present with norepinephrine (NE) in all sympathetic nerves innervating the cardiovascular system (Ekblad *et al*, 1984) and in brain (O'Donohue *et al*, 1985). Together with the gastrointestinal peptides, peptide YY (PYY, approximately 70% homology) and pancreatic polypeptide (PP, 50 % homology) (Lundberg *et al*, 1982), NPY forms a family of pancreatic peptides. NPY is one of the best conserved peptides in mammalian species (Larhammar *et al*, 1987) and has multiple biological actions indicating that it is physiologically important. NPY is believed to act as a sympathetic co-transmitter mediating vasoconstriction independently of catecholamines, and as a modulator of autonomic cardiovascular responses (reviews: O'Donohue *et al*, 1985; McDonald, 1988; Waeber *et al*, 1988). NPY is also present in epinephrine-containing chromaffin cells of the adrenal medulla (Allen *et al*, 1983), and, under some conditions, may act as an adrenomedullary hormone (Corder *et al*, 1985). Finally, our data (Myers *et al*, 1988) indicate the extraneuronal presence of NPY in rat platelets where it may serve autocrine and paracrine functions in platelet-vascular interactions.

Thus, at least in rats, there are three potential sources for circulating NPY - the sympathetic nerves, the adrenal medulla and platelets. The latter source probably largely contributes to high basal circulating NPY levels in rats. In normal humans who have no platelet-derived NPY (Myers *et al*, 1991), as well as in some other species such as pigs (Lundberg *et al*, 1986), cats (Theodorsson-Norheim *et al*, 1985) and guinea pigs (Dahlof *et al*, 1986), circulating plasma NPY levels are several-fold lower than in rats, and probably mainly represent NPY released neuronally.

The cardiovascular effects of NPY are multiple. In resistance vessels (Ohlen *et al*, 1990) and in some small arteries such as coronary and cerebral arteries (Maturi *et al*, 1989; Edvinsson, 1985), NPY causes vasoconstriction by an action mediated by Y1 receptors (Wahlestest *et al*, 1986) and systemically evokes pressor responses. In many blood vessels (e.g., pulmonary artery) NPY amplifies other vasoconstrictor actions, especially that of NE (Wahlestedt *et al*, 1990). Although not all blood vessels are responsive to NPY-induced vasoconstriction, those pre-exposed to high circulating levels of NE become markedly hypersensitive (Wahlestedt *et al*, 1990). NPY also mediates sympatho-inhibitory (Wahlestedt *et al*, 1986) and histamine-releasing effects which may lead to hypotension (Shen *et al*, 1991). Studies performed on isolated (Schoups *et al*, 1986; Haass *et al*, 1989) or *in situ* perfused organs (Lundberg *et al*, 1986) have established that NPY is released via calcium-dependent exocytosis during stimulation of sympathetic nerves. The release of NPY has also been shown *in vivo* during prolonged and intense stress (Kaijser *et al*, 1990; Morris *et al*, 1986). In humans, plasma NPY-ir levels are increased by exercise, cold pressor test (Morris *et al*, 1986) and hypoglycemia (Takahashi *et al*, 1988), but not by an orthostatic test or isometric handgrip (Pernow *et al*, 1986), suggesting that only more intense sympathetic activation enhances the release of NPY from the nerves and perhaps also from the adrenal medulla. Whether adrenomedullary NPY contributes to circulating levels of the peptide is not clear. Although stimulation of splanchnic nerves (Allen *et al*, 1985) and hemorrhage (Briand *et al*, 1990) increase adrenal plasma NPY levels, the output of the peptide into the circulation actually decreases because of the concomitant reduction in adrenal blood flow. In rats, which have large amounts of NPY in platelets, the contribution of neuronal and adrenomedullary pools to circulating plasma NPY levels are even more difficult to establish.

We have previously shown in rats that some stressors such as exposure to cold water (Zukowska and Vaz, 1989) elevate plasma NPY-ir whereas others such as immobilization do not (Zukowska-Grojec et al, 1987). We argued then that differential plasma NPY-ir responses are due to differential activation of the sympathetic nerves by various stressors, since those which primarily activate the adrenal medulla (e.g., immobilization) fail to raise plasma NPY levels. However, direct evidence for contributions of the sympathetic-adrenomedullary sources to plasma NPY is still sparse. Although established in other species, in vivo release of NPY from sympathetic nerves in rats has never been reported, since rats have been considered a poor model where "artifactual release" of platelet-derived NPY confuses the issue.

In the present study, we have examined the contributions of neuronal and extra-neuronal pools to plasma NPY levels. First, we attempted to assess purely neuronal release of NPY in response to sympathetic nerve stimulation in pithed rats, and then, platelet release of NPY by infusion of collagen in vivo. Next, we developed an approach to minimize artifactual ex vivo platelet aggregation (and NPY release) and studied the contribution of sympathetic nerves and adrenal medulla to plasma NPY responses to stress.

MATERIALS and METHODS

Pithed Rat Model

Rats (albino male, 300-400 g) were anesthetized, vagotomized, pithed by a steel rod, paralyzed and artificially respired as previously described (Zukowska-Grojec et al, 1987). Electrical stimulation of sympathetic outflow was performed at 40 V, 1 msec pulse duration, at varying frequencies, patterns (continuous or in bursts delivering the same amount of pulses, and stimulation periods repeated 1-4 times) and durations (1 to 10 minutes). In another group of pithed rats, platelet release of NPY in vivo was studied by infusing collagen i.v. at 2 mg/kg (calculated equivalent to an ED_{50} for platelet NPY release in *vitro* (Myers *et al*, 1988), and at a lethal dose of 5-10 mg/kg.

Conscious Rat Model

Chronically cannulated rats (albino male, 300-400 g) in which cannulas were flushed daily with heparinized saline (300 units/kg/day) were subjected to two acute stress paradigms: cold water (4°C) exposure (Zukowska-Grojec *et al*, 1988) and immobilization (Zukowska-Grojec *et al*, 1987). There were two groups of rats subjected to cold stress: Group I - in which arterial blood samples were collected in 2.25 mg EDTA and 0.5 tripsin-inhibitory unit of aprotinin per 0.5 ml of blood (Zukowska-Grojec *et al*, 1987, 1988), without special care taken to prevent platelet aggregation during collection; and Group II - in which a stringent technique of blood collection was adopted to minimize or eliminate *ex vivo* platelet aggregation. In Group II, blood was collected by free flow from a short arterial catheter into a tube containing acid-citrate-dextrose solution (ACD: 0.065 M citric acid, 0.085 M sodium citrate, 2% dextrose) in a 9:1 ratio. This method of collection has been chosen from several others including collecting in higher concentrations of aprotinin, and a platelet-stabilization mixture used in the study of platelet factor 4 which contains EDTA, adenosine, theophylline and prostacyclin. Only ACD proved to be effective in completely preventing collagen-induced aggregation of rat platelets *in vitro*. Similar blockade of *in vitro* platelet aggregation was also obtained with a large dose of heparin (1000 units/kg) injected intravenously, and this method was used in one group of cold-stressed rats to assess the *in vivo* contribution of platelet aggregation.

In immobilized rats, blood samples were collected as in cold-stressed rats from Group I. Both cold water stress (Group I) and immobilization were performed in intact rats and in rats pretreated with a ganglionic blocker, chlorisondamine (3 mg/kg, i.v.).

Plasma NPY-immunoreactivity.

NPY-ir was measured by an RIA as previously described (Zukowska-Grojec *et al*, 1987) using NPY specific antiserum against porcine NPY (Peninsula Laboratories, Belmont, CA), porcine and rat NPY as standards (the antiserum recognizes 100% of porcine and 50% of rat NPY), and ^{125}I-NPY (Amersham) in plasma samples lyophilized and resuspended in RIA buffer.

RESULTS

Plasma NPY Responses to Sympathetic Nerve Stimulation in Pithed Rats

Figure 1 shows plasma NPY-ir levels obtained during sympathetic stimulation in three groups of pithed rats (Panels A,B,C). Panels A and B represent NPY-ir levels measured in blood collected by a previously used standard method (Zukowska-Grojec *et al*, 1987, 1988), whereas Panel C shows stimulation-induced NPY-ir responses as measured in blood stabilized against platelet aggregation by ACD. Sympathetic nerve stimulation tended to evoke frequency-dependent plasma NPY-ir responses but the increase in NPY was significant only at the highest frequency of 3 Hz (p < 0.05, Figure 1A); this increase was prevented by ganglionic blockade with chlorisondamine. In Group B (Figure 1B), no significant differences were observed between different patterns of nerve stimulation. Continuous stimulation at 1 Hz for 10 minutes (total of 600 pulses) increased plasma NPY-ir levels similarly to bursts of either 10 Hz or 30 Hz delivering the same number of pulses (Figure 1B). Plasma NPY-ir levels in blood samples not stabilized against platelet activation, presented in Panels A and B, showed large variations.

Basal plasma NPY-ir levels were similar in blood collected by a standard method and in platelet-stabilized blood, but there was less variability in the latter group. In platelet-stabilized blood, plasma NPY-ir responses to stimulation at 3 Hz, 40 V for 2 minutes (repeated 4 times) showed a significant 2.5-fold increase over baseline levels (p < 0.05).

Plasma NPY-ir Responses to Collagen *In Vivo*

Infusion of collagen to pithed rats caused a dose dependent increase in plasma NPY-ir levels (Figure 2). Collagen evoked a 2-fold (p < 0.05) rise in plasma NPY-ir at 2 mg/kg injected over 1 minute, and a 5-fold (p < 0.02) increase at a lethal dose of 5-10 mg/kg.

Plasma NPY-ir Levels in Cold Water Stress

Plasma NPY-ir levels increased by 2.5-fold at 30 minutes of cold water stress, as measured in blood samples not stabilized for platelet

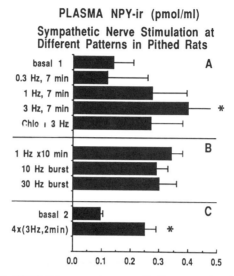

FIGURE 1.   Plasma NPY-ir responses to sympathetic nerve stimulation of different patterns in pithed rats. Each panel represents groups of rats: in A and B - blood samples were collected without, and in C - with platelet stabilization.   Chlo - chlorisondamine, 3 mg/kg, i.v.   * $p < 0.05$ as compared to baseline by paired Student's t-test.

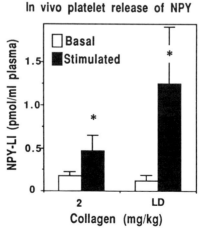

FIGURE 2.   Plasma NPY-like immunoreactivity (NPY-LI) responses to intravenous injection of collagen in pithed rats. LD - a lethal dose of collagen (5-10 mg/kg). *$p < 0.05$ as compared to baseline by paired Student's t-test.

aggregation (Figure 3A). This response was similarly prevented by either ganglionic blockade with chlorisondamine (3 mg/kg) or an antiaggregatory dose of heparin (1000 units/kg). When blood was collected in ACD, plasma NPY-ir levels did not show any significant increase at 30 minutes but did at 60 minutes of cold water stress ($p < 0.05$, Figure 3B), and this effect was smaller than in Group II.

Plasma NPY-ir during Immobilization in Adrenal Demedullated Rats

Immobilization failed to increase plasma NPY-ir levels in sham-operated rats (Figure 4). However, in adrenal demedullated rats (DEM) rats, plasma NPY-ir levels increased by 50% ($p < 0.05$, Figure 4) at 60 minutes of immobilization; this response was prevented by ganglionic blockade with chlorisondamine (Figure 4). Plasma NPY-ir levels in these rats were measured in blood samples collected without platelet stabilization, which again contributed to high resting values and large variations.

DISCUSSION

In postganglionic sympathetic nerves, NPY is apparently present in large dense core vesicles, similar to or the same as those containing norepinephrine (NE) (Fried et al, 1985). Neuronal release of NPY has been shown previously in vivo (Dahlof et al, 1986) and in vitro in sympathetically innervated organs of several species (Lundberg et al, 1989; Haass et al, 1989), and now, our studies confirm that it can also be evoked in vivo in rats. Previously, evidence for purely sympathetic release of NPY in rats was lacking due to high platelet NPY content and release in this species (Myers et al, 1988). Platelet NPY release may occur both in vivo as well as ex vivo during blood collection and modify the magnitude and cause of variations in NPY responses. Thus, measurement of NPY in rats requires the same precautions as is being used for any other platelet-derived factor (Levine and Krentz, 1977). Stabilization of platelets during blood collection by ACD reduced large intra- and inter-individual variability in plasma NPY-ir levels. In pithed rats, plasma NPY-ir responses to sympathetic nerve stimulation were dependent on frequency and duration (occurring at frequencies greater than 1 Hz, and at longer than 1 minute duration of stimulation) but were independent of continuous or bursting

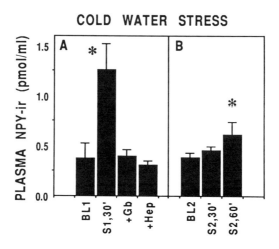

FIGURE 3.   Plasma NPY-ir responses to cold water stress (S) in conscious rats.
Each panel represents different group of rats in which blood samples were
collected without (A) and with (B) platelet stabilization. Gb - ganglionic blockade
with chlorisodamine (3 mg/kg); Hep - heparin 1000 units/kg, i.v. * $p < 0.05$ as
compared to baseline (BL) by paired Student's t-test.

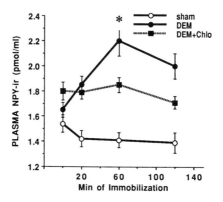

FIGURE 4.   Plasma NPY-ir responses to immobilization stress in sham-operated
(sham) and adrenal demedullated (DEM) rats. Chlo - chlorisondamine pretreatment
(3mg/kg).

stimulation patterns.

The potential contribution of *in vivo* platelet release of NPY is indicated by the ability of collagen to raise circulating NPY-ir levels by 3-5-fold, at doses equivalent to those which cause platelet aggregation *in vitro* (Myers *et al*, 1988). Since these effects of collagen, were observed in pithed rats which are devoid of any circulatory reflexes, they are unlikely to be secondary to hemodynamic effects of collagen but rather reflect its aggregatory action. Thus, platelet aggregation *in vivo* may be potentially an important and sizable pool of NPY being released directly into the circulation. Platelets may be a sole source of NPY (collagen) or may be activated secondary to the activation of the sympatho-adrenomedullary system via stimulation of pro-aggregatory alpha-adrenergic receptors (Kerry *et al*, 1984). This appears to be the case during cold water stress which increases circulating NPY-ir levels primarily by activation of sympathetic nerves; platelet aggregation *ex vivo*, and probably *in vivo*, can markedly amplify this response. Our recent studies using stringent blood collection procedures to prevent *ex vivo* platelet aggregation underscore the importance of the platelet component. Considering this, previous data on plasma NPY-ir levels in rats (Morris *et al*, 1987; Castagne *et al*, 1987) should be critically re-evaluated in view of the possible platelet involvement. This is of importance since rats are commonly used models for studying stress and hypertension and plasma NPY levels could be incorrectly inferred to reflect activation of the sympathetic-adrenomedullary system, whereas they may be platelet-dependent, and at times, artifactual.

The contribution of the adrenal medulla to circulating plasma NPY-ir levels appears to be negligible in normal (e.g., sham-operated) rats. Our findings with immobilization stress agree with those of Morris *et al* (1987) who also found that adrenal demedullation fails to affect resting NPY-ir levels but augmented plasma NPY responses to hemorrhage. The nature of this augmentation remains unclear. Possibly, the absence of the adrenal medulla causes greater activation of the sympathetic nerves in response to stress. Alternatively, since circulating plasma NPY levels depend on release as well as inactivation processes, higher circulating NPY-ir levels in adrenal demedullated rats may reflect decreased clearance of NPY.

Plasma NPY-ir responses to stress vary depending on stress paradigms (Zukowska-Grojec *et al*, 1987, 1988). Since sympathetic nerves and platelets are two major sources of NPY in normal rats, changes in circulating NPY levels will depend on the degree of their activation during stress. Provided that artifactual *ex vivo* platelet aggregation is appropriately prevented, plasma NPY levels may reflect neuronal release. However, intense and generalized sympathetic nerve activation is required for significant effects. This principle applies to rats as well as to other species (Pernow *et al*, 1986; Briand *et al*, 1990) and suggests that spillover of NPY into the circulation may be effectively prevented by several processes occurring between the sympathetic neuro-effector junction and the circulation such as enzymatic degradation, specific and non-specific binding of the peptide and reduction of local blood flow. The presence of a non-adrenergic component of pressor responses to sympathetic nerve stimulation (McGrath *et al*, 1982) combined with the evidence of neuronal NPY release *in vitro* (Haass *et al*, 1989) suggest, however, that NPY may be released intrajunctionally (although it may not spill over into the plasma).

The physiological importance of NPY in cardiovascular responses to stress may be enhanced by known NPY interactions with the adrenergic system. We have previously shown that NPY-induced vasopressor action is markedly augmented by high circulating catecholamines, and in addition, NPY potentiates adrenergic vasoconstriction (Wahlestedt *et al*, 1990). Since all stressors increase plasma catecholamine levels (McCarty *et al*, 1988) creating a hyperadrenergic state, they may predispose to NPY mediated vasoconstriction in such important vascular beds as coronary and cerebral vascular beds, resulting in stress-related cardiovascular pathology. Those types of stressors (high intensity and long duration) which produce high circulating NPY levels could be particularly detrimental to the cardiovascular system.

ACKNOWLEDGEMENTS

This work has been supported in part by U.S. Public Health Service grants HL40718 and HL43160.

REFERENCES

Allen, J.M., Adrian, T.E., Polak, J.M., and Bloom, S.R. (1983). Neuropeptide Y (NPY) in the adrenal gland. *Journal of the Autonomic Nervous System* **9**,

559-563.
Allen, J.M., Bircham, M.M., Bloom, S.R., and Edwards, A.V. (1984). Release of neuropeptide Y in response to splanchnic nerve stimulation in the conscious calf. *Journal of Physiology* **357**, 401-408.
Briand, R., Yamaguchi, N., Gagne, J., Nadeau, R., and de Champlain, J.(1990). Alpha-2 adrenoceptor modulation of catecholamine and neuropeptide Y responses during hemorrhagic hypotension in anesthetized dogs. *Journal of the Autonomic Nervous System* **30**, 11-122.
Castagne, V., Corder, R., Gaillard, R., and Mormede, P.(1987). Stress-induced changes of circulating neuropeptide Y in the rat: comparison with catecholamines. Regulatory Peptides, 19, 55-63.
Corder, R., Lowry, P.J., Emson, P.C., and Gaillard, R.C. (1985). Chromatographic characterization of the circulating neuropeptide Y immunoreactivity from patients with phaechromocytoma. *Regulatory Peptides* 10, 91-97.
Dahlof, C., Dahlof, P., and Lundberg, J.M. (1986). Alpha-2-adrenoceptor-mediated inhibition of nerve stimulation-evoked release of neuropeptide Y (NPY)-like immunoreactivity in the pithed guinea-pig. *European Journal of Pharmacology* **131**, 279-283.
Edvinsson, L. (1985). Characterization of the contractile effect of neuropeptide Y in feline cerebral arteries. *Acta Physiologica Scandanavica* **125**, 33-41.
Ekblad, E., Edvisson, L., Wahlestedt, C., Uddman, R., Hakanson, R., and Sundler, F. (1984). Neuropeptide Y coexists and cooperates with noradrenaline in perivascular nerve fibers. *Regulatory Peptides* **8**, 225-235.
Fried, G., Terenius, L., Hokfelt, T., and Goldstein, M. (1985). Evidence for differential localization of noradrenaline and neuropeptide Y (NPY) in neuronal storage vesicles isolated from rat vas deferens. *Journal of Neuroscience* **5**, 450-458.
Haass, M., Hock, M., Richart, G., and Schömig, A. (1989). Neuropeptide Y differentiates between exocytotic and non-exocytotic noradrenaline release in guinea-pig heart. *Naunyn Schmiedeberg's Archives of Pharmacology* **340**, 509-515.
Kaijser, L., Pernow, J., Berglund, B., and Lundberg, J.M. (1990). Neuropeptide Y is released together with noradrenaline from the human heart during exercise and hypoxia. *Clinical Physiology* 10, 179-188.
Kerry, R., Scrutton, M.C., and Wallis, R.B. (1984). Mammalian platelet adrenoceptors. *British Journal of Pharmacology* **81**, 91-102.
Larhammar, D., Ericsson, A. and Persson, H. (1987). Structure and expression of the rat neuropeptide Y gene. *Proceedings of the National Academy of Sciences* (USA) **84**, 2068-2072.
Levine, S.P., and Krentz, L.S. (1977). Development of a radioimmunoassay for human platelet factor 4. *Thrombosis Research* 11, 673-686.
Lundberg, J.M., and Tatemoto, K.(1982). Pancreatic polypeptide family (APP, BPP, NPY and PYY) in relation to sympathetic vasoconstriction resistant to alpha-adrenoceptor blockade. *Acta Physiologica Scandanavica* **116**, 393-402.
Lundberg, J.M., Rudehill, A., Sollevi, A., Thoeodorsson-Norheim, E., and Hamberger, B. (1986). Frequency- and reserpine-dependent chemical coding of sympathetic transmission: differential release of noradrenaline and

neuropeptide Y from pig spleen. *Neuroscience Letters* **63**, 96-100.

Maturi, A.F., Greene, R., Speir, E., Burrus, C., Dorsey, L.M.A., Markle, D.R., Maxwell, M., Schmidt, W., Goldstein, S.R., and Patterson, R.E. (1989). Neuropeptide Y: A peptide found in human coronary arteries constricts primarily small coronary arteries to produce myocardial ischemia in dogs. *Journal of Clinical Investigation* **83**, 1217-1224.

McCarty, R., Horvath, K., and Konarska, M. (1988).   Chronic stress and sympatho-adrenomedullary responsiveness. *Social Science and Medicine* **26**, 333-341.

McDonald, J.K. (1988).   NPY and related substances.   *CRC Critical Reviews in Neurobiology* **4**, 97-135.

McGrath, J.C. (1982).   Evidence for more than one type of postjunctional alpha-adrenoceptors. *Biochemical Pharmacology* **31**, 467-484.

Morris, M., Kapoor, V., and Chalmers, J. (1987).   Plasma neuropeptide Y concentration is increased after hemorrhage in conscious rats: relative contributions of sympathetic nerves and the adrenal medulla. *Journal of the Cardiovascular Pharmacology* **9**, 541-545.

Morris, M.J., Russell, A.E., Kapoor, V., Elliott, J.M., West, M.J., Wing, L.M.H., and Chalmers, J.P. (1986).   Increases in plasma neuropeptide Y concentrations during sympathetic activation in man. *Journal of Autonomic Nervous System* **17**, 143-49.

Myers, A.K., Abi-Younes, S., and Zukowska-Grojec, Z. (1991).   Re-evaluation of the effects of neuropeptide Y on aggregation of human platelets. *Life Sciences* **49**, 545-551.

Myers, A.K., Farhat, M.Y., Vaz, C.A., Keiser, H.R., and Zukowska-Grojec, Z. (1988).   Release of immunoreactive neuropeptide Y by rat platelets. *Biochemical and Biophysical Research Communications* **155**, 118-122.

O'Donohue, T.L., Chronwall, B. M., Pruss, R.M., Mezey, E., Kiss, J. Z., Eiden, L.E., Massari, V.J., Tessel, R.E., Pickel, V.M., DiMaggio, D.A.., Hotchkiss, A.J., Crowley, W.R., and Zukowska-Grojec, Z. (1985).   Neuropeptide Y and peptide YY neuronal and endocrine systems. *Peptides* **6**, 755-768.

Ohlen, A., Persson, M.G., Lindbom, L., Gustafsson, L.E., and Hedquist, P. (1990). Nerve-induced non-adrenergic vasoconstriction and vasodilatation in skeletal muscle. *American Journal of Physiology* **258**, H1334-H1338.

Pernow, J., Lundberg, J.M., Kaijser, L., Hjemdahl, P., Theodorsson-Norheim, E., Martinsson, A., and Pernow, B. (1986).   Plasma neuropeptide Y-like immunoreactivity and catecholamines during various degrees of sympathetic activation in man. *Clinical Physiology* **6**, 561-578.

Schoups, A.A., Saxena, V.K., Tombeur, K., and De Potter, W.P. (1988).   Facilitation of the release of norepinephrine and neuropeptide Y by the alpha-2 adrenoceptor blocking agents idazoxan and hydergine in the dog spleen. *Life Sciences* **42**, 517-523.

Shen, G.H., Grundemar, L., Zukowska-Grojec, Z., Hakanson, R, and Wahlestedt, C. (1991).   C-terminal neuropeptide Y fragments are mast cell-dependent vasodepressor agents. *European Journal of Pharmacology*, in press.

Takahashi, K., Mouri, T., Murakami, O., Itoi, K., Sone, M., Ohneda, M., Nozuki, M., and Yoshinaga, K. (1988). Increases of neuropeptide Y-like immunoreactivity in plasma during insulin-induced hypoglycemia in man. *Peptides* **9**, 433-435.

Theodorsson-Norheim, E., Hemsen, A., and Lundberg, J.M. (1985). Radioimmunoassay for neuropeptide Y (NPY): chromatographic characterization of immunoreactivity in plasma and tissue extracts. Scandinavian Journal of Clinical Investigation, 45, 355-365.

Waeber, B., Aubert, J.-F., Corder, R., Evoquez, D., Nussberger, J., Gaillard, R., and Brunner, H.R. (1988). Cardiovascular effects of neuropeptide Y. *American Journal of Hypertension* **1**, 193-199.

Wahlestedt, C., Hakanson, R., Vaz, A.C., and Zukowska-Grojec, Z. (1990). Norepinephrine and neuropeptide Y: vasoconstrictor cooperation *in vivo* and *in vitro*. *American Journal of Physiology* **258**, R736-R742.

Wahlestedt, C., Yanaihara, N., and Hakanson, R. (1986). Evidence for different pre- and post-junctional receptors for neuropeptide Y and related peptides. *Regulatory Peptides* **13**, 307-318.

Zukowska-Grojec, Z., and Vaz, C.A. (1988). Role of neuropeptide Y (NPY) in cardiovascular responses to stress. *Synapse* **2**, 293-298.

Zukowska-Grojec, Z., Konarska, M., and McCarty, R. (1987). Differential plasma catecholamine and neuropeptide Y responses to acute stress in rats. *Life Sciences* **42**, 1615-1624.

Stress: Neuroendocrine and Molecular Approaches
Edited by R. Kvetnansky, R. McCarty and J. Axelrod

1992 Gordon and Breach Science
Publishers S.A., New York, USA.
Photocopying permitted by license only.

RELATIONSHIP BETWEEN NEUROPEPTIDE Y AND NOREPINEPHRINE RELEASE FROM CARDIAC SYMPATHETIC NERVES

M. Haass, G. Richardt and A. Schömig

Department of Cardiology, University of Heidelberg
Heidelberg, Germany

INTRODUCTION

Neuropeptide Y (NPY), which was originally isolated from porcine hypothalamus (Tatemoto, 1982), is co-localized with catecholamines in central and peripheral sympathetic nerve fibers and the adrenal medulla (for review, refer to Haass and Schömig, 1989). Upon sympathetic activation, systemic plasma concentrations of NPY have been found to increase significantly in man (Figure 1). Under experimental conditions, stimulation of sympathetic efferents is accompanied by a marked overflow of NPY from a variety of organs with dense sympathetic innervation, such as cat and pig spleen (Haass and Schömig, 1989).

The sympathetic co-transmitter NPY interferes both pre- and postsynaptically with sympathetic transmission (Edvinsson *et al*, 1987). Furthermore, NPY exerts direct vasoconstrictor effects on a variety of vascular beds including the coronary circulation. Within the heart, high concentrations of NPY are observed in cardiac perivascular sympathetic nerve fibers (Wharton and Polak, 1990). This brief review focuses on our most recent investigations on a characterization of the relationship of NPY to norepinephrine release from the guinea pig perfused heart. In the coronary-venous overflow, NPY and norepinephrine were determined by polyclonal radioimmunoassay

211

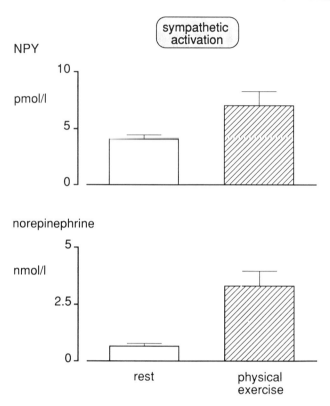

FIGURE 1.   Plasma concentrations of neuropeptide Y (NPY) and norepinephrine
at rest and after 12 minutes of graded bicycle exercise in 8 healthy volunteers.
Physical exercise significantly ($p < 0.05$) increased the plasma concentrations of
both transmitters.   The respective blood samples were withdrawn from an
antecubital vein and values are presented as means ± SEM.

(Haass *et al*, 1989a) and by high pressure liquid chromatography in
combination with electrochemical detection (Schömig *et al*, 1987),
respectively.

DISCUSSION

Exocytotic Co-release of NPY and Norepinephrine

Electrical stimulation of the left stellate ganglion evoked a frequency-dependent co-release of NPY and norepinephrine from the guinea pig perfused heart (Haass *et al*, 1989a). Likewise, nicotine (Figure 2), veratridine and potassium depolarization induced a marked overflow of both transmitters (Richardt *et al*, 1988; Haass *et al*, 1989b, 1990c). Most likely due to its higher molecular weight, the occurrence of NPY in the coronary-venous overflow was slightly delayed compared to norepinephrine (Haass *et al*, 1989a). As evidence for the sympathetic origin of NPY, the molar ratio of evoked transmitter overflow was consistently 1 (NPY) to 600-1000 (norepinephrine) (Haass *et al*, 1990b). Electrical stimulation, nicotine, veratridine, and high potassium concentrations are known to evoke calcium-dependent exocytotic transmitter release by depolarization of sympathetic nerve fibers (Winkler, 1988).

Calcium. Consistent with an exocytotic release mechanism, the stimulation-induced co-release of both transmitters from the guinea pig heart was critically dependent on the presence of extracellular calcium (Haass *et al*, 1990a). Likewise, calcium-free perfusion prevented the overflow of NPY and norepinephrine evoked by either nicotine (Figure 2), veratridine or potassium depolarization (Haass *et al*, 1989, 1990c).

Calcium channels. Using various ligands for calcium channels, calcium entry through neuronal (N-type) calcium channels was found to be a prerequisite for the co-release of NPY and norepinephrine evoked by either electrical stimulation (Haass *et al*, 1990 or by nicotine (Haass *et al*, 1991). Transmitter overflow was unaffected by the L-type calcium channel blocker felodipine, a dihydropyridine derivative. However, blockade of N-type calcium channels by either *w*-conotoxin or by cadmium chloride markedly reduced NPY and norepinephrine release (Figure 2).

Protein kinase C. Consistent with a modulatory role of the calcium sensitive protein kinase C for the evoked co-release of both transmitters (for review, refer to Winkler, 1988), inhibition of the enzyme by polymyxin B suppressed (Figure 2) and activation of the enzyme by phorbol 12-myristate 13-acetate markedly enhanced the overflow of NPY and norepinephrine (Haass *et al*, 1990a, 1991). A nonspecific effect of the phorbol ester was ruled out, since 4α-phorbol 12,13-didecanoate, a phorbol ester not interfering with protein kinase C, did not affect transmitter overflow (Haass *et al*, 1990a, 1991).

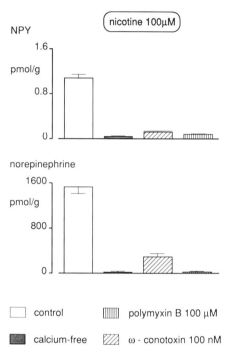

FIGURE 2. Overflow of neuropeptide Y (NPY) and norepinephrine from the guinea pig isolated perfused heart evoked by nicotine (100 µM). Calcium-free perfusion, blockade of N-type calcium channels (by *w*-conotoxin), and inhibition of protein kinase C (by polymyxin B) significantly ($p < 0.05$) reduced nicotine-induced transmitter overflow. If not otherwise indicated, all hearts were perfused with a calcium-containing (1.85 mM) Krebs-Henseleit-buffer at a constant flow rate of 5 ml per minute and gram heart weight. The concentrations of both transmitters in the coronary-venous overflow are expressed as overflow per gram heart weight (n = 4-8 in each group). Values are means ± SEM.

Presynaptic receptors. The stimulation-evoked co-release of NPY and norepinephrine was characterized further through its modulation by inhibitory presynaptic adenosine A1 and stimulatory presynaptic angiotensin II receptors (Haass *et al*, 1989a, 1990b). Additionally, transmitter release was modulated by presynaptic autoreceptors, including alpha$_2$-adrenoceptors and NPY receptors, via negative feedback mechanism (Haass *et al*, 1989a).

Neuronal amine re-uptake. In contrast to the parallel changes of NPY and norepinephrine overflow reported so far, blockade of neuronal amine re-uptake (uptake$_1$) by desipramine or nisoxetine enhanced norepinephrine but reduced NPY overflow in the guinea pig perfused heart (Haass *et al*, 1989a, 1989b). Similar results were obtained in the guinea pig perfused spleen (Lundberg *et al*, 1989). This dissociation is due to a combination of two effects elicited by uptake$_1$ blockade: by inhibiting norepinephrine re-uptake, blockade of uptake$_1$ enhances the concentrations of the amine in the synaptic cleft. Enhanced intrasynaptic concentrations of norepinephrine, however, stimulate inhibitory presynaptic alpha$_2$-adrenoceptors, tending to reduce evoked overflow of NPY and norepinephrine (Haass *et al*, 1989b).

Differential Release of NPY and Norepinephrine Due to Nonexocytotic Mechanisms

Norepinephrine can be released by exocytosis as well as by a nonexocytotic release mechanism. Nonexocytotic norepinephrine release depends on two major preconditions. First, vesicular uptake of norepinephrine into sympathetic storage vesicles has to be disturbed, resulting in increased axoplasmic amine concentrations. Second, the neuronal amine carrier (uptake$_1$) has to reverse its normal transport direction allowing for outward transport of free axoplasmic norepinephrine (Graefe, 1989). Nonexocytotic norepinephrine release can be induced by the indirectly acting sympathomimetic agent tyramine (Thoa *et al*, 1975) and by energy depletion of nerve endings (Schömig *et al*, 1987).

Calcium. In the guinea pig perfused heart, tyramine induced a concentration-dependent overflow of norepinephrine (Haass *et al*, 1989b), which was not accompanied by any substantial release of NPY (Figure 3). This finding is not limited to cardiac sympathetic nerves, since similar results were obtained in the guinea pig perfused spleen (Lundberg *et al*, 1989). In contrast to nicotine-evoked transmitter overflow, tyramine-induced norepinephrine release was independent of the presence of extracellular calcium (Figure 3). Under calcium-free conditions, anoxic and substrate-free perfusion also evoked a marked overflow of norepinephrine without a concomitant overflow of NPY (Haass *et al*, 1989b). Likewise, no NPY was detectable in the coronary-venous effluent after short-term myocardial ischemia (Franco-Cereceda *et al*, 1989).

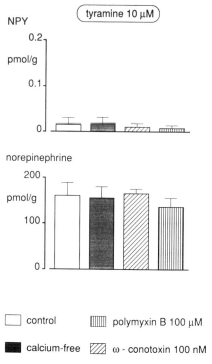

FIGURE 3. Overflow of neuropeptide Y (NPY) and norepinephrine from the guinea pig isolated perfused heart evoked by tyramine (10 μM). Tyramine-induced norepinephrine overflow was neither affected by calcium-free perfusion, nor by blockade of N-type calcium channels (*w*-conotoxin) nor by inhibition of protein kinase C (by polymyxin B). No major overflow of NPY was observed under all four conditions. For further details see Figure 2 (n = 4-6 in each group). Values are means ± SEM.

Calcium channels and protein kinase C. In contrast to exocytotic norepinephrine release, tyramine-evoked transmitter overflow was neither modulated by blockade of N-type calcium channels nor by inhibition of protein kinase C (Schömig and Haass, 1990) (Figure 3).

Neuronal amine re-uptake. Since nonexocytotic norepinephrine release from heart is a carrier-meditated process, inhibition of the neuronal uptake$_1$-carrier by desipramine prevented norepinephrine overflow induced by tyramine and by anoxia plus glucose-free perfusion (Haass *et al*, 1989b).

CONCLUSIONS

A concomitant release of NPY and norepinephrine is observed upon depolarization of cardiac sympathetic nerve fibers. Consistent with an exocytotic release mechanism, the co-release of both transmitters critically depends on the presence of calcium. The co-release of NPY and norepinephrine is further characterized by its requirement of calcium influx through N-type calcium channels, its dependence on the activation of protein kinase C, and its parallel presynaptic modulation. In contrast to calcium-dependent exocytotic transmitter release, nonexocytotic, carrier-mediated norepinephrine release is not accompanied by overflow of NPY. Nonexocytotic norepinephrine release is independent of the presence of extracellular calcium and also occurs after inhibition of N-type calcium channels and protein kinase C. Since the relationship of both transmitters depends on the release mechanism, simultaneous determination of NPY and norepinephrine allows a differentiation between exocytotic and nonexocytotic norepinephrine release.

ACKNOWLEDGEMENTS

This research was supported by a grant from the German Research Foundation (SFB 320 -- Cardiac function and it's regulation).

REFERENCES

Edvinsson, L., Hakanson, R., Wahlestedt, C., and Uddman, R. (1987). Effects of neuropeptide Y on the cardiovascular system. *Trends in Pharmacological Sciences* **8**, 231-235.

Franco-Cereceda, A., Saria, A., and Lundberg, J.M. (1989). Differential release of calcitonin gene-related peptide and neuropeptide Y from the isolated heart by capsaicin, ischaemia, nicotine, bradykinin and ouabain. *Acta Physiologica Scandinavica* **135**, 173-187.

Graefe, K.H. (1989). On the mechanism of nonexocytotic release of noradrenaline from noradrenergic neurones. In: J. Brachmann, and A. Schömig (Eds.), "Adrenergic System and Ventricular Arrhythmias in Myocardial Infarction," pp. 44-52. Berlin: Springer-Verlag.

Haass, M., and Schömig, A. (1989). Neuropeptide Y and sympathetic transmission. In: J. Brachmann and A. Schömig (Eds.), "Adrenergic System and Ventricular Arrhythmias in Myocardial Infarction," pp. 21-33. Berlin: Springer-Verlag.

Haass, M., Cheng, B., Richardt, G., Lang, R.E., and Schömig, A. (1989a). Characterization and presynaptic modulation of stimulation-evoked exocytotic co-release of noradrenaline and neuropeptide Y in guinea pig heart. *Naunyn-Schmiedeberg's Archives of Pharmacology* 339, 71-78.

Haass, M., Hock, M., Richardt, G., and Schömig, A. (1989b). Neuropeptide Y differentiates between exocytotic and nonexocytotic noradrenaline release in guinea pig heart. *Naunyn-Schmiedeberg's Archives of Pharmacology* 340, 509-515.

Haass, M., Förster, C., Kranzhöfer, R., Richardt, G., and Schömig, A (1990a). Role of calcium channels and protein kinase C for release of norepinephrine and neuropeptide Y. *American Journal of Physiology* 259, R925-R930.

Haass, M., Richardt, G., Lang, R.E., and Schömig, A. (1990b). Common features of neuropeptide Y and noradrenaline release in guinea pig heart. *Annals of the New York Academy of Sciences* 611, 450-452.

Haass, M., Hock, M., Richardt, G., and Schömig, A. (1990c). Acidosis suppresses exocytotic noradrenaline and neuropeptide Y release in the energy-depleted heart. *Circulation* 82 (Suppl.III), III-454 (abstract).

Haass, M., Richardt, G., Brenn, Th., Schömig, E., and Schömig, A. (1991). Nicotine-induced release of noradrenaline and neuropeptide Y in guinea-pig heart: Role of calcium channels and protein kinase C. *Naunyn-Schmiedeberg's Archives of Pharmacology* 344, 527-531.

Lundberg, J.M., Rudehill, A., Sollevi, A., and Hamberger, B. (1989). Evidence for co-transmitter role of neuropeptide Y in pig spleen. *British Journal of Pharmacology* 96, 675-687.

Richardt, G., Haass, M., Brenn, Th, Neeb, S., Hock, M., Lang, R.E., and Schömig, A. (1988). Nicotine-induced release of noradrenaline and neuropeptide Y in guinea pig heart. *Klinische Wochenschrift* 66 (Suppl. XI): 21-27.

Schömig, A., Fischer, S., Kurz, Th., Richardt, G., and Schömig E. (1987). Nonexocytotic release of endogenous noradrenaline in the ischemic and anoxic rat heart: mechanism and metabolic requirements. *Circulation Research* 60, 194-200.

Schömig, A., and Haass, M. (1990). Neuropeptide Y -- a marker for noradrenaline exocytosis. *Annals of the New York Academy of Sciences* 611, 447-449.

Tatemoto, K., Carlquist, M., and Mutt, W. (1982). Neuropeptide Y: a novel brain peptide with structural similarities to peptide YY and pancreatic polypeptide. *Nature* 296, 659-660.

Thoa, N.B., Wooten, G.F., Axelrod, J., and Kopin I.J. (1975). On the mechanism of release of norepinephrine from sympathetic nerves by depolarizing agents and sympathomimetic drugs. *Molecular Pharmacology* 11, 10-18.

Wharton, J., and Polak, J.M. (1990). Neuropeptide tyrosine in the cardiovascular system. *Annals of the New York Academy of Sciences* 611, 133-144.

Winkler, H. (1988). Occurrence and mechanism of exocytosis in adrenal medulla and sympathetic nerve. In: U. Trendelenburg and N. Weiner (Eds.), "Catecholamines I-Handbook of Experimental Pharmacology, Vol. 90/I," pp. 43-118. Berlin: Springer-Verlag.

PART THREE

MOLECULAR GENETICS OF STRESS HORMONES

Stress: Neuroendocrine and Molecular Approaches
Edited by R. Kvetnansky, R. McCarty and J. Axelrod

1992 Gordon and Breach Science
Publishers S.A., New York, USA.
Photocopying permitted by license only.

CHRONIC STRESS ELEVATES ENKEPHALIN EXPRESSION IN THE PARAVENTRICULAR AND SUPRAOPTIC NUCLEI AND MODIFIES OPIOID RECEPTOR BINDING IN THE MEDIAN EMINENCE AND POSTERIOR PITUITARY OF RATS

W.S. Young, III[1] and S.L. Lightman[2]

[1]Laboratory of Cell Biology
National Institute of Mental Health, Bethesda, Maryland, USA
[2]Neuroendocrinology Unit, Charing Cross and Westminster Medical
School, Charing Cross Hospital, Fulham Palace Road
London W6 8RF, United Kingdom

INTRODUCTION

The magnocellular neurons of the paraventricular (PVN) and supraoptic (SON) nuclei of the hypothalamus express the genes for the nonapeptides, vasopressin (VP) and oxytocin (OT), that are transported to the posterior pituitary and released into the circulation in response to hyperosmolality, parturition, and lactation (Robinson, 1986). The magnocellular neurons also express dynorphin (with VP) and enkephalin (especially with OT) (Watson *et al*, 1982; Weber *et al*, 1982; Vanderhaeghen *et al*, 1983; Sherman *et al*, 1986; Lightman and Young, 1987) genes and these opioids are found in the posterior pituitary nerve endings (Coulter *et al*, 1981; Martin and Voigt, 1981; Whitnall *et al*, 1983; Adachi *et al*, 1985; Panula and Lindberg, 1987). There are complex interactions between opioids and the magnocellular hypothalamic-neurohypophysial systems. Opiate drugs and opioids influence the activity of PVN and SON neurons (Muehlethaler et al, 1980; Pittman et al, 1980; Wakerley et al, 1983; Wuarin et al, 1988; Blanco et al, 1989; Leng et al, 1989). It is likely that dynorphin released from axon terminals of the magnocellular neurons of the PVN and SON acts locally, through actions at k-opioid receptors (Bunn et al, 1985; Gerstberger and Barden, 1986; Herkenham et al, 1986) to inhibit release of VP and OT from the pituitary (Clarke et al, 1979; Iversen et al, 1980; Lightman et al, 1982; Maysinger et al, 1984; Bondy et al, 1988; Falke, 1988; Zhao et al, 1988; Summy-Long et al, 1990). The role of enkephalin gene products in regulating VP or OT release is less well understood.

Recently we examined the hypothalamic expression of the enkephalin gene in response to acute stress. We noted that certain stressors such as naloxone-precipitated morphine withdrawal and single intraperitoneal injections of bacterial cell wall and hypertonic saline preparations (Lightman and Young, 1987; Sternberg et al, 1989), but not all stressors (Harbuz and Lightman, 1989), led to elevated expression of the enkephalin gene in the parvocellular neurons of the PVN. These neurons contain corticotropin-releasing factor (Antoni et al, 1983; Swanson et al, 1983) and increase expression of VP after adrenalectomy (Tramu et al, 1983) and, probably, after acute stress (Lightman and Young, 1987). We also observed that repeated administration of hypertonic saline led to even greater expression of enkephalin in the PVN (Lightman and Young, 1989), at least in part, because enkephalin expression also appeared in the magnocellular neurons of the PVN and SON. Herein we describe the influence of chronic stress on the increase in and site of expression of the enkephalin and dynorphin genes, as well as the effects on opioid receptor binding in the median eminence and posterior pituitary.

MATERIALS AND METHODS

Animals

All studies were performed on female Sprague-Dawley rats (150-200g) that were housed in a 12 hour light/dark, temperature, and humidity-controlled environment and that had free access to food and water. All studies started at 0900 hours and stress responses were induced by i.p. injection of 1.5M saline (HS; 1.8ml/kg body weight), whereas control animals received an equivalent volume of 0.15M saline. The rats were immediately returned to their home cages and after 4 hours (at 1300 hours) they were decapitated. Chronic stress was induced by daily injections of 1.5M saline for up to 12 days. Four hours after injection on the 2nd (28 hours), 3rd (52 hours), 5th (124 hours), or 12th day (292 hours), groups of rats were decapitated and their brains were removed. There were two control groups: in the first, a single injection of 1.5M saline was given and the animals were killed 4 hours later. In the second group, drinking water was replaced with 2% saline for 12 days at which time they were killed. In addition, some animals had their adrenals removed through a dorsal incision under equithesin anesthesia (4.25g chloral hydrate, 2.12g magnesium sulfate, and 0.97g pentobarbitone in 39.6ml propylene glycol, 20 ml ethanol and 30 ml water). On return to their cages, these animals were given free access to both drinking water and 0.15M saline and were sacrificed at 2, 5, and 12 days postoperatively.

Hybridization Histochemistry

Upon sacrifice, the brains were removed and coronal slices containing the hypothalamus were frozen on dry ice. The frozen slices were stored at -80°C until sections were cut 1.8-1.9mm caudal to the bregma (Paxinos and Watson, 1986), thaw-mounted onto twice gelatin-coated slides, and returned to -80°C until processed for hybridization histochemistry. The cryostat sections were brought to room temperature and processed as previously described (Young, 1990). The sections were 12 μm in thickness except for some cut at 4 or 8 μm to look for co-localization in adjacent sections. Briefly, the sections were fixed in 4% formaldehyde, treated with acetic anhydride, dehydrated and delipidated in graded ethanols and chloroform, and rehydrated. The sequences and specificities of the enkephalin, dynorphin and OT oligodeoxynucleotide nucleotide probes have already been described (Young et al, 1986ab). They were 3'-end-labeled using terminal deoxynucleotidyl transferase (Boehringer-Mannheim) and ^{35}S-dATP (> 1000Ci/mmol, New England Nuclear). The specific activities of the ^{35}S-labeled probes were 0.5-1 x 10^4 Ci/mmol. The sections were apposed to Kodak X-Omat AR film for 12 days. Quantitative analysis (Young, 1990), using the Image software (W. Rasband, NIH), was performed on sections that were hybridized with the same batch of enkephalin probe and processed simultaneously. Some sections were then dipped into 50% Kodak NTB3 emulsion and exposed for 2 days (OT) or for 5-8 weeks (enkephalin).

Receptor Autoradiography

Fresh frozen sections (12 μ) from rats injected with isotonic or hypertonic saline (see above) were used for *in vitro* receptor autoradiography using the μ-, δ-, and k-opioid receptor-preferring ligands as previously described (McLean *et al*, 1987). Briefly, μ-receptors were detected using 2nM [^{3}H]D-ala^2-N-methyl-phe^4,-glyol^5-enkephalin ("DAGO") and δ-receptors using 5nM [^{3}H]D-ala^2-D-leu^5-enkephalin ("DADL") in the presence of 30nM oxymorphone and 2μM GTP. The sections for k-receptors were first pretreated with 1μM each of 2-(p-ethoxybenzyl)-1-diethylaminoethyl- 5-isothiocyanato-benzimidizole ("BIT") and N-phenyl-N-[1-(2-(p-isothiocyanato) phenylethyl)-4-piperidinyl] propan-amide ("FIT") to alkylate μ- and δ-receptors, respectively (Rice *et al*, 1983). Those sections were then incubated with 2nM [^{3}H]bremazocine. Non-specific binding was obtained in the presence of 1μM (μ and δ) or 10μM (k) naloxone. After washing, all sections were apposed to tritium-sensitive film for 16 weeks. Analysis was also performed using the Image software and statistical comparisons were made using Student's t-test (two-tailed).

RESULTS

Analysis of enkephalin expression in the PVN and SON revealed a steady increase in levels of probe hybridized per PVN and SON section (Figure 1). By 12 days, the levels of enkephalin probe hybridized rose from undetectable (after a 12 day film exposure) to 16 and 8 x 10^3 copies per PVN and SON section, respectively. In contrast, 12 days of drinking 2% (0.33M) saline led to levels of only 3 and 2 x 10^3 copies per respective section. Dynorphin expression after 12 daily injections of hypertonic saline increased by 115% (n=11, p=0.001 by Student's two-tailed t-test) and 170% (n=11, p=0.003), respectively, in the PVN and SON.

When enkephalin expression was examined at higher resolution after dipping the sections in nuclear emulsion, it was apparent that 4 hours after the HS injection, the principle increase in the PVN was in

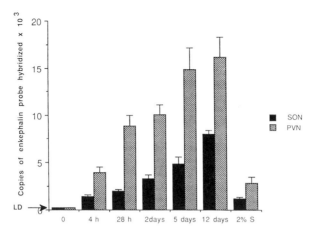

FIGURE 1.   Labeling by the ³⁵S-ENK probe of sections through the PVN and SON after intraperitoneal injections of 1.5M saline. Levels of hybridized probe increase after daily injections beyond the undetectable levels present in control sections. Levels in the sections from rats given only 2% saline to drink are above background but much lower than after 12 days of HS injections. LD = limit of detection.

the parvocellular region (Figure 2D). However, increased numbers of grains were also present over the magnocellular region of both the PVN (Figure 2D) and SON (Figure 3D). The large increases in expression after 12 days of injections were due, for the most part, to greater expression in the magnocellular regions (Figures 2E, 3E). Chronic saline led to small increases in magnocellular expression (Figures 2F, 3F), whereas sections from controls (Figures 2B, 3B) and adrenalectomized (Figures 2C, 3C) rats showed only occasional neurons labeled with the enkephalin probe.

In an attempt to determine which neurons expressed the enkephalin gene, we prepared 4 and 8 µm alternating sections. Cell sections labeled with the enkephalin probe were consistently observed sandwiched between two sections from the same cell labeled with the OT probe (Figure 4). However, as suspected from Figures 2 and 3, enkephalin expression was also detected in unlabeled, non-OT (i.e., VP) magnocellular cells.

Kappa and μ (albeit at lower levels) receptors were depressed 20% (p = 0.009, n = 11) and 30% (p = 0.024, n = 11), respectively, in the neurohypophysis by the hypertonic saline treatment (δ receptors were too low to measure). Mu receptors were also reduced (13%, p = 0.008,

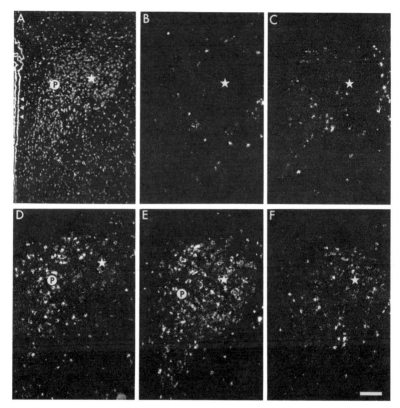

FIGURE 2. Expression of the enkephalin gene in the PVN in sections from control (B), adrenalectomized (C), single HS injection (D), 12 days of HS injections (E), and 12 days of 2% saline (F) animals. Panel A is a cell stain produced with ethidium bromide (Young *et al*, 1984). P is in the medial parvocellular region, star is in the posterior magnocellular core. Bar = 100μm. Exposure was for 5 weeks.

n = 1) in the median eminence, whereas k and δ receptors were increased (14%, p = 0.036, n = 11) and unchanged there, respectively (Figures 5).

DISCUSSION

Enkephalin is the most widely distributed of the endogenous opioid peptides in the central nervous system and there is considerable evidence for region-specific regulation of proenkephalin mRNA.

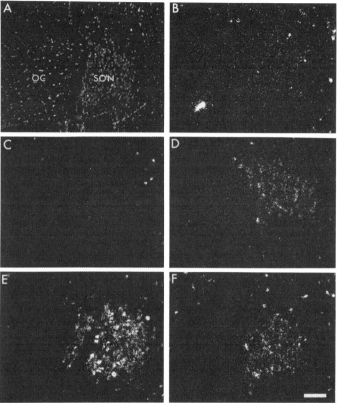

FIGURE 3.   Expression of the enkephalin gene in the SON in sections from control (B), adrenalectomized (C), single HS injection (D), 12 days of HS injections (E), and 12 days of 2% saline (F) animals.   Panel A is a cell stain produced with ethidium bromide.   oc = optic chiasm   Bar = 100µm.   Exposure was for 5 weeks.

Within the hypothalamus itself, enkephalin neurons in the ventromedial nucleus, for example, are clearly estrogen responsive, whereas those in the preoptic area are not (Lauber *et al*, 1990). Parvocellular enkephalin neurons in the PVN are the only hypothalamic enkephalin neurons which have been shown to respond to stress (Lightman and Young, 1987, 1988; Sternberg *et al*, 1989). Methionine-enkephalin co-exists with corticotropin-releasing factor and vasopressin in some of these parvocellular PVN neurons and corticotropin-releasing factor mRNA in these cells is clearly responsive to glucocorticoid feedback (Tramu *et al*, 1983; Lightman and Young, 1989; Harbuz *et al*, 1990).

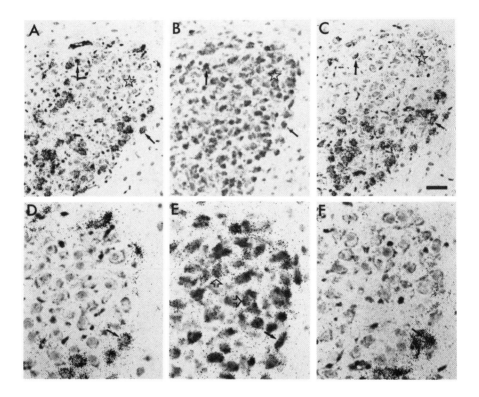

FIGURE 4. Expression of the OT (A,C,D,F) and enkephalin (B,E) genes in adjacent thin sections. The lower panels are higher magnifications of the posterior magnocellular regions indicated by the stars. The solid arrows indicate some pairs of labeled OT neuronal sections (4μm) that flank internal sections (8μm) of the same neurons that are labeled with the enkephalin probe. Note that many neurons (some examples shown by open arrows) in the posterior magnocellular core contain enkephalin but not OT transcripts. Bar = 50μm for A-C, 25μm for D-F. Exposure was 2 days for OT and 8 weeks for enkephalin.

In addition to the enkephalin cells in the parvocellular PVN that project to the median eminence, methionine-enkephalin has also been demonstrated by immunohistochemistry to co-exist with oxytocin in axonal projections from the magnocellular PVN to the neurohypophysis (Coulter et al, 1981; Martin and Voigt, 1981; Adachi et al, 1985) and the enkephalin content of the neurohypophysis can clearly respond to changes in plasma osmolality (Zamir et al, 1985; Jessop et al, 1990). Immunocytochemical results at the level of

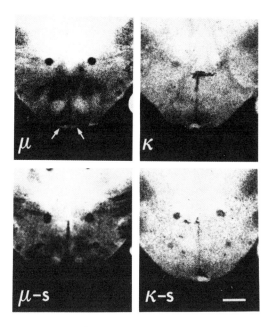

FIGURE 5.   Receptor autoradiography reveals μ- and k-opioid receptors in the median eminence (between arrows) from control and chronically stressed (s; 12 days of HS injections) rat hypothalami. Film exposures were for 16 weeks. Bar equals 1mm.

the cell bodies, however, have been controversial—some groups not finding enkephalin in the system (Micevych and Elde, 1980; Weber *et al*, 1982) and some finding enkephalin co-localized with oxytocin (Vanderhaeghen *et al*, 1983; Adachi *et al*, 1985) or with both oxytocin and vasopressin (Coulter *et al*, 1981; Martin and Voigt, 1981). This is understandable in light of hybridization histochemical studies carried out in the normal physiological state which have shown only minimal levels of enkephalin transcripts in these magnocellular neurons, either in the PVN or SON (Harlan *et al*, 1987; Lightman and Young, 1987). There was also very little change following osmotic stimulation (Lightman and Young, 1987). This contrasts sharply with the very marked increases seen with another opioid peptide, dynorphin, which co-exists with vasopressin in the PVN and SON magnocellular cells (Sherman *et al*, 1986; Lightman and Young, 1987).

Our data clearly demonstrate that enkephalin is expressed in both vasopressin and oxytocin magnocellular cells and the remarkable increase in enkephalin transcripts following chronic stress confirms that the enkephalin gene in these cells can be activated by physiological stimuli. It is important to note that the response was not simply due to the osmotic load since animals drinking 2% saline for 12 days did not show this great increase. The decrease in k-opioid binding in the posterior pituitary is consistent with the effect of hyperosmolality (Brady and Herkenham, 1987), as the increases in dynorphin expression are below what we have observed previously after 12 days of drinking 2% saline (Lightman and Young, 1987). We do not know if the slight increase in enkephalin expression after drinking 2% saline could cause a decrease in μ-opioid binding in the posterior pituitary.

The role of this elevation in enkephalin transcripts is uncertain. We presume that the increased levels reflect increased translation of the prepropeptide and subsequent processing of methionine- and leucine-enkephalin for secretion. Although an obvious site of action would be the neurohypophysis where its release could have a local regulatory effect on oxytocin or vasopressin secretion, neural lobe opiate receptors are predominantly kappa (Herkenham et al, 1986) and would preferentially bind dynorphin.

Magnocellular enkephalin may also be released directly from magnocellular dendrites and collaterals as has been demonstrated for oxytocin (Moos et al, 1989) and there is good evidence for a profound effect of the δ-opiate receptor agonist [D-Ala^{2}D-Leu^5]enkephalin on the electrical activity of these same magnocellular cells (Wakerley et al, 1983). It is also possible that enkephalin may be released from magnocellular neuronal axons at the level of the median eminence as has been suggested for vasopressin and oxytocin (Antoni et al, 1990). If this is the case, the local secretion of enkephalin at this site could effect the release of multiple releasing factors and anterior pituitary hormones (Carter and Lightman, 1986). Chronic stress certainly results in different neuroendocrine effects from acute stress, including adaptation of the hypothalamic pituitary-adrenal axis (Sakellaris and Vernikos-Danellis, 1975; Dallman and Wilkinson, 1978) and naltrexone -reversible inhibition of luteinizing hormone release (Briski et al, 1984). Our results of opioid binding in the median eminence certainly suggest that this region is, indeed, dynamically involved in responding to opioids during chronic stress.

ACKNOWLEDGEMENTS

We thank Ms. E. Shepard for technical assistance and Mr. R. Dreyfuss for photographic assistance. We also thank Drs. L. Brady and R.

Rothman for generous contribution of the ligands used for the receptor autoradiographic studies.

REFERENCES

Adachi, T., Hisano, S., and Daikoku, S. (1985). Intragranular colocalization of immunoreactive methionine-enkephalin and oxytocin within the nerve terminals in the posterior pituitary. *Journal of Histochemistry and Cytochemistry* **33**, 891-899.

Antoni, F.A., Fink, G., and Sheward, W.J. (1990). Corticotropin-releasing peptides in rat hypophysial portal blood after paraventricular lesions: a marked reduction in the concentration of corticotrophin-releasing factor-41, but no change in vasopressin. *Journal of Endocrinology* **125**, 175-183.

Antoni, F.A., Palkovits, M., Makara, G.B., Linton, E.A., Lowry, P.J., and Kiss, J.Z. (1983). Immunoreactive corticotropin-releasing hormone in the hypothalamoinfundibular tract. *Neuroendocrinology* **36**, 415-423.

Blanco, E., Carretero, J., Sànchez, F., Riesco, J. M., and Vàzquez, R. (1989). Sex-specific effects of met-enkephalin treatment on vasopressin immunoreactivity in the rat supraoptic nucleus. *Neuropeptides* **13**, 115-120.

Bondy, C.A., Gainer, H., and Russell, J.T. (1988). Dynorphin A inhibits and naloxone increases the electrically stimulated release of oxytocin but not vasopressin from the terminals of the neural lobe. *Endocrinology* **122**, 1321-1327.

Brady, L.S., and Herkenham, M. (1987). Dehydration reduces k-opiate receptor binding in the neurohypophysis of the rat. *Brain Research* **425**, 212-217.

Briski, K.P., Quigley, K., and Meites, J. (1984). Endogenous opiate involvement in acute and chronic stress-induced changes in plasma LH concentrations in the male rat. *Life Sciences* **34**, 2485-2493.

Bunn, S.J., Hanley, M.R., and Wilkin, G.P. (1985). Evidence for a kappa-opioid receptor on pituitary astrocytes: an autoradiographic study. *Neuroscience Letters* **55**, 317-323.

Carter, D., and Lightman, S. L. (1986). Opiates. In: S.L. Lightman and B.J. Everitt (Eds.), "Neuroendocrinology," pp. 103-153. St. Louis, Missouri: Blackwell Scientific Publications.

Clarke, G., Wood, P., Merrick, L., and Lincoln, D. W. (1979). Opiate inhibition of peptide release from the neurohumoral terminals of hypothalamic neurones. *Nature* **282**, 746-748.

Coulter, H.D., Elde, R.P., and Unverzagt, S.L. (1981). Co-localization of neurophysin- and enkephalin-like immunoreactivity in cat pituitary. *Peptides* **2**, 51-55.

Dallman, M.F., and Wilkinson, C.W. (1978). Feedback and facilitation in the adrenocortical system and the endocrine responses to repeated stimuli. In: I. Assenmacher and D.S. Farner (Eds.), "Environmental Endocrinology," pp. 252-259. Heidelberg: Springer Verlag.

Falke, N. (1988). Dynorphin (1-8) inhibits stimulated release of oxytocin but not vasopressin from isolated neurosecretory endings of the rat neurohypophysis. *Neuropeptides* 11, 163-167.

Gerstberger, R., and Barden, N. (1986). Dynorphin 1-8 binds to opiate kappa receptors in the neurohypophysis. *Neuroendocrinology* 42, 376-382.

Harbuz, M.S., and Lightman, S.L. (1989). Responses of hypothalamic and pituitary mRNA to physical and psychological stress in the rat. *Journal of Endocrinology* 122, 705-711.

Harbuz, M.S., Nicholson, S.A., Gillham, B., and Lightman, S.L. (1990). Stress responsiveness of hypothalamic corticotrophin-releasing factor and pituitary pro-opiomelanocortin mRNAs following high-dose glucocorticoid treatment and withdrawal in the rat. *Journal of Endocrinology* 127, 407-415.

Harlan, R.E., Shivers, B.D., Romano, G.J., Howells, R.D., and Pfaff, D.W. (1987). Localization of preproenkephalin mRNA in the rat brain and spinal cord by *in situ* hybridization. *Journal of Comparative Neurology* 258, 159-184.

Herkenham, M., Rice, K.C., Jacoson, A.E., and Rothman, R.B. (1986). Opiate receptors in rat pituitary are confined to the neural lobe and are exclusively kappa. *Brain Research* 382, 365-371.

Iversen, L.L., Iversen, S.D., and Bloom, F.E. (1980). Opiate receptors influence vasopressin release from nerve terminals in rat neurohypophysis. *Nature* 284, 350-351.

Jessop, D., Sidhu, R., and Lightman, S.L. (1990). Osmotic regulation of methionine enkephalin in the posterior pituitary of the rat. *Brain Research* 516, 41-45.

Lauber, A.H., Romano, G.J., Mobbs, C.V., Howells, R.D., and Pfaff, D.W. (1990). Estradiol induction of proenkephalin messenger RNA in hypothalamus: dose-response and relation to reproductive behavior in the female rat. *Molecular Brain Research* 8, 47-54.

Leng, G., Russell, J.A., and Grossman, R. (1989). Sensitivity of magnocellular oxytocin neurones to opioid antagonist in rats treated chronically with intracerebroventricular (i.c.v.) morphine. *Brain Research* 484, 290-296.

Lightman, S.L., Iversen, L.L., and Forsling, M.L. (1982). Dopamine and [D-Ala2, D-Leu5]enkephalin inhibit the electrically stimulated neurohypophyseal release of vasopressin *in vitro*: evidence for calcium-dependent opiate action. *Journal of Neuroscience* 2, 78-81.

Lightman, S.L., and Young, W.S., III. (1987). Changes in hypothalamic preproenkephalin A mRNA following stress and opiate withdrawal. *Nature* 328, 643-645.

Lightman, S.L., and Young, W.S., III. (1987). Vasopressin, oxytocin, dynorphin, enkephalin, and corticotrophin releasing factor mRNA stimulation in the rat. *Journal of Physiology* (London) 394, 23-39.

Lightman, S.L., and Young, W.S., III. (1988). Corticotrophin-releasing factor, vasopressin and pro-opiomelanocortin mRNA responses to stress and opiates in the rat. *Journal of Physiology* (London) 403, 511-523.

Lightman, S.L., and Young, W.S., III. (1989). Influence of steroids on the hypothalamic CRF and enkephalin messenger ribonucleic acid responses to stress. *Proceedings of the National Academy of Sciences* (USA) 86, 4306-4310.

Martin, R., and Voigt, K.H. (1981). Enkephalins co-exist with oxytocin and vasopressin in nerve terminals of rat neurohypophysis. *Nature* 289, 502-404.

Maysinger, D., Vermes, I., Tilders, F., Seizinger, B.R., Gramsch, C., Höllt, V., and Herz, A. (1984). Differential effects of various opioid peptides on vasopressin and oxytocin release from the pituitary *in vitro. Archives of Pharmacology* **328**, 191-195.

McLean, S., Rothman, R.B., Jacobson, A.E., Rice, K.C., and Herkenham, M. (1987). Distribution of opiate receptor subtypes and enkephalin and dynorphin immunoreactivity in the hippocampus of squirrel, guinea pig, rat, and hamster. *Journal of Comparative Neurology* **255**, 497-510.

Micevych, P., and Elde, R. (1980). Relationship between enkephalinergic neurons and the vasopressin-oxytocin neuroendocrine system of the cat: an immunohistochemical study. *Journal of Comparative Neurology* **190**, 135-146.

Moos, F., Poulain, D.A., Rodriguez, F., Guernè, Y., Vincent, J.-D., and Richard, P. (1989). Release of oxytocin within the supraoptic nucleus during milk ejection reflex in rats. *Experimental Brain Research* **76**, 593-602.

Muehlethaler, M., Gaehwiler, B.H., and Dreifuss, J.J. (1980). Enkephalin-induced inhibition of hypothalamic paraventricular neurons. *Brain Research* **197**, 264-268.

Panula, P., and Lindberg, I. (1987). Enkephalins in the rat pituitary gland: immunohistochemical and biochemical observations. *Endocrinology* **121**, 48-58.

Paxinos, G., and Watson, C. (1986). "The Rat Brain in Stereotaxic Coordinates." New York: Academic Press.

Pittman, Q.J., Hatton, J.D., and Bloom, F.E. (1980). Morphine and opioid peptides reduce paraventricular neuronal activity: studies on the rat hypothalamic slice preparation. *Proceedings of the National Academy of Sciences* (USA) **77**, 5527-5531.

Rice, K.C., Jacobson, A.E., Burke, T.R., Bajwa, B.S., Streaty, R.A., and Klee, W.A. (1983). Irreversible ligands with high selectivity toward δ or μ opiate receptors. *Science* **220**, 314-316.

Robinson, I.C.A.F. (1986). The magnocellular and parvocellular OT and AVP systems. In: S.L. Lightman and B.J. Everitt (Eds.), "Neuroendocrinology," pp. 154-176. Boston: Blackwell Scientific Publications.

Sakellaris, P.C., and Vernikos-Danellis, J. (1975). Increased rat of response of the pituitary-adrenal system in rats adapted to chronic stress. *Endocrinology* **97**, 597.

Sherman, T.G., Civelli, O., Douglass, J., Herbert, E., and Watson, S.J. (1986). Coordinate expression of hypothalamic pro-dynorphin and pro-vasopressin mRNAs with osmotic stimulation. *Neuroendocrinology* **44**, 222-228.

Sternberg, E.M., Young, W.S., III, Bernardini, R., Calogero, A.E., Chrousos, G.P., Gold, P.W., and Wilder, R.L. (1989). A central nervous system defect in corticotropin releasing hormone biosynthesis is associated with susceptibility to streptococcal cell wall arthritis in Lewis rats. *Proceedings of the National Academy of Sciences* (USA) **86**, 4771-4775.

Summy-Long, J.Y., Rosella-Dampman, L.M., McLemore G.L., and Koehler, E. (1990). Kappa opiate receptors inhibit release of oxytocin from the magnocellular system during dehydration. *Neuroendocrinology* **51**, 376-384.

Swanson, L.W., Sawchenko, P.E., Rivier, J., and Vale, W.W. (1983). Organization of ovine corticotropin- releasing factor immunoreactive cells and fibers in the rat brain: an immunohistochemical study. *Neuroendocrinology* **36**, 165-186.

Tramu, G., Croix, C., and Pillez, A. (1983). Ability of CRF immunoreactive neurons of the paraventricular nucleus to produce a vasopressin-like material. *Neuroendocrinology* **37**, 467-469.

Vanderhaeghen, J.J., Lotstra, F., Liston, D.R., and Rossier, J. (1983). Proenkephalin, [Met]enkephalin, and oxytocin immunoreactivities are co-localized in bovine hypothalamic magnocellular neurons. *Proceedings of the National Academy of Sciences* (USA) **80**, 5193-5143.

Wakerley, J.B., Noble, R., and Clarke, G. (1983). Effects of morphine and D-ala,D-leu enkephalin on the electrical activity of supraoptic neurosecretory cells *in vitro*. *Neuroscience* **10**, 73-81.

Watson, S.J., Akil, H., Fischli, W., Goldstein, A., Zimmerman, E., Nilaver, G., and van Wimersma Greidanus, T.B. (1982). Dynorphin and vasopressin: common localization in magnocellular neurons. *Science* **216**, 85-87.

Weber, E., Roth, K.A., Evans, C.J., Chang, J.-K., and Barchas, J.D. (1982). Immunohistochemical localization of dynorphin(1-8) in hypothalamic magnocellular neurons: evidence for absence of proenkephalin. *Life Sciences* **31**, 1761-1764.

Whitnall, M.H., Gainer, H., Cox, B.M., and Molineaux, C.J. (1983). Dynorphin-A-(1-8) is contained within vasopressin neurosecretory vesicles in rat pituitary. *Science* **222**, 1137-1139.

Wuarin, J.-P., Dubois-Dauphin, M., Raggenbass, M., and Dreifuss, J.J. (1988). Effect of opioid peptides on the paraventricular nucleus of the guinea pig hypothalamus is mediated by μ-type receptors. *Brain Research* **445**, 289-296.

Young, W.S., III. (1990). *In situ* hybridization histochemistry. In: A. Björklund, T. Hökfelt, F.G. Wouterlood, and A. van den Pol (Eds.), "Methods for the Analysis of Neuronal Microcircuits and Synaptic Interactions," pp. 481-512. New York: Elsevier.

Young, W.S., III, Alheid, G.F., and Heimer, L. (1984). The ventral pallidal projection to the mediodorsal thalamus: a study with fluorescent retrograde tracers and immunohistofluorescence. *Journal of Neuroscience* **4**, 1626-1638.

Young, W.S., III, Bonner, T.I., and Brann, M.R. (1986a). Mesencephalic dopamine neurons regulate the expression of neuropeptide mRNAs in the rat forebrain. *Proceedings of the National Academy of Sciences* (USA) **83**, 9827-9831.

Young, W.S., III, Mezey, E., and Siegel, R.E. (1986b). Vasopressin and oxytocin mRNAs in adrenalectomized and Brattleboro rats: analysis by quantitative *in situ* hybridization histochemistry. *Molecular Brain Research* **1**, 231-241.

Zamir, N., Zamir, D., Eiden, L., Palkovits, M., Brownstein, M.J., Eskay, R.L., Weber, E., Faden, A.I., and Feuerstein, G. (1985). Methionine- and leucine-enkephalin in rat neurohypophysis: different responses to osmotic stimuli and T_2 toxin. *Science* **226**, 606-608.

Zhao, B., Chapman, C., and Bicknell, R.J. (1988). Opioid-noradrenergic interactions in the neurohypophysis. I. Differential opioid receptor regulation of oxytocin, vasopressin, and noradrenaline release. *Neuroendocrinology* **48**, 16-24.

Stress: Neuroendocrine and Molecular Approaches
Edited by R. Kvetnansky, R. McCarty and J. Axelrod

REGULATION OF HYPOTHALAMIC CRF mRNA

S. L. Lightman[1], M. S. Harbuz[1] and W. S. Young, III[2]

[1]Neuroendocrinology Unit, Charing Cross and Westminster Medical School, Charing Cross Hospital, London W6 8RF, United Kingdom
[2]Laboratory of Cell Biology, National Institute of Mental Health, Bethesda, Maryland, USA

INTRODUCTION

Corticotropin-releasing factor (CRF) was isolated from the hypothalamus in 1955 (Guillemin and Rosenburg, 1955; Saffran and Schally, 1955) and the primary sequence of the human (Shibahara *et al*, 1983) and rat (Jingami *et al*, 1985; Thompson *et al*, 1987) genes described between 1983 and 1987. The regulation of hypothalamic CRF mRNA has therefore only been studied for a few years and is relatively poorly understood. CRF transcripts are first detectable by *in situ* hybridization in the rat hypothalamus on day 17 of gestation (Grino *et al*, 1989) and adult intensity of hybridization is seen by postpartum day 7.

PARVOCELLULAR PARAVENTRICULAR CRF mRNA

Glucocorticoid Regulation of CRF mRNA

Changes in CRF mRNA in response to various stimuli are relatively modest in comparison to the large changes that can be seen for instance in anterior pituitary pro-opiomelanocortin (POMC) mRNA. The clearest demonstration of specific regulation of CRF mRNA comes from studies in which glucocorticoid feedback is altered. Jingami *et al* (1985) first described a significant increase in CRF mRNA following adrenalectomy which could be reversed by the administration of dexamethasone. We have shown that CRF mRNA inhibition by circulating corticosterone is clearly concentration-dependent (Figure 1). This has also been reported by Swanson and Simmons (1989), whose laboratory has demonstrated that diurnal changes in CRF mRNA are a factor in the negative feedback control of ACTH from circulating glucocorticoids (Watts and Swanson, 1989). All these glucocorticoid-dependent changes in CRF mRNA are relatively small, and appear at least in part to be due to activation of glucocorticoid receptors in the paraventricular nucleus, since glucocorticoid receptors have been detected in these neurons (Agnati *et al*, 1985) and local administration of glucocorticoid reduces ipsilateral CRF-41 and CRF mRNA (Kovacs and Mezey, 1987; Sawchenko, 1987; Kovacs and Makara, 1988; Harbuz and Lightman, 1989a). It is clear, however, that extrahypothalamic type I receptors must also be involved indirectly by a trans-synaptic mechanism (Dallman *et al*, 1989a; Herman *et al*, 1990).

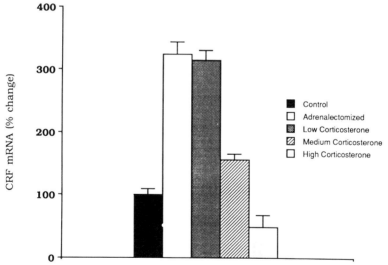

FIGURE 1.   CRF mRNA in the hypothalamic paraventricular nucleus in normal control, sham-adrenalectomized animals (assigned a value of 100%), adrenalectomized animals on a 0.9% saline, and adrenalectomized animals given low (2.5 mg/l), medium (25 mg/l), or high (125 mg/l) corticosterone replacement therapy in 0.9% saline drinking water for one week after surgery.

Neural Regulation of CRF mRNA

Hypothalamic CRF mRNA is not simply regulated by changes in steroid feedback. Indeed, far from simply responding to changes in circulating steroids, CRF-41 itself is the main regulator of circulating corticosteroids via the intermediary of ACTH secretion. Even in the absence of glucocorticoids following adrenalectomy, or during constant steroid replacement with pellets, there is still a diurnal rhythm of ACTH secretion (Dallman *et al*, 1989b). The regulation of CRF neurons in the hypothalamic PVN must therefore also result from input from neural pathways within the CNS, and the neuromorphology of these afferent pathways and their neurotransmitters has been summarized by Palkovits (1987).

What evidence do we have that these pathways can affect CRF mRNA? We have approached this question by defining stimuli that can alter CRF mRNA through a non-glucocorticoid dependent mechanism. The most obvious approach was to activate a normal physiological stimulus to ACTH stimulation--a stress response. This we have now done by performing a large number of stress paradigms and these studies provide convincing evidence for a rapid effect of both physical and psychological stimuli (Lightman and Young, 1988; Harbuz and Lightman, 1989b) on the accumulation of CRF mRNA in the paraventricular nucleus (Figure 2). Since these same stimuli actually result in a rapid increase in plasma corticosterone, this increase in CRF mRNA occurs in spite of concurrent increases in glucocorticoid feedback inhibition.

The relationship between glucocorticoid feedback and stress-induced activation of CRF gene transcription is clearly a very important one. We were interested to know how changes in glucocorticoid levels might affect CRF mRNA responses to stress. This was investigated either by removing the source of circulating glucocorticoids by adrenalectomy, or the addition of exogenous steroids. Although adrenalectomy resulted in a large increase in CRF mRNA which peaked within 3-5 days, stressful stimulation still resulted in a further increase in CRF transcripts to a similar extent to that seen in control animals (Lightman and Young, 1989a). Furthermore, glucocorticoid administration which markedly reduces basal CRF mRNA does not prevent a normal stress-induced increase in CRF transcripts (Lightman and Young, 1989a) even after 14 days of very high dose oral steroid medication (Harbuz *et al*, 1990). It is clear,

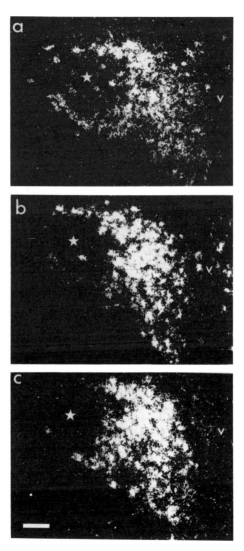

FIGURE 2.   Dark-field photomicrograph of the distribution of autoradiographic grains representing CRF probe hybridized in cells of the paraventricular nucleus of the hypothalamus. The parvocellular neurons are found between the magnocellular core of the nucleus (*) and the third ventricle (V). A, control animals; B, 24 hours after naloxone- precipitated opiate withdrawal stress; and C, 4 hours after i.p. hypertonic saline stress.   Exposure was for 3 weeks.   Bar equals 100μm. (from Lightman, S.L., and Young, W.S., III, 1988).

therefore, that the activated glucocorticoid receptor alone cannot prevent the stimulatory effect of stress-stimulated trans-acting factors. The molecular mechanisms involved in this stress-mediated response are as yet unknown, although a functional cyclic AMP responsive element has been detected in the 5' flanking sequence of the rat CRF gene (Seasholtz et al, 1988) and forskolin and 8-bromo-cAMP rapidly increase CRF mRNA in a mouse corticotroph cell line transfected with a human CRF genomic fragment (Dorin et al, 1989).

CRF mRNA Non-responsiveness During Lactation

The CRF mRNA response to stress is clearly a very robust one. It cannot be attenuated by changes in circulating corticosteroids and is even present in anesthetized animals (Harbuz et al, 1991). Is it a consistent response in all circumstances? We have had a particular interest in a physiologically important condition which is associated with a marked reduction in the response of the hypothalamo-pituitary-adrenal axis to stress (Stern et al, 1973). This is the state of lactation. We therefore investigated whether this lack of hormone responsiveness was associated with any change in CRF mRNA responsiveness at the hypothalamic level.

Remarkably, we found that lactation was associated with an abolition of the CRF mRNA response to stress (Lightman and Young, 1989b), and although removal of the litters did not result in any return of responsiveness after 18 hours, by 48 hours the CRF mRNA response to stress had returned to normal. These changes in CRF mRNA are not due to increased feedback sensitivity to glucocorticoids (Lightman et al, 1990), and represent an impressive plasticity of hypothalamic function.

Chronic Immunological Activation of the Hypothalamic-Pituitary-Adrenal Axis

So far we have discussed responses to single acute stressors. The question now arises whether chronic stressors result in similar or different effects on CRF mRNA. Multiple daily stressors in the form of i.p. injections of hypertonic saline result in CRF mRNA levels little different from those found after one single injection (Lightman and Young, 1989a). When we looked at a model of chronic stress resulting from the development of an immunologically mediated inflammatory arthritis, however, we had some very surprising results (Harbuz *et al*, submitted). As the arthritis developed there was the expected increase in corticosterone, ACTH and adrenal weight and pituitary POMC mRNA. CRF mRNA, however, actually fell to about 55% of control levels (Figure 3). We therefore have a situation in which there was increased pituitary-adrenal activity at a time of diminished CRF mRNA. Since it was clearly important to check whether this fall in mRNA was associated with a change in translation to the CRF-41 peptide, we also performed portal sampling on these animals. This reinforced our mRNA results since the development of arthritis was also associated with a fall in hypophyseal portal CRF-41 peptide (Figure 3), and interestingly an increase in portal arginine vasopressin.

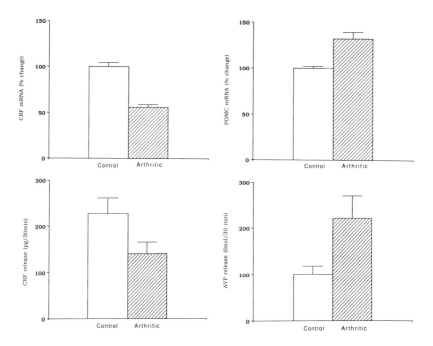

FIGURE 3.   The effect of a single injection (id) of *Mycobacterium butyricum* adjuvant (hatched bars) or vehicle (open bars) on (a) hybridization of probe to CRF mRNA in the hypothalamic paraventricular nucleus, (b) Hybridization of probe to POMC mRNA in the anterior pituitary, (c) CRF-41 peptide release in 30 minutes into the hypophyseal portal blood (HPB) and (d) AVP release in 30 minutes into the HPB. All results at 28 days after induction of arthritis.

In this model of chronic inflammatory stress our data certainly suggest that activation of the pituitary-adrenal axis is through portal vasopressin, while CRF mRNA is actually suppressed. The mechanisms involved are unclear but may involve cytokines activated by the immunological process in these animals. We certainly have evidence in these animals that interleukin-2 increases POMC mRNA but not CRF mRNA, whereas interleukin-1β activates both (Figure 4). The cause of the fall in CRF mRNA, however, is a mystery, particularly since this suppression also occurs in adrenalectomized animals with no glucocorticoid negative feedback. Further studies in this area should be very rewarding for our understanding of CRF mRNA regulation during immune mediated inflammatory arthritis. Moreover, it may also have potential pathophysiological importance in view of the evidence that a defect in CRF mRNA regulation may be an important etiological factor in the development of streptococcal cell wall arthritis in the Lewis strain of rats (Sternberg *et al*, 1989).

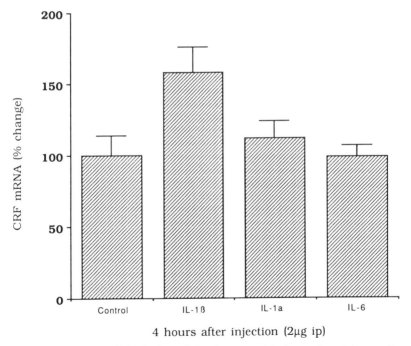

4 hours after injection (2μg ip)

FIGURE 4. CRF mRNA in the hypothalamic paraventricular nucleus 4 hours after a single injection (2μg,ip) of either interleukin-1-α, interleukin-1β or interleukin-6. Changes in CRF mRNA are expressed as the percentage change from control.

REGULATION OF HYPOTHALAMIC MAGNOCELLULAR
CRF mRNA

There is one other physiological stimulus that reduces parvocellular paraventricular CRF mRNA. During our studies on the hypothalamic response to increased osmolality, we noticed an increase in supraoptic nucleus but not paraventricular nucleus CRF mRNA (Young, 1986; Lightman and Young, 1987). On looking at our slides in more detail, it became apparent that there was indeed an increase in CRF mRNA-- but only in the magnocellular cells which project to the posterior pituitary, while there was a decrease of similar magnitude in the parvocellular cells which project to the median eminence (Figure 5). Again, it was important to check that these changes in message correlated with changes in translated peptide, and we found that these animals had a more than two-fold increase in neurointermediate lobe CRF-41 (Jessop *et al*, 1989) and markedly decreased plasma ACTH levels, even after adrenalectomy (Jessop *et al*, 1990).

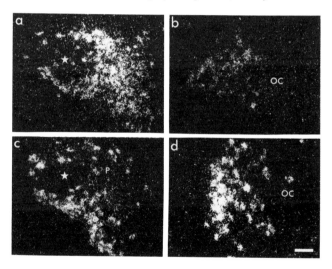

FIGURE 5. Cells containing CRF mRNA in the paraventricular nucleus (A and C) and supraoptic nucleus (B and D) in control (A and B) and salt-loaded (C and D, 12 day) animal sections. There was an increase in CRF transcripts in magnocellular neurons in the PVN and SON while a decrease occured in the parvocellular neurons (P). The star is placed over the core of vasopressin magnocellular neurons and OC is optic chiasm. Bar equals 100 μm. (from Lightman, S.L., and Young, W.S. III, 1987).

CONCLUSIONS

In this brief chapter, we have sought to illustrate different mechanisms involved in the regulation of hypothalamic CRF mRNA. Corticosteroid feedback at several sites in the central nervous system, endogenous diurnal rhythm generators, stress-activated afferent pathways to the paraventricular nucleus, together with as yet unspecified regulatory mechanisms activated by lactation, changes in plasma osmolality and chronic inflammation all contribute to the regulation of CRF mRNA. We are still a long way from understanding the complexities of the regulation of this fascinating compound.

REFERENCES

Agnati,, L.F., Fuxe, K., Yu, Z.Y., Härfstrand, A., Okret, S., Wikström, A.C., Goldstein, M., Zoli, M., Vale, W., and Gustafsson, J-A. (1985). Morphometrical analysis of the distribution of corticotropin-releasing factor, glucocorticoid receptor and phenylethanolamine-N-methyl transferase immunoreactive structures in the paraventricular nucleus of the rat. *Neuroscience Letters* 54, 147-152.

Dallman, M.F., Akana, S.F., Levin, N., Jacobson, L., Cascio, C.S., Darlington, D.N., Suemaru, S., and Scribner, K. (1989b). Corticosterone (B) replacement in adrenalectomized rats: insight into the regulation of ACTH secretion. In: F.C. Rose (Ed.), "The Control of the Hypothalamo-Pituitary-Adrenocortical Axis." Connecticut, USA: International Universities Press.

Dallman, M.F., Levin, N., Cascio, C.S., Akana, S.F., Jacobsen, L., and Kuhn, R.W. (1989a). Pharmacological evidence that the inhibition of diurnal adrenocorticotropin secretion by corticosteroids is mediated via type I corticosterone preferring receptors. *Endocrinology* 124, 2844-2850.

Dorin, R.I., Takahashi, H., Nakai, Y., Fukuta, J., Naitoh, Y., and Imura, H. (1989). Regulation of human corticotropin-releasing hormone gene expression by 3', 5', -cyclic adenosine monophosphate in a transformed mouse corticotroph cell line. *Molecular Endocrinology* 3, 1537-1544.

Grino, M., Young, W.S. III, and Burgunder, J-M. (1989). Ontogeny of expression of the corticotropin-releasing factor gene in the hypothalamic paraventricular nucleus and of the proopiomelanocortin gene in rat pituitary. *Endocrinology* 124, 60-68.

Guillemin, R., and Rosenberg, B. (1955). Humoral hypothalamic control of anterior pituitary: a study with combined tissue cultures. *Endocrinology* 57, 599-607.

Harbuz, M.S., and Lightman, S.L. (1989a). Glucocorticoid inhibition of stress-induced changes in hypothalamic corticotrophin-releasing factor messenger RNA and proenkephalin A messenger RNA. *Neuropeptides* 14, 17-20.

Harbuz, M.S., and Lightman, S.L. (1989b). Responses of hypothalamic and pituitary mRNA to physical and psychological stress in the rat. *Journal of*

Endocrinology **122**, 705-711.

Harbuz, M.S., Nicholson, S.A., Gillham, B., and Lightman, S.L. (1990). Stress responsiveness of hypothalamic corticotrophin-releasing factor and pituitary pro-opiomelanocortin mRNAs following high dose glucocorticoid treatment and withdrawal in the rat. *Journal of Endocrinology* **127**, 407-415.

Harbuz, M.S., Russell, J.A., Sumner, B.E.H., Kawata, M., and Lightman, S.L. (1991). Rapid changes in the content of proenkephalin A and corticotrophin releasing hormone mRNAs in the paraventricular nucleus during morphine withdrawal in urethane-anaesthetized rats. *Molecular Brain Research* **9**, 285-291.

Harbuz, M.S., Rees, R.G., Eckland, D.J.A., Jessop, D.S., and Lightman, S.L. (Submitted). Paradoxical responses of hypothalamic CRF mRNA and CRF-41 peptide and adenohypophyseal POMC mRNA during chronic inflammatory stress.

Herman, J.P., Wiegand, S.J., and Watson, S.J. (1990). Regulation of basal corticotropin-releasing hormone and arginine vasopressin messenger ribonucleic acid expression in the paraventricular nucleus: effects of selective hypothalamic deafferentations. *Endocrinology* **127**, 2408-2417.

Jessop, D.S., Eckland, D.J.A., Todd, K., and Lightman, S.L. (1989). Osmotic regulation of hypothalamo-neurointermediate lobe corticotrophin-releasing factor-41 in the rat. *Journal of Endocrinology* **120**, 119-124.

Jessop, D.S., Chowdrey, H.S., and Lightman, S.L. (1990). Inhibition of rat corticotropin-releasing factor and adrenocorticotropin secretion by an osmotic stimulus. *Brain Research* **523**, 1-4.

Jingami, H., Matsukura, S., Numa, S., and Imura, H. (1985). Effects of adrenalectomy and dexamethasone administration on the level of prepro-corticotropin-releasing factor messenger ribonucleic acid (mRNA) in the hypothalamus and adrenocorticotropin/ß lipotropin precursor mRNA in the pituitary in rats. *Endocrinology* **117**, 1314-1320.

Jingami, H., Mizuno, N., Takahashi, H., Shibahara, S., Furutani, Y., Imura, H., and Numa, S. (1985). Cloning and sequence analysis of cDNA for rat corticotropin-releasing factor precursor. *FEBS Letters* **191**, 63-66.

Kovacs, K.J., and Mezey, E. (1987). Dexamethasone inhibits corticotropin-releasing factor gene expression in the rat paraventricular nucleus. *Neuroendocrinology* **46**, 365-368.

Kovacs, K.H., and Makara, G.B. (1988). Corticosterone and dexamethasone act at different brain sites to inhibit adrenalectomy-induced adrenocorticotropin hypersecretion. *Brain Research* **474**, 205-210.

Lightman, S.L., and Young, W.S., III (1987). Vasopressin, oxytocin, dynorphin, enkephalin and corticotrophin-releasing factor mRNA stimulation in the rat. *Journal of Physiology* **394**, 23-40.

Lightman, S.L., and Young, W.S., III (1988). Corticotrophin-releasing factor, vasopressin and proopiomelanocortin mRNA responses to stress and opiates in the rat. *Journal of Physiology* **403**, 511-523.

Lightman, S.L., and Young, W.S., III (1989a). Influence of steroids on the hypothalamic corticotropin-releasing factor and preproenkephalin mRNA responses to stress. *Proceedings of the National Academy of Sciences* (USA) **86**, 4306-4310.

Lightman, S.L., and Young, W.S., III (1989b). Lactation inhibits stress-mediated secretion of corticosterone and oxytocin and hypothalamic accumulation of corticotropin-releasing factor and enkephalin messenger ribonucleic acids. *Endocrinology* 124, 2358-2364.

Lightman, S.L., Walker, C-D., Scribner, K., Akana, S.F., Cascio, C.S., and Dallman, M.F. (1990). Lactating rats have a nearly normal diurnal shift in ACTH feedback sensitivity to corticosterone despite lacking the basal diurnal rhythm and stress-response. *Program of the 72nd Annual Meeting of the Endocrine Society*, Atlanta, GA, p.122 (Abstract).

Palkovits, M. (1987). Anatomy of neural pathways affecting CRH secretion. *Annals of the New York Academy of Sciences* 512, 139-148.

Saffran, M., and Schally, A.V. (1955). Release of corticotrophin by anterior pituitary tissue *in vitro*. *Canadian Journal of Biochemistry and Physiology* 33, 408-415.

Sawchenko, P.E. (1987). Evidence for a local site of action for glucocorticoids in inhibiting CRF and vasopressin expression in the paraventricular nucleus. *Brain Research* 403, 213-224.

Seasholtz, A.F., Thompson, R.C., and Douglass, J.O. (1988). Identification of a cyclic adenosine monophosphate-responsive element in the rat corticotropin-releasing hormone gene. *Molecular Endocrinology* 2, 1311-1319.

Shibahara, S., Morimoto, Y., Furutani, Y., Notake, M., Takahashi, H., Shimizu, S., Horikawa, S., and Numa, S. (1983). Isolation and sequence analysis of the human corticotropin-releasing factor precursor gene. *EMBO Journal* 2, 775-779.

Stern, J.M., Goldman, L., and Levine, S. (1973). Pituitary-adrenal responsiveness during lactation in rats. *Neuroendocrinology* 12, 179-191.

Sternberg, E.M., Young, W.S. III, Bernardini, R., Calogero, A.E., Chrousos, G.P., Gold, P.W., and Wilder, R.L. (1989). A central nervous system defect in corticotropin releasing hormone biosynthesis is associated with susceptibility to streptococcal cell wall arthritis in Lewis rats. *Proceedings of the National Academy of Sciences* (USA) 86, 4771-4775.

Swanson, L.W., and Simmons, D.M. (1989). Differential steroid hormone and neural influences on peptide mRNA levels in CRH cells of the paraventricular nucleus: A hybridization histochemical study in the rat. *Journal of Comparative Neurology* 285, 413-435.

Thompson, R.C., Seasholtz, A.F., and Herbert, E. (1987). Rat corticotropin-releasing hormone gene: sequence and tissue-specific expression. *Molecular Endocrinology* 1, 363-370.

Watts, A.G., and Swanson, L.W. (1989). Diurnal variations in the content of preprocorticotropin-releasing hormone messenger ribonucleic acids in the hypothalamic paraventricular nucleus of rats of both sexes as measured by *in situ* hybridization. *Endocrinology* 125, 1734-1738.

Young, W.S., III (1986). Corticotropin-releasing factor mRNA in the hypothalamus is affected differently by drinking saline and by dehydration. *FEBS Letters* 208, 158-162.

Stress: Neuroendocrine and Molecular Approaches
Edited by R. Kvetnansky, R. McCarty and J. Axelrod

1992 Gordon and Breach Science
Publishers S.A., New York, USA.
Photocopying permitted by license only.

CHANGES IN mRNA LEVELS OF POMC, CRH AND STEROID HORMONE RECEPTORS IN RATS EXPOSED TO ACUTE AND REPEATED IMMOBILIZATION STRESS

E. Mamalaki[1], R. Kvetnansky[2], L. S. Brady[1]
and P. W. Gold[1]

[1]Section on Functional Neuroanatomy, CNE
National Institute of Mental Health and
[2]National Institute of Neurological Disorders and Stroke
National Institutes of Health, Bethesda, Maryland, USA

INTRODUCTION

Activation of the pituitary-adrenal system in response to acute stress, as confirmed by elevated ACTH and corticosterone plasma levels, is well established. Corticotropin releasing hormone (CRH) produced in the parvocellular neurons of the paraventricular nucleus (PVN) appears to play a principle role in the stress response (Meyerhoff *et al*, 1987; Rivier *et al*, 1982). However, the function of the hypothalamic-pituitary-adrenal axis (HPA) during chronic stress, as well as the underlying mechanisms, are still unclear.

Although adrenal steroids exert an inhibitory feedback on ACTH and CRH production and secretion (Beyer *et al*, 1988; Keller-Wood and Dallman, 1984; Young *et al*, 1986), it appears that responses of the system during chronic stress are not attenuated. In the presence of prior stress-induced elevations in corticosteroid levels, the response of corticotrophs to subsequent stimuli has been reported to be

247

increased in magnitude (Keller-Wood and Dallman, 1984; Vernikos *et al*, 1982; Young *et al*, 1990). Several stressors, chronically applied, result in increased hypothalamic CRH mRNA (Imaki *et al*, 1991; Herman *et al*, 1989); however, reports on CRH content have been conflicting (de Goeij *et al*, 1991; Haas and George, 1988; Herman *et al*, 1989; Kiss *et al*, 1989).

Glucocorticoids exert their feedback effects on CRH and ACTH secretion via type I (mineralocorticoid receptor, MR) and type II (glucocorticoid receptor, GR) receptors. The two types of receptors have different anatomical distributions in the central nervous system (CNS) and different affinities for corticosterone; consequently they mediate different actions of glucocorticoids on the CNS (Reul and de Kloet, 1985). The hippocampus, which has long been considered to have an inhibitory influence on the hypothalamic-pituitary-adrenocortical (HPA) axis (Sapolsky *et al*, 1984a), contains both types of receptors (Van Eekelen *et al*, 1988). Other potential sites of glucocorticoid feedback on HPA function are pituitary (Sapolsky and McEwen, 1985; Sheppard *et al*, 1990) and hypothalamus (Kovács and Mezey, 1987; Sapolsky *et al*, 1990). The relative importance of the regulation of the two types of receptors in the control of HPA axis has yet to be defined, although there is evidence that both are critical in the regulation of stress-induced HPA activation (Ratka *et al*, 1989; Sapolsky *et al*, 1990).

In the following study, we used immobilization (IMO) (Kvetnansky and Mikulaj, 1970), a well-established stressor, as a model to investigate the effects of acute and repeated stress on the mRNA expression of POMC in the pituitary and of CRH in the PVN. We also assessed the effects of immobilization stress on MR and GR mRNA expression in the anterior pituitary, PVN, and hippocampus.

MATERIALS AND METHODS

Male Sprague-Dawley rats (280-320 g, Taconic Farms, Germantown, NY) were housed 3-4 per cage under light-controlled conditions (lights on from 0700-1900 hours) with a room temperature of 23 ± 2°C. Food and water were available *ad libitum*. Rats were randomly separated into control and experimental groups and experiments began 7 days after arrival of the animals.

Experimental animals were exposed to forced immobilization (IMO) as previously described (Kvetnansky and Mikulaj, 1970). Rats were immobilized for 2 hours in the morning, during the nadir of the circadian rhythm. In the acute stress experiment (1 x IMO), rats were exposed to the stressor once and sacrificed immediately thereafter. In the chronic experiments, rats were immobilized once daily for 6 days and sacrificed 24 hours after the 6th IMO (ADAPTED CONTROL) or were given one final stress session the 7th day and sacrificed immediately (7 x IMO). Control animals (CONTROL) were habituated by daily transfer to the experimental room. All groups were sacrificed by decapitation on the same day.

Brains were rapidly removed after decapitation, frozen by immersion in 2-methyl butane at -30° C, and stored at -70° C prior to sectioning. Frozen sections (14 μm-thick) were cut coronally through the hypothalamus, dorsal hippocampus, and pituitary. Sections were thaw-mounted onto gelatin-coated slides, dried, and stored at -40° C prior to processing for *in situ* hybridization histochemistry.

Synthetic 48-base oligodeoxyribonucleotide probes (generously donated by W. Scott Young, Laboratory of Cell Biology, National Institute of Mental Health) were directed against CRH bases 496-543 (Jingami *et al*, 1985a) and POMC bases encoding amino acids 96-111 (Drouin *et al*, 1985). The probes were labeled at the 3'-end using [α-^{35}S]dATP (specific activity > 1000 Ci/mmol, New England Nuclear, Boston, MA), terminal deoxynucleotidyl transferase (25 units/μl, Boehringer-Mannheim Biochemicals, Indianapolis, IN), and tailing buffer (Bethesda Research Laboratory, Bethesda, MD).

Ribonucleotide probes (plasmids generously donated by Jeffrey L. Arriza, The Salk Institute) were directed against bases 81-528 of the rat GR sequence (Miesfeld *et al*, 1986) and 513 bases encoding the carboxy-terminal 25 amino acids and a portion of the 3'-untranslated region of the rat MR sequence (Arriza *et al*, 1988). Transcription of the probes was carried out using the Riboprobe System (Promega Biotech, Madison, WI) in the presence of [α-^{35}S]UTP (specific activity 1000-1500 Ci/mmol, New England Nuclear).

The *in situ* hybridization protocols were performed as described previously for oligonucleotide (Young *et al*, 1986) and ribonucleotide (Whitfield *et al*, 1990) probes. Briefly, sections were incubated overnight at 37°C with 5 x 10^5 cpm of labeled CRH or POMC probe, washed in four 15 minute rinses of 2x SSC (1x SSC is 0.15 M sodium chloride, 0.015 M sodium citrate, pH 7.2) containing 50% formamide at 40°C, two 30 minute rinses of 1x SSC at 25°C, briefly rinsed in distilled water, 70% ethanol, and dried. Sections were incubated overnight at 53°C with 5 x 10^5 cpm of labeled MR or GR probe, treated with RNase A (20 μg/ml, Boehringer-Mannheim) for 30 minutes at 25°C, washed sequentially for 60 minutes in 2x SSC at 50°C, 60 minutes in 0.2x SSC at 55°C, and 60 minutes in 0.2x SSC at 60°C, briefly rinsed in a graded series of ethanol containing 0.3M ammonium acetate, and dried.

Slides and ^{35}S-impregnated brain paste standards of known radioactivity and wet weight were placed in x-ray cassettes, apposed to film (Hyperfilm-βMax, Amersham) for the following durations for each probe: POMC in the intermediate pituitary lobe - 12 minutes; POMC in the anterior pituitary lobe - 6 hours; CRH in the PVN - 3 days; GR in the PVN - 4 days; GR in the pituitary and hippocampus - 33 hours; MR in the PVN - 6 weeks. MR in the pituitary and hippocampus - 7 days. Films were developed (D19, Eastman Kodak Co.) for 5 minutes at 20°C.

To determine the anatomical localization of probe at the cellular level, sections were dipped in NTB-2 nuclear emulsion (Kodak) as described previously (Herkenham, 1988; Herkenham and Pert, 1982), exposed for 1 day to 12 weeks, developed (D19) for 2 minutes at 16°C, and counterstained with cresyl violet.

Autoradiographic film images of brain sections and ^{35}S-standards were digitized on a Macintosh II computer-based image analysis system with IMAGE software (Wayne Rasband, Research Services Branch, National Institute of Mental Health). A threshold function was applied to each brain section to select and measure light transmittance of probe hybridized in PVN, pituitary and dorsal hippocampus. A second-order polynomial calibration curve was constructed using the transmittance values of ^{35}S-brain paste standards containing known amounts of radioactivity. Transmittance measurements for each probe were made on 2 consecutive sections from each brain region per rat and were converted to disintegrations per minute per milligram wet weight of tissue using the calibration curve. The average value for each animal in experimental or control groups was used to calculate group means (n = 6-8 per group). Statistical significance between brain regions in control and experimental groups was determined by one-way analysis of variance (ANOVA), followed by Fisher's protected least significant difference (PLSD) test.

RESULTS

Effects of Acute and Repeated Stress on POMC mRNA Levels in the Pituitary

POMC mRNA expression in the anterior pituitary lobe was increased by 119% after two hours of IMO. During repeated IMO it remained significantly increased, although less pronounced, in both the adapted control and repeatedly stressed rats (by 58% and 68%, respectively); no further increase was evident in response to the 7th IMO, compared to the adapted control (Table 1, Figure 1). In the intermediate pituitary lobe, POMC mRNA levels were increased only after acute IMO (Table 1).

TABLE 1. Expression of mRNA in rat brain and plasma levels of corticosterone after acute and repeated immobilization stress.

	Control	Acute Stress	Adapted Contol	Repeated Stress
POMC mRNA				
Anterior pituitary	100 ± 22	$219\pm37^{**}$	$158\pm36^{**}$	$168\pm31^{**}$
Intermediate pituitary	100 ± 16	$173\pm19^{**}$	110 ± 23	108 ± 34
Posterior pituitary	100 ± 16	$173\pm19^{**}$	110 ± 23	108 ± 34
CRH mRNA				
PVN	100 ± 8	$115\pm8^{*}$	$114\pm15^{*}$	$123\pm10^{**,\#}$
GR mRNA				
Anterior Pituitary	100 ± 15	$123\pm16^{*}$	113 ± 16	109 ± 7
PVN	100 ± 9	$124\pm14^{**}$	111 ± 11	106 ± 9
Hippocampus	100 ± 22	96 ± 16	89 ± 8	95 ± 6
MR mRNA				
Anterior Pituitary	100 ± 5	99 ± 3	$112\pm4^{**}$	100 ± 9
PVN	100 ± 20	$169\pm42^{**}$	$140\pm24^{*}$	$161\pm31^{**}$
Hippocampus	100 ± 15	$125\pm4^{**}$	112 ± 5	107 ± 9
Plasma Corticosterone	8 ± 2	$478\pm38^{**}$	$116\pm39^{**}$	$456\pm41^{**,\#}$

Values for mRNA are means and SD (n = 6-7) expressed as percentage of control values. Values for corticosterone are means and SEM (n = 6-8) expressed in ng/ml. *P < 0.05 relative to control; **P < 0.01 relative to control; $^{\#}$ P < 0.05 relative to adapted control.

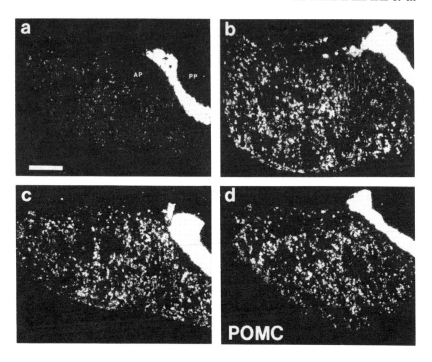

FIGURE 1.   Dark-field photomicrographs of POMC mRNA hybridized in the pituitary of control (a), acutely stressed (b), adapted control (c), and repeatedly stressed (d) rats.   In control animals, POMC message is uniformly, although sparsely localized over the anterior pituitary (AP) corticotrophs. In all experimental groups there is an increase in POMC mRNA expression and in the number of POMC-labeled cells. There is no POMC mRNA hybridized in the posterior pituitary (PP).   Because exposure time of emulsion-coated tissue was optimized for the anterior pituitary, silver grains over the intermediate lobe (star) are densely packed (saturation) and consequently no differences are seen between groups. Bar equals 410 μm.

Effects of Acute and Repeated Stress on CRH mRNA Levels in the PVN.

Acute IMO induced a small but statistically significant increase (15%) in CRH mRNA levels in the parvocellular region of the PVN. CRH mRNA levels were still increased (14%) 24 hours after the 6th IMO and were elevated an additional 10% in response to the 7th IMO (Table 1, Figure 2).

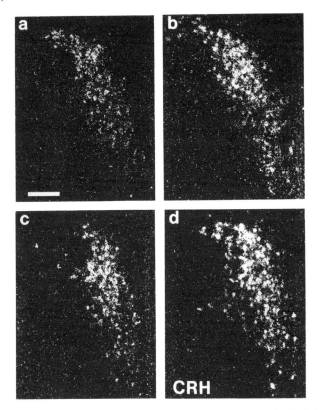

FIGURE 2. Dark-field photomicrographs of CRH mRNA hybridized in paraventricular nucleus of control (a), acutely stressed (b), adapted control (c), and repeatedly stressed (d) rats. In control animals, CRH message is densely and uniformly localized over the dorsomedial parvocellular region of the PVN (a). After acute or repeated stress, CRH mRNA is observed in a larger number of neurons which are uniformly distributed within the parvocellular region of the PVN (b-d). The stress-induced increase in CRH message is more pronounced at the end of the 7th IMO (d). Bar equals 127 μm.

Effects of Acute and Repeated Stress on GR and MR mRNA Levels in the Pituitary and Different Brain Areas

After acute IMO, GR mRNA levels were increased in the anterior pituitary lobe by 23% and in the PVN by 12%. No significant change in hippocampal GR mRNA expression was seen with any experimental condition (Table 1).

MR mRNA levels were significantly increased in the PVN in the acute IMO (70%) as well as in the adapted control (40%) and repeatedly stressed (61%) groups (Table 1). No change in MR mRNA levels was seen in the anterior pituitary. In the hippocampus, MR mRNA was increased by 25% only after acute IMO (Table 1).

Plasma Corticosterone

Plasma corticosterone was increased 60-fold at the end of 2 hours of immobilization stress. During repeated stress, corticosterone was increased 14-fold in the adapted control group and 57-fold in the repeatedly stressed group relative to controls (Table 1).

DISCUSSION

In the present study, 2 hours of IMO induced a significant increase in mRNA expression of CRH in the medial parvocellular PVN, accompanied by a significant elevation in anterior pituitary POMC mRNA levels, in agreement with previous studies (Suda *et al*, 1988; Lightman and Young, 1988; Harbuz and Lightman, 1989). Our results support a rapid activation (two hours after the beginning of the stressor) of the HPA axis at the level of mRNA expression in response to IMO stress.

During repeated IMO, CRH mRNA levels in the PVN remained increased in the adapted control group. Furthermore, CRH neurons appear to preserve the capability to respond additionally to a subsequent IMO in the 7 x IMO group. Several other studies have demonstrated increases in CRH mRNA in the PVN following chronic stress in rats (Herman *et al*, 1989; Imaki *et al*, 1991). These data, coupled with decreases in PVN CRH content (Herman *et al*, 1989; Kiss *et al*, 1989) suggest that CRH neurons in the PVN, when exposed to chronic stress, are in a state of accelerated synthesis and release of CRH.

The chronic changes in CRH mRNA levels in the PVN were accompanied by significant elevations of pituitary POMC mRNA levels in the anterior pituitary lobe in agreement with previous studies (Shiomi *et al*, 1986; Höllt *et al*, 1986). It might be expected that the subsequent release of corticosterone would limit the corticotroph response to chronic stress through negative feedback. However, there is abundant evidence that some stimuli that activate the HPA axis result in decreased sensitivity of the corticotrophs to glucocorticoid negative feedback (Young *et al*, 1990; Keller-Wood and Dallman, 1984; Vernikos *et al*, 1982). The mechanism by which chronic stress induces a rise in POMC mRNA levels is unknown. Continuous administration of CRH to rats results in increased accumulation of POMC mRNA in corticotrophs (Höllt and Haarman, 1984) and persistent stimulation of ACTH secretion (Rivier and Vale, 1983a; Bruhn *et al*, 1984). Other stimulatory factors have been shown to play a physiological role in stress-induced ACTH secretion (Rivier and Vale, 1983b; Plotsky *et al*, 1985); these factors could interfere with the regulation of POMC mRNA expression as well.

The increase in POMC mRNA levels in the anterior pituitary was less pronounced during chronic IMO, when compared with acute IMO values, and the corticotrophs did not respond to the 7th IMO, as did the CRH neurons. These findings suggest that although corticotrophs remain highly activated during repeated stress, they nevertheless show partial desensitization. It is possible that glucocorticoids exert their negative feedback effects at the level of mRNA expression more dominantly on pituitary corticotrophs than on hypothalamic CRH neurons (Jingami *et al*, 1985b). A down-regulation of CRH receptors in the anterior pituitary, however, cannot be excluded (Hauger *et al*, 1988)

POMC mRNA levels were also increased in the intermediate pituitary lobe after acute IMO, whereas no changes were seen during repeated stress, suggesting that the two pituitary lobes respond in different ways to chronic stress. Previous research has also shown differential regulation of POMC mRNA expression in the anterior and intermediate pituitary lobes (Höllt and Haarman, 1984). There is evidence that nerve fibers in the intermediate lobe contain CRH (Saavedra *et al*, 1984). Furthermore, acute administration of CRH causes a release of POMC-derived peptides from cultured rat intermediate pituitary cells *in vitro* (Vale *et al*, 1983) suggesting a direct action of CRH on POMC mRNA regulation in the intermediate lobe *in vivo*.

The failure of elevated plasma corticosterone concentrations during chronic IMO to inhibit activation of the HPA may indicate a down-regulation of glucocorticoid receptors (Sapolsky *et al*, 1984b). However, in the present study, no significant changes in mRNA expression of either MR or GR was seen during chronic IMO, with the exception of MR mRNA in the PVN, which was increased. In a recent study (Chao *et al*, 1989), adrenalectomy and corticosterone administration have been shown to alter corticosterone binding in the hippocampus without changing GR or MR mRNA expression, suggesting that post-transcriptional mechanisms may be involved in regulation of these receptors Accordingly, glucocorticoids have been shown to affect the turnover of GR protein without affecting the rate of GR synthesis (McIntyre and Samuels, 1985). The physiological role of increased expression of MR mRNA in the PVN remains unclear.

In conclusion, immobilization in rats induces a rapid and significant activation of the HPA axis, as reflected in increased CRH mRNA and POMC mRNA levels. Repeated exposure of animals to the stressor increases CRH and POMC mRNA expression. GR and MR mRNA levels are differentially regulated in the hippocampus, PVN, and pituitary by acute IMO but are not altered during chronic IMO. The physiological role of increased expression of MR mRNA in the PVN remains unclear. Further studies are required to define better the stress-induced up-regulation of the HPA axis and the underlying mechanisms.

REFERENCES

Arriza, J.L., Simerly, R.B., Swanson, L.W., and Evans, R.M. (1988). The neuronal mineralocorticoid receptor as a mediator of glucocorticoid response. *Neuron* **1**, 887-900.

Beyer, H.S., Matta, S.G., and Sharp, B.M. (1988). Regulation of the messenger ribonucleic acid for corticotropin-releasing factor in the paraventricular nucleus and other brain sites of the rat. *Endocrinology* **123**, 2117-2123.

Bruhn, T.O., Sutton, R.E., Rivier, C.L., and Vale, W.W. (1984). Corticotropin-releasing factor regulates proopiomelanocortin messenger ribonucleic acid levels *in vivo*. *Neuroendocrinology* **39**, 170-175.

Chao, H.M., Choo, P.H., and McEwen, B.S. (1989). Glucocorticoid and mineralocorticoid receptor mRNA expression in rat brain. *Neuroendocrinology* **50**, 365-371.

de Goeij, D.C.E., Kvetnansky, R., Whitnall, M.H., Jezova, D., Berkenbosch, F., and Tilders, F.J.H. (1991). Repeated stress-induced activation of corticotropin-releasing factor neurons enhances vasopressin stores and colocalization with corticotropin-releasing factor in the median eminence of rats. *Neuroendocrinology* **53**, 150-159.

Drouin, J., Chamberland, M., Charron, J., Jeannotte, L., and Nemer, M. (1985). Structure of the rat pro-opiomelanocortin (POMC) gene. *FEBS Letters* **193**, 54-58.

Haas, D.A., and George, S.R. (1988). Effect of dopaminergic and α-adrenergic modulation on corticotropin-releasing factor immunoreactivity in rat hypothalamus. *Canadian Journal of Physiology and Pharmacology* 66, 754-761.

Harbuz, M.S., and Lightman, S.L. (1989). Responses of hypothalamic and pituitary mRNA to physical and psychological stress in the rat. *Journal of Endocrinology* 122, 705-711.

Hauger, R.L., Millan, M.A., Lorang, M., Harwood, J.P., and Aguilera, G. (1988). Corticotropin- releasing factor receptors and pituitary- adrenal responses during immobilization stress. *Endocrinology* 123, 396-405.

Herkenham, M. (1988). Influence of tissue treatment on receptor autoradiography. In: F.W. Van Leeuwen, R.M. Buijs, C.W. Pool, and O. Pach (Eds.), "Molecular Neuroanatomy", pp. 111-120. New York: Elsevier Science Publishers.

Herkenham, M., and Pert, C.B. (1982). Light microscopic localization of brain opiate receptors: A general autoradiographic method which preserves tissue quality. *Journal of Neuroscience* 2, 1129-1149.

Herman, J.P., Schafer, M.K.-H., Sladek, C.D., Day, R., Young, E.A., Akil, H., and Watson, S.J. (1989). Chronic electroconvulsive shock treatment elicits up-regulation of CRF and AVP in select populations of neuroendocrine neurons. *Brain Research* 501, 235-246.

Höllt, V., Przewlocki, R., Haarmann, I., Almeida, O.F.X., Kley, N., Millan, M.J., and Herz, A. (1986). Stress-induced alterations in the levels of messenger RNA coding for pro-opiomelanocortin and prolactin in rat pituitary. *Neuroendocrinology* 43, 277-282.

Höllt, V., and Haarman, I. (1984). Corticotropin- releasing factor differentially regulates pro-opiomelanocortin messenger ribonucleic acid levels in anterior as compared to intermediate pituitary lobes of rats. *Biochemical and Biophysical Research Communications* 124, 407-415.

Imaki, T., Nahan, J.-L., Rivier, C., Sawchenko, P.E., and Vale, W. (1991). Differential regulation of corticotropin-releasing factor mRNA in rat brain regions by glucocorticoids and stress. *Journal of Neuroscience* 1, 585-599.

Jingami, H., Mizuno, N., Takahashi, H., Shibahara, S., Furutani, Y., Imura, H., and Numa, S. (1985a). Cloning and sequence analysis of cDNA for rat corticotropin-releasing factor precursor. *FEBS Letters* 191, 63-66.

Jingami, H., Matsukara, S., Numa, S., and Imura, H. (1985b). Effects of adrenalectomy and dexamethasone administration on the level of prepro-corticotropin-releasing factor messenger ribonucleic acid (mRNA) in the hypothalamus and adrenocorticotropin / ß-lipotropin precursor mRNA in the pituitary in rats. *Endocrinology* 117, 1314-1320.

Keller-Wood, M.E., and Dallman, M.F. (1984). Corticosteroid inhibition of ACTH secretion. *Endocrine Reviews* 5, 1-24.

Kiss, A., Mikulajova, M., Kvetnansky, R., Rokaeus, A., Skirboll, L.R., and Palkovits, M. (1989). Effect of repeated immobilization stress on vasopressin, corticotropin-releasing factor (CRF) and galanin immunoreactivity in the median eminence of the rat. In: G.R. Van Loon, R. Kvetnansky, R. McCarty, and J. Axelrod (Eds.), "Stress: Neurochemical and Humoral Mechanisms," pp. 461-472. New York: Gordon and Breach.

Kovacs, K.J., and Mezey, E. (1987). Dexamethasone inhibits corticotropin-releasing factor gene expression in the rat paraventricular nucleus. *Neuroendocrinology* **46**, 365-368.

Kvetnansky, R., and Mikulaj, L. (1970). Adrenal and urinary catecholamines in rats during adaptation to repeated immobilization stress. *Endocrinology* **87**, 738-743.

Lightman, S.L., and Young III, W.S. (1988). Corticotropin-releasing factor, vasopressin and pro-opiomelanocortin mRNA responses to stress and opiates in the rat. *Journal of Physiology* **403**, 511-523.

McEwen, B.S., Chao, H., Spencer, R., Brinton, R.E., MacIsaac, L., and Harrelson, A. (1987). Corticosteroid receptors in brain: Relationship of receptors to effects in stress and aging. *Annals of the New York Academy of Sciences* **512**, 394-401.

McIntyre, W.R., and Samuels, H.H. (1985). Triamcinolone acetonide regulates glucocorticoid-receptor levels by decreasing the half-life of the activated nuclear- receptor form. *Journal of Biological Chemistry* **260**, 418-427.

Meyerhoff, J.L., Mougey, E.H., and Kant, G.J. (1987). Paraventricular lesions abolish the stress-induced rise in pituitary cyclic adenosine monophosphate and attenuate the increases in plasma levels of pro-opiomelanocortin-derived peptides and prolactin. *Neuroendocrinology* **46**, 222-230.

Miesfeld, R., Rusconi, S., Godowski, P.J., Maler, B.A., Okret, S., Wikström, A.C., Gustafsson, J.-Å., and Yamamoto, K.R. (1986). Genetic complementation of a glucocorticoid receptor deficiency by expression of cloned receptor cDNA. *Cell* **46**, 389-399.

Plotsky, P.M., Bruhn, T.O., and Vale, W. (1985). Evidence for multifactor regulation of the adrenocorticotropin secretory response to hemodynamic stimuli. *Endocrinology* **116**, 633-639.

Ratka, A., Sutanto, W., Bloemers, M., and de Kloet, E.R. (1989). On the role of brain mineralocorticoid (type I) and glucocorticoid (type II) receptors in neuroendocrine regulation. *Neuroendocrinology* **50**, 117-123.

Reul, J.H., and de Kloet, E.R. (1985). Two receptor systems for corticosterone in brain: Microdistribution and differential occupation. *Endocrinology* **117**, 2505-2511.

Rivier, C., and Vale, W. (1983a). Influence of the frequency of ovine corticotropin-releasing factor administration on adrenocorticotropin and corticosterone secretion in the rat. *Endocrinology* **113**, 1422-1426.

Rivier, C., and Vale, W. (1983b). Modulation of stress-induced ACTH release by corticotropin- releasing factor, catecholamines, and vasopressin. *Nature* **305**, 325-327.

Rivier, C., Rivier, J., and Vale, W. (1982). Inhibition of adrenocorticotropin hormone secretion in the rat by immunoneutralization of corticotropin-releasing factor. *Science* **218**, 377-379.

Saavedra, J.M., Rougeot, C., Culman, J., Israel, A., Niwa, M., Tonon, M.C., and Vaudry, H. (1984). Decreased corticotropin-releasing factor immunoreactivity in rat intermediate and posterior pituitary after stalk section. *Neuroendocrinology* **39**, 93-95.

Sapolsky, R.M., Armanini, M.P., Packan, D.R., Sutton, S.W., and Plotsky, P.M.

(1990). Glucocorticoid feedback inhibition of adrenocorticotropic hormone secretagogue release. *Neuroendocrinology* **51**, 328-336.

Sapolsky, R.M., and McEwen, B.S. (1985). Down- regulation of neural corticosteroid receptors by corticosterone and dexamethasone. *Brain Research* **339**, 287-292.

Sapolsky, R.M., Krey, L.C., and McEwen, B.S. (1984a). Glucocorticoid-sensitive hippocampal neurons are involved in terminating the adrenocortical stress response. *Proceedings of the National Academy of Sciences* (USA) **81**, 6174-6177.

Sapolsky, R.M., Krey, L.C., and McEwen, B.S. (1984b). Stress down-regulates corticosterone receptors in a site-specific manner in the brain. *Endocrinology* **114**, 287-292.

Sheppard, K.E., Roberts, J.L., and Blum, M. (1990). Differential regulation of type II corticosteroid receptor messenger ribonucleic acid expression in the rat anterior pituitary and hippocampus. *Endocrinology* **127**, 431-439.

Shiomi, H., Watson, S.J., Kelsley, J.E., and Akil, H. (1986). Pretranslational and post-translational mechanisms for regulating ß-endorphin-adrenocorticotropin of the anterior pituitary lobe. *Endocrinology* **119**, 1793-1799.

Suda, T., Tozawa, F., Yamada, M., Ushiyama, T., Tomori, N., Sumitomo, T., Nakagami, Y., Demura, H., and Shizume, K. (1988). Insulin- induced hypoglycemia increases corticotropin-releasing factor messenger ribonucleic acid levels in rat hypothalamus. *Endocrinology* **123**, 1371-1375.

Vale, W., Vaughan, J., Smith, M., Yamamoto, G., Rivier, J., and Rivier, C. (1983). Effects of synthetic ovine corticotropin-releasing factor, glucocorticoids, catecholamines, neurohypophysial peptides, and other substances on cultured corticotropic cells. *Endocrinology* **113**, 1121-1131.

Van Eekelen, J.A., Jiang, W., de Kloet, E.R., and Bohn, M.C. (1988). Distribution of the mineralocorticoid and the glucocorticoid receptor mRNAs in the rat hippocampus. *Journal of Neuroscience Research* **21**, 88-94.

Vernikos, J., Dallman, M.F., Bonner, C., Katzen, A., and Shinsako, J. (1982). Pituitary-adrenal function in rats chronically exposed to cold. *Endocrinology* **110**, 413.

Whitfield, H.J., Brady, L.S., Smith, M.A., Mamalaki, E., Fox, R.J., and Herkenham, M. (1990). Optimization of cRNA probe *in situ* hybridization methodology for localization of glucocorticoid receptor mRNA in rat brain: A detailed protocol. *Cellular and Molecular Neurobiology* **10**, 145-157.

Young, E.A., Akana, S., and Dallman, M.F. (1990). Decreased sensitivity to glucocorticoid fast feedback in chronically stressed rats. *Neuroendocrinology* **51**, 536-542.

Young III, W.S., Mezey, E., and Seigel, R.E. (1986). Quantitative *in situ* hybridization histochemistry reveals increased levels of corticotropin-releasing factor mRNA after adrenalectomy in rats. *Neuroscience Letters* **70**, 198-203.

Stress: Neuroendocrine and Molecular Approaches
Edited by R. Kvetnansky, R. McCarty and J. Axelrod

1992 Gordon and Breach Science
Publishers S.A., New York, USA.

ONTOGENY OF EXPRESSION AND GLUCOCORTICOID REGULATION OF CRF AND AVP GENES IN RAT HYPOTHALAMUS: RELATIONSHIP TO THE STRESS NON-RESPONSIVE PERIOD

M. Grino[1], L. Muret[1], A. Priou[1], J.-M. Burgunder[2]
and C. Oliver[1]

[1]Laboratoire de Neuroendocrinologie Experimentale, INSERM U 297,
Marseille 13326, France and Neuromorphologisches Labor
[2]Neurologische Klinik und Poliklinik der Universität,
Inselspital, CH-3010, Berne, Switzerland

INTRODUCTION

The secretion of glucocorticoids is a central feature of the response to stressful stimuli. Glucocorticoids play an important role in adaptation to stress-induced changes. They mobilize readily usable energy substrates in blood, increase cardiovascular tone and inhibit immune and inflammatory responses. Besides their role in stress, they have also been shown to play a role in the regulation of gene expression, metabolic processes and brain function (Munck *et al*, 1984).

In adult rats, secretion of corticosterone by the adrenal cortex depends on pituitary release of ACTH. Corticotropin-releasing factor (CRF) and arginine vasopressin (AVP) are believed to be the major stimulators of ACTH secretion under both basal and stressful conditions (Linton *et al*, 1985). CRF is mainly synthesized in the parvocellular portion of the paraventricular nucleus (PVN) while AVP is synthesized in both the parvocellular and the magnocellular portions of the PVN and in the supraoptic nucleus (SON). These differences are of functional significance. Indeed, parvocellular neurons are known to project to the median eminence and to release their peptide products into the hypophysial portal blood. The AVP precursor synthesized in the magnocellular portion of the PVN and in the SON is transported axonally to the posterior pituitary. The subsequently matured AVP is secreted into the general circulation and acts on the kidneys (Zimmerman, 1985).

Circulating AVP levels are relatively low and therefore are thought to be insufficient to participate in the regulation of ACTH secretion from the anterior pituitary (Linton et al, 1985). Nevertheless, it has been shown that magnocellular neurons in passage through the external and internal zones of the median eminence can release AVP (Holmes et al, 1986). As a consequence, magnocellular AVP could be involved in regulation of anterior pituitary ACTH secretion. In addition, glucocorticoids have a negative feedback effect on CRF and AVP, but only in the parvocellular portion of the PVN. In adult rats, adrenalectomy (ADX) induces an increase in levels of parvocellular CRF and AVP mRNAs (Young et al, 1986ab). ADX also increases the concentration of the mRNA coding for one of the enzymes involved in the biosynthesis of CRF and AVP (peptidyl-glycine α-amidating monooxygenase), indicating that glucocorticoids regulate the biosynthesis of CRF and AVP (Grino et al, 1990). Both CRF and AVP seem to be necessary for a complete response of the pituitary to stressors. Passive immunization against CRF and/or AVP reduces significantly the ACTH response to restraint or formalin stress (Linton et al, 1985). There is an increase in the secretion rate of CRF and AVP into rat and sheep hypophysial portal blood during stress (Guillaume et al, 1989; Caraty et al, 1988, 1990). In addition, both CRF and AVP gene expression is increased in adult rat PVN during stress (Lightman and Young, 1987).

During the perinatal period in rats, the pattern of pituitary-adrenal secretion is peculiar. Indeed, beginning in the late fetal period, the secretion of corticosterone is very high, then decreases progressively around birth. In the early postnatal period, basal corticosterone release is greatly reduced. From postnatal day 2 until the second week of life, rats fail to respond or respond only weakly to various stressors that usually induce a large increase in ACTH and corticosterone secretion in adults. This period has been called the "stress non-responsive period" (SNRP). The adaptive value of glucocorticoid hypersecretion during the late fetal period is as yet unexplained. Low circulating glucocorticoid levels during the SNRP are believed to be essential for normal brain and behavioral development. Indeed, rats treated with glucocorticoids during the first week of life have permanently reduced brain weights; neuronal, glial and myelin alterations; as well as behavioral changes (reviewed in Sapolsky and Meaney, 1986).

To study the regulation of the hypothalamo-pituitary-adrenal (HPA) axis in developing rats we examined, using quantitative *in situ* hybridization, the ontogeny of CRF and AVP gene expression in the hypothalamus.

CRF AND AVP GENE EXPRESSION IN FETUSES

CRF mRNA was first detected on day 17 of gestation (E17), in the putative PVN. From E18 to E20, although the PVN was still not identifiable, the grain density over the same region labeled in the E17 brain increased. On E21, while the morphology around the third ventricle and the surface covered with grains were similar to those on E20, the density showed a marked diminution (Grino *et al*, 1989a) (Table 1).

AVP mRNA was first detected by Laurent *et al* (1989) in the PVN at E18. Consistent with the above mentioned report, we detected AVP mRNA on E20 (the first age used for AVP hybridization) in both the putative parvocellular and magnocellular portions of the PVN. Levels of AVP mRNA were increased at E21 (Burgunder and Grino, 1991) (Table 1) . Comparable patterns of development have been reported for immunoreactive CRF (I-CRF) (Chatelain *et al*, 1988) and I-AVP (Rundle and Funder, 1988) using immunocytochemistry or radioimmunoassay, suggesting that the increase in CRF and AVP mRNA levels is followed by increased biosynthesis of the corresponding peptides. These observations indicate that the hypothalamus may be involved in the regulation of fetal ACTH secretion during late pregnancy. Indeed, Boudouresque *et al* (1988) have demonstrated that endogenous CRF begins to play a physiological role in the regulation of fetal ACTH and corticosterone secretion as early as E17. Passive immunization of pregnant rats with an antiserum to CRF significantly reduces the levels of both hormones in the fetal circulation at this stage. However, fetal plasma ACTH and

TABLE 1. CRF and AVP mRNA levels in the PVN during development.

AGE	CRF (copies of probe/mm^3 x 10^2)	AVP (copies of probe/mm^{3}x10^2)
E20	5.50 ± 0.40	2.89 ± 0.48
E21	2.05 ± 0.10	4.37 ± 0.31
P1	1.48 ± 0.06	7.35 ± 0.67
P3	2.20 ± 0.38	9.35 ± 1.50
P7	4.60 ± 0.40	9.92 ± 0.13
P14	3.80 ± 0.10	10.10 ± 0.64

Each value is the mean ± SE from 5 animals per age. (Data are from Grino *et al*, 1989b, and Burgunder and Grino, 1991).

corticosterone levels were only partially reduced, although the amount of CRF antiserum in fetal plasma was likely to bind all the endogenous circulating CRF. Besides CRF, other factors such as AVP may play a physiological role in the regulation of fetal ACTH secretion.

CRF AND AVP GENE EXPRESSION IN NEONATES

As noted above, neonatal rats exhibit a reduced capacity to secrete ACTH and corticosterone in response to stress beginning on postnatal day 2 (P2) and continuing through the first 10 days of life. Several hypotheses have been advanced to explain this phenomenon; one is that the pituitary-adrenal axis is immature. During the SNRP, administration of synthetic CRF stimulates ACTH secretion, although with lower potency than in adult rats (Walker *et al*, 1986a). Another possibility is a decreased ability of CRF to be released and to reach the pituitary. Walker *et al* (1986a) have reported that by the third day of life, a urethane injection is followed by an increase in plasma ACTH that is partially blocked by passive immunization against CRF. This indicates that under certain conditions, CRF can be secreted and can stimulate ACTH secretion in young rats. Nevertheless, we do not know if CRF is secreted normally in response to stressors. Finally, an increased glucocorticoid negative feedback in the hypothalamus and/or pituitary may occur. This could be due to a decrement in plasma transcortin-binding capacity, which in turn leads to an increase in the concentration of unbound corticosterone and, hence, an increase in tissue to serum ratios of glucocorticoids (Koch, 1969).   Walker *et al* (1986b) have reported that corticotroph sensitivity to corticosterone is increased in young rats.

On the day after birth, hypothalamic CRF mRNA levels decreased compared to levels on E21.   Labeling intensity was low on P3, increased from P3 to P7 and remained stable until P14 (Grino *et al*, 1989a) (Table 1).   Similar findings have been reported by other authors (Baram and Schultz, 1989).   In contrast, hypothalamic AVP mRNA levels increased steadily from E21 to P3 and remained stable from P3 to P14 (Burgunder and Grino, 1991) (Table 1).   Different patterns of development have been reported for concentrations of I-CRF and I-AVP in the hypothalamus.   Indeed, I-CRF increases steadily from E17 up to P21 (Chatelain *et al*, 1988), while I-AVP shows a small increase between E19 and P3 and a large increase between P14 and P40 (Rundle and Funder, 1988).   These different mRNA/peptide ratios could be due to differences in peptide synthesis or storage, or alternatively to developmental variations in the post translational processing of prohormones to their mature products.   In addition, differences in peptide release could explain this observation. Indeed, it has been established that, after ADX, which induces a large increase in CRF and AVP mRNA levels (Young *et al*, 1986a,b) and in CRF and AVP secretion (Plotsky and Sawchenko, 1987; Eckland *et al*, 1988), I-CRF and I-AVP in the median eminence are decreased dramatically (Bugnon *et al*, 1983).

CONCLUSIONS

Taken together, these data suggest that, during the immediate postnatal period, there is decreased CRF synthesis and release, while AVP synthesis and release increase steadily.   Our findings showing that the expression of proopiomelanocortin (POMC, the precursor for ACTH) mRNA in the anterior pituitary is decreased during the same time (Grino *et al*, 1989a) supports this hypothesis.   Indeed CRF has a strong stimulatory effect on POMC gene expression while AVP may decrease POMC gene transcription (Levin *et al*, 1989).   These differences in levels of CRF and AVP gene transcription can, at least in part, account for the SNRP.   Indeed most of the stressors (such as ether or traumatic stress) which do not stimulate ACTH secretion during the SNRP seem to involve increased CRF secretion and/or synthesis (Antoni, 1986). Our finding that CRF biosynthesis is decreased during the SNRP could explain these results. However, during insulin-induced hypoglycemia (IIH) in adult rats, CRF seems to play a tonic role, with AVP being the dynamic mediator of ACTH secretion (Plotsky *et al*, 1985).

Recent findings from our laboratory, indicate that 8-day-old rats are able to respond to IIH. In addition, passive immunization against AVP is able to antagonize partially the stimulatory effect of IIH on ACTH secretion (L. Muret, paper in preparation). The factors responsible for decreased CRF gene expression during the SNRP are still unknown. Walker *et al* (1986b) have suggested that glucocorticoid negative feedback is increased during the SNRP. This could be due to increased circulating free corticosterone, ontogenic changes in the concentration of corticosterone-binding receptors, or enhanced postreceptor transduction mechanisms (Olpe and McEwen, 1976; Clayton *et al*, 1977; Meaney *et al*, 1985).

To test this hypothesis, we studied the effects of surgical ADX on levels of CRF and AVP mRNAs in the PVN in 7-day-old rats, i.e., within the SNRP. Fourteen day old rats (which have an intact stress response) were used as controls. We found that 48 hours after ADX in 14 day old rats, a significant increase in both CRF and AVP mRNA levels occurred in the parvocellular PVN, which was comparable to that found in adult rats. In 7 day old rats, ADX induced a significant stimulation of AVP mRNA levels in the parvocellular PVN, while CRF mRNA levels were unchanged (Figure 1). This phenomenon was accompanied by a reduced post-ADX increase in anterior pituitary POMC mRNA levels in 7 day old rats compared to 14 day old animals, which showed an adult-like response. In addition, chronic treatment of ADX 7 day old rats with CRF induced an increase in anterior pituitary POMC mRNA concentrations comparable to that in ADX untreated 14-day-old rats (Grino *et al*, 1989b).

These observations suggest that in rats regulation of AVP-producing cells is mature very early, while the regulation of CRF synthesizing neurons is immature during the first ten days of life. The discrepancy between regulation by glucocorticoids of AVP and CRF gene expression during development suggests that endogenous corticosterone may act by different pathways on the CRF and AVP genes in the PVN. Indeed, Herman *et al* (1990) have recently demonstrated that, in adult rats, the glucocorticoid negative feedback

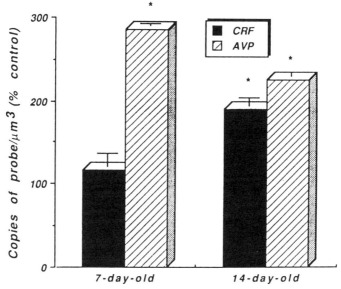

FIGURE 1.   Effect of ADX on CRF and AVP gene transcription in the parvocellular portion of the PVN.   Hybridization signals were measured in brain slices obtained from 7 or 14 day old rats 48 hours after ADX.   Results (expressed as % of the hybridization signal measured in sham-operated controls are the means ± SE from values obtained from 5 animals in each group.   [* P<0.05 versus control. (Data are from Grino *et al*, 1989b and Burgunder and Grino, 1991)].

on CRF neurons is dependent on neuronal inputs to the PVN. On the other hand, local glucocorticoid feedback is clearly able to restrict the expression of the AVP gene in the parvocellular portion of the PVN. Therefore, these observations support the hypothesis that innervation of the PVN is not mature during the SNRP (Grino *et al*, 1989b).

The factors that regulate CRF neurons in adult rats are still not completely elucidated. Serotonin and/or central biogenic amines have been proposed to stimulate hypothalamic CRF secretion and/or synthesis (Fuller, 1981; Grino *et al*, 1989c). While there is a dense serotoninergic innervation of the hypothalamus, and in particular of the PVN, very early (E19; Lidov and Molliver, 1982), the catecholaminergic innervation of the PVN follows a different pattern of development. Phenylethanolamine N-methyltransferase-positive fibers appear on postnatal day 1 and increase progressively from days 1-7. A marked increase occurs from days 7-14, at which time an adult-like pattern is established (Katchaturian and Sladeck, 1980).

Similarly, the number of α- and ß-adrenergic receptors in brain increases rapidly during the second week of life (Pitman *et al*, 1980; Deskin *et al*, 1981). Several reports (Kaneko and Hiroshige, 1978; Feldman *et al*, 1983; Sawchenko, 1988) indicate that the catecholaminergic innervation of the hypothalamus can mediate, at least in part, the effects of ADX on CRF neurons. All these observations suggest that an immaturity of the catecholaminergic innervation of the hypothalamus could be responsible for the immaturity of the regulation of CRF neurons during the SNRP. Further studies, investigating more precisely the ontogeny of the innervation of hypothalamic CRF neurons, will be helpful in understanding the regulation of the HPA axis in developing rats.

ACKNOWLEDGMENTS

We thank R.-M. Saura (INSERM U 297) and T. Lauterburg (Neuromorphologisches Labor) for their technical assistance, and R. Quérat (INSERM U 297) for typing the manuscript.

REFERENCES

Antoni, F.A. (1986). Hypothalamic control of adrenocorticotropin secretion: advances since the discovery of 41-residue corticotropin-releasing factor. *Endocrine Reviews* 7, 351-378.

Baram, T.Z., and Schultz, L. (1989). Corticotropin-releasing hormone: pre- and postnatal gene expression in the developing rat brain. *Neuroscience Abstracts.*

Boudouresque, F., Guillaume, V., Grino, M., Strbak, V., Chautard, T., Conte-Devolx, B., and Oliver, C. (1988). Maturation of pituitary-adrenal function in rat fetuses. *Neuroendocrinology* 48, 417-422.

Bugnon, C., Fellman, D., and Gouget, A. (1983). Changes in corticoliberin and vasopressin-like immunoreactivities in the zona externa of the median eminence in adrenalectomized rats, an immunocytochemical study. *Neuroscience Letters* 37, 43-49.

Burgunder, J.-M., and Grino, M. (1991). Ontogeny of expression and glucocorticoid regulation of the arginine vasopressin gene in the rat hypothalamic paraventricular nucleus. *73rd Annual Meeting of the Endocrine Society,* Abstract 1634.

Caraty, A., Grino, M., Locatelli, A., and Oliver, C. (1988). Secretion of corticotropin-releasing factor (CRF) and vasopressin (AVP) into the hypophysial portal blood of conscious, unrestrained rams. *Biochemical and Biophysical Research Communications* 155, 841-849.

Caraty, A., Grino, M., Locatelli, A., Guillaume, V., Boudouresque, F., Conte-Devolx, B., and Oliver, C. (1990). Insulin-induced hypoglycemia stimulates corticotropin-releasing factor and arginine vasopressin secretion into hypophysial portal blood of conscious, unrestrained rams. *Journal of Clinical Investigation* **85**, 1716-1721.

Chatelain, A., Boudouresque, F., Chautard, T., Dupouy, J.-P., and Oliver, C. (1988). Corticotrophin-releasing factor immunoreactivity in the hypothalamus of the rat during the perinatal period. *Journal of Endocrinology* **119**, 59-64.

Clayton, C.J., Grosser, B.I., and Stevens, W. (1977). The ontogeny of corticosterone and dexamethasone receptors in rat brain. *Brain Research* **134**, 445-453.

Deskin, R., Seidler, F.J., Whitmore, W.L., and Slotkin, T.A. (1981). Development of α-noradrenergic and dopaminergic receptor systems depends on maturation of their presynaptic nerve terminals in the rat brain. *Journal of Neurochemistry* **36**, 1683-1690.

Eckland, D.J.A., Todd, K., and Lightman, S.L. (1988). Immunoreactive vasopressin and oxytocin in hypothalamo-hypophysial blood of the Brattleboro and Long-Evans rat: effect of adrenalectomy and dexamethasone. *Journal of Endocrinology* **117**, 27-34.

Feldman, S., Siegel, R.A., Weidenfeld, J., Conforti, N., and Melamed, E. (1983). Role of medial forebrain bundle catecholaminergic fibers in the modulation of glucocorticoid feedback effects. *Brain Research* **260**, 297-300.

Fuller, R.W. (1981). Serotoninergic stimulation of pituitary-adrenocortical function in rats. *Neuroendocrinology* **32**, 118-127.

Grino, M., Young, W.S. III, and Burgunder, J.-M. (1989a). Ontogeny of expression of the corticotropin-releasing factor gene in the hypothalamic paraventricular nucleus and of the proopiomelanocortin gene in rat pituitary. *Endocrinology* **124**, 60-69.

Grino, M., Burgunder, J.-M., Eskay, R.L., and Eiden, L.E. (1989b). Onset of glucocorticoid responsiveness of anterior pituitary corticotrophs during development is scheduled by corticotropin-releasing factor. *Endocrinology* **124**, 2686-2689.

Grino, M., Guillaume, V., Conte-Devolx, B., Szafarczyk, A., Joanny, P., and Oliver, C. (1989c). Role of central catecholamines in the control of CRF and ACTH secretion. In: G.R. Van Loon, R. Kvetnansky, R. McCarty and J. Axelrod (Eds.), "Stress: Neurochemical and Humoral Mechanisms," pp. 355-367. New York: Gordon and Breach.

Grino, M., Guillaume, V., Boudouresque, F., Conte-Devolx, B., Maltese, J.-Y., and Oliver, C. (1990). Glucocorticoids modulate peptidyl-glycine α-amidating monooxygenase (PAM) gene expression in the rat hypothalamic paraventricular nucleus. *Molecular Endocrinology* **4**, 1613-1619.

Guillaume, V., Grino, M., Conte-Devolx, B., Boudouresque, F., and Oliver, C. (1989). Corticotropin-releasing factor secretion increases in rat hypophysial portal blood during insulin-induced hypoglycemia. *Neuroendocrinology* **49**, 676-679.

Herman, J.P., Wiegand, S.J., and Watson, S.J. (1990). Regulation of basal corticotropin-releasing hormone and arginine vasopressin messenger ribonucleic acid expression in the paraventricular nucleus: effects of selective hypothalamic deafferentations. *Endocrinology* **127**, 2408-2417.

Holmes, M.C., Antoni, F.A., Aguilera, G., and Catt, K.J. (1986). Magnocellular axons in passage through the median eminence release vasopressin. *Nature* **319**, 326-329.

Kaneko, M., and Hiroshige, T. (1978). Site of fast, rate-sensitive feedback inhibition of adrenocorticotropin secretion during stress. *American Journal of Physiology* **234**, R46-R51.

Katchaturian, H., and Sladeck Jr, J.R. (1980). Simultaneous monoamine histofluorescence and neuropeptide immunocytochemistry. III. Ontogeny of catecholamine varicosities and neurophysin neurons in the rat supraoptic and paraventricular nuclei. *Peptides* **1**, 77-95.

Koch, B. (1969). Fraction libre de la corticosterone plasmatique et response hypophyso-surrenalienne au stress durant la periode post-natale chez le rat. *Hormone and Metabolic Research* **1**, 301-308.

Laurent, F.M., Hindelang, C., Klein, M.J., Stoeckel, M.E., and Felix, J.M. (1989). Expression of the oxytocin and vasopressin genes in the rat hypothalamus during development. *Developmental Brain Research* **46**, 145-154.

Levin, N., Blum, M., and Roberts, J.L. (1989). Modulation of basal and corticotropin-releasing factor-stimulated proopiomelanocortin gene expression by vasopressin in rat anterior pituitary. *Endocrinology* **125**, 2957-2966.

Lidov, H.G.W., and Molliver, M.E. (1982). An immunohistochemical study of serotonin neuron development in the rat: ascending pathways and terminal fields. *Brain Research Bulletin* **8**, 389-430.

Lightman, S.L., and Young, W.S. III. (1987). Vasopressin, oxytocin, dynorphin, enkephalin and corticotrophin-releasing factor mRNA stimulation in the rat. *Journal of Physiology* **394**, 23-39.

Linton, E.A., Tilders, F.J.H., Hodgkinson, S., Berkenbosch, F., Vermes, I., and Lowry, P.J. (1985). Stress-induced secretion of adrenocorticotropin in rats is inhibited by administration of antisera to ovine corticotropin-releasing factor and vasopressin. *Endocrinology* **116**, 966-970.

Meaney, M.J., Sapolsky, R.M., and McEwen, B.S. (1985). The development of the glucocorticoid receptor system in the rat limbic brain. I. Ontogeny and autoregulation. *Developmental Brain Research* **18**, 159-164.

Munck, A., Guyre, P.M., and Holbrook, N.J. (1984). Physiological functions of glucocorticoids in stress and their relation to pharmacological actions. *Endocrine Reviews* **5**, 25-44.

Olpe, H.R., and McEwen, B.S. (1976). Glucocorticoid binding to receptor-like proteins in rat brain and pituitary: ontogenic and experimentally induced changes. *Brain Research* **105**, 121-128.

Pittman, R.N., Minneman, K.P., and Molinoff, P.B. (1980). Ontogeny of β_1- and β_2-adrenergic receptors in rat cerebellum and cerebral cortex. *Brain Research* **188**, 357-368.

Plotsky, P.M., Bruhn, T.O., and Vale, W. (1985). Hypophysiotropic regulation of adrenocorticotropin secretion in response to insulin-induced hypoglycemia. *Endocrinology* **117**, 323-329.

Plotsky, P.M., and Sawchenko, P.E. (1987). Hypophysial-portal plasma levels, median eminence content, and immunohistochemical staining of corticotropin-releasing factor, arginine vasopressin, and oxytocin after pharmacological adrenalectomy. *Endocrinology* **120**, 1361-1369.

Rundle, S.E., and Funder, J.W. (1988). Ontogeny of corticotropin-releasing factor and arginine vasopressin in the rat. *Neuroendocrinology* **47**, 374-378.

Sapolsky, R.M., and Meaney, M.J. (1986). Maturation of adrenocortical stress response: neuroendocrine control mechanism and the stress hyporesponsive period. *Brain Research Reviews* **11**, 65-76.

Sawchenko, P.E. (1988). Effects of catecholamine-depleting medullary knife cuts on corticotropin-releasing factor and vasopressin immunoreactivity in the hypothalamus of normal and steroid-manipulated rats. *Neuroendocrinology* **48**, 459-470.

Walker, C.D., Perrin, M., Vale, W., and Rivier, C. (1986a). Ontogeny of the stress response in the rat: role of the pituitary and the hypothalamus. *Endocrinology* **118**, 1445-1451.

Walker, C.D., Sapolsky, R.M., Meaney, M.J., Vale, W.W., and Rivier, C.L. (1986b). Increased pituitary sensitivity to glucocorticoid feedback during the stress non responsive period in the neonatal rat. *Endocrinology* **119**, 1816-1821.

Young, W.S. III, Mezey, E., and Siegel, R.E. (1986a). Quantitative *in situ* hybridization histochemistry reveals increased levels of corticotropin-releasing factor mRNA after adrenalectomy in rats. *Neuroscience Letters* **70**, 198-204.

Young, W.S. III, Mezey, E., and Siegel, R.E. (1986b). Vasopressin and oxytocin mRNAs in adrenalectomized and Brattleboro rats: analysis by quantitative hybridization histochemistry. *Molecular Brain Research* **1**, 231-241.

Zimmerman, E.A. (1985). Anatomy of vasopressin-producing cells. *Frontiers in Hormone Research* **13**, 1-21.

Stress: Neuroendocrine and Molecular Approaches
Edited by R. Kvetnansky, R. McCarty and J. Axelrod

1992 Gordon and Breach Science
Publishers S.A., New York, USA.
Photocopying permitted by license only.

CHRONIC STRESS AND ADRENALECTOMY BOTH INDUCE VASOPRESSIN SYNTHESIS AND STORAGE IN CORTICOTROPIN RELEASING FACTOR NEURONS IN THE PARAVENTRICULAR NUCLEUS

D. C. E. de Goeij[1], D. Jezova[2] and F. J. H. Tilders[1]

[1]Department of Pharmacology, Medical Faculty, Free University,
Amsterdam, The Netherlands
[2]Institute of Experimental Endocrinology, Slovak Academy of Sciences,
Bratislava, Czecho-Slovakia

INTRODUCTION

There is general agreement that corticotropin-releasing factor (CRF) plays a crucial role in the control of ACTH release by anterior pituitary corticotrophs (Linton *et al*, 1985; Tilders *et al*, 1985; Antoni, 1986; Rivier *et al*, 1986). CRF involved in this process is produced by a circumscribed group of parvocellular neurons in the paraventricular nucleus of the hypothalamus (PVN) and all of these neurons project to the external zone of the median eminence (ZEME). In addition to CRF, these neurons are capable of producing arginine vasopressin (AVP), which strongly potentiates the ACTH-releasing effects of CRF (Gillies *et al*, 1982; Rivier *et al*, 1983; Whitnall, 1988).

Following bilateral adrenalectomy (ADX), synthesis of both CRF and AVP increases, as reflected in an elevation of CRF and AVP immunoreactivity (CRF$_i$, AVP$_i$) in these parvocellular cell bodies (Roth *et al*, 1982; Antoni *et al*, 1983; Merchenthaler *et al*, 1983; Tramu *et al*, 1983; Kiss *et al*, 1984; Sawchenko *et al*, 1984; Liposits *et al*, 1985).

273

In addition, ADX enhances CRF_i and AVP_i in nerve terminals in the ZEME (Zimmerman *et al*, 1977; Roth *et al*, 1982; Silverman *et al*, 1982; Hisano *et al*, 1987; Alonso *et al*, 1988; Berkenbosch *et al*, 1988). The enhancement of CRF_i and AVP_i in perikarya is accompanied by increases in amounts of mRNA transcripts of both the CRF gene and the AVP gene in the parvocellular area (Wolfson *et al*, 1985; Davis *et al*, 1986; Young *et al*, 1986ab; Schafer *et al*, 1987; Beyer *et al*, 1988; Lightman *et al*, 1989; Swanson *et al*, 1989; Imaki *et al*, 1991). In fact, after ADX nearly all CRF neurons in this part of the PVN start to produce AVP, which is associated with a near complete co-localization of CRF and AVP in nerve terminals in the ZEME (Tramu *et al*, 1983; Kiss *et al*, 1984; Sawchenko *et al*, 1984; Hisano *et al*, 1987; Sawchenko, 1987; Whitnall *et al*, 1987; Whitnall, 1988).

Several authors have suggested that this increase in CRF and AVP synthesis is due to a lack of glucocorticoid feedback, since corticosterone replacement or dexamethasone administration prevents these effects of ADX (Stillman *et al*, 1977; Jingami *et al*, 1985; Kovacs *et al*, 1986, 1987; Sawchenko, 1987; Whitnall *et al*, 1987; Beyer *et al*, 1988; Swanson *et al*, 1989; Imaki *et al*, 1991). Accordingly, it is anticipated that AVP and CRF will be suppressed by repeated or chronic stress, conditions which are known to enhance glucocorticoid levels.

In contrast to this hypothesis, Silverman *et al* (1989) showed increased CRF_i in the PVN after chronic stress. Recently, we reported that repeated daily exposure of rats to immobilization stress had no effect on CRF_i content in the ZEME, but increased AVP_i content (de Goeij *et al*, 1991). Furthermore, the co-localization of AVP in CRF-positive swellings in the ZEME also increased significantly.

In addition, we found similar effects in the ZEME of male rats exposed to chronic psychosocial stress. This was achieved by housing 6 male rats together with a dominant male and 4 females in a large (3.0 x 1.85 m) enriched environment. After three weeks of living in such a hierarchically controlled colony, CRF_i stores in the ZEME of subordinate rats remained unaltered, whereas AVP_i increased approximately two-fold. This elevation of AVP_i correlated with the intensity of the social interactions (de Goeij *et al*, in preparation). Furthermore, repeated daily activation of the hypothalamic-pituitary-adrenal system by insulin (for 11 consecutive days) resulted in an increased AVP/CRF ratio in the ZEME. Based on the results obtained from these three different models we concluded that the increase in AVP storage and co-localization in CRF terminals in the ZEME is a typical response to chronic or repeated stress.

In contrast to an absence of CRF changes in nerve terminals after repeated/chronic stress, several authors have demonstrated increased CRF mRNA in parvocellular neurons which could be attenuated by dexamethasone treatment (Lightman et al, 1988; Suda et al, 1988; Herman et al, 1989; Harbuz et al, 1989; Lightman et al, 1989; Imaki et al, 1991). Lightman and Young (1988) also found an increase in AVP mRNA in the parvocellular area of the PVN after hypertonic saline. However, the presence of massive amounts of AVP mRNA in magnocellular neurons, some of which are scattered throughout the parvocellular PVN, makes it difficult to measure changes in parvocellular neurons selectively.

The fact that we found a clear increase in AVP content in the ZEME without an effect on CRF content might be explained by a preferential secretion of CRF, resulting in a gradual accumulation of AVP in nerve terminals. Alternatively, the increase in AVP synthesis in CRF cell bodies may even be more pronounced than that of CRF. Therefore, we investigated the effect of repeated daily immobilization stress on the co-localization of AVP_i in CRF_i containing parvocellular neurons by means of a double staining procedure.

Although AVP was found in approximately 25-50% of CRF terminals in the ZEME of untreated Wistar or Sprague-Dawley rats, CRF and AVP cannot be detected in the cell bodies of such animals (Whitnall, 1988; de Goeij et al, 1991). To facilitate visualization of CRF and AVP in parvocellular neurons, rats were treated with colchicine. This drug blocks axonal transport and leads to accumulation of peptides in the cell bodies. However, it should be noted that high doses of colchicine may activate the hypothalamo-pituitary-adrenal system (Berkenbosch et al, 1988). In order to run detailed CRF/AVP co-localization studies, a double staining procedure was performed on 10 μm sections by using a rat monoclonal antibody to CRF (Van Oers et al, 1989) and a rabbit polyclonal antibody to AVP (refer to de Goeij et al, 1991). CRF and AVP staining was completed by use of a rhodamine-labeled goat anti rat IgG antiserum (Cappel) and an Fitc-labelled goat anti rabbit IgG antiserum (Tago), respectively. The number of CRF_i- and AVP_i- labeled cells were counted in the medial parvocellular part of the PVN, a region that is known to project massively to the external lamina of the median eminence (Lechan et al, 1980; Swanson et al, 1980, 1983). Only cells

TABLE 1: Effect of repeated daily immobilization (IMMO) or handling (HANDL) on the number of CRF and AVP positive parvocellular neurons in the PVN.

TREATMENT	CRF	AVP	%CO-LOCALIZATION
HANDL	94.7 ± 14.5	11.5 ± 3.7	11.5 ± 6.2
IMMO	165.0 ± 19.5*	53.0 ± 13.8*	31.0 ± 12.6*

Groups of rats (n=6) were subjected to once daily handling or once daily immobilization (150 minutes) for 16 consecutive days. Rats were decapitated 24 hours after i.c.v. colchicine (50 μg/ 10 μl). Data are expressed as the means ± SEM of CRF_i- and AVP_i-positive neurons per section and as percentage ± SEM of co-localization of AVP_i in CRF_i neurons. Data were analyzed by two-way analysis of variance using SPSS/PC+ (SPSS Inc., Chicago, III) followed by a two-tailed t-test. * Significance was defined at the 0.05 level.

with a nucleus in the section plane were counted.

As can be seen in Table 1, repeated immobilization stress induced a significant increase in the number of CRF_i- and AVP_i-positive parvocellular neurons. More importantly, the fraction of CRF-positive cells that also contained AVP increased approximately threefold (p<0.05). Thus, although CRF synthesis appeared to increase after repeated stress, the increase in AVP synthesis in CRF neurons was much more pronounced.

The fact that CRF mRNA, CRF_i and AVP_i in parvocellular cell bodies all increase after chronic stress despite elevated corticosterone levels, poses fundamental questions concerning the mechanisms by which CRF and AVP synthesis is regulated during chronic stress.

HYPOTHESES

Several hypotheses will be considered to explain the similarity in PVN responses after ADX and repeated/chronic stress.

Hypothesis A

CRF neurons exhibit glucocorticoid receptor mRNA and glucocorticoid receptors (GRs) (Agnati *et al*, 1985; Swanson *et al*, 1989) and therefore may be considered as targets for circulating corticosterone. Prolonged periods or repeated episodes of high plasma corticosterone levels may lead to down-regulation of GRs as demonstrated by Sapolsky *et al* (1984) in other brain structures, including hippocampus, after chronic stress. This in turn may lead to attenuation of corticosterone effects on parvocellular neurons in spite of elevated corticosterone levels.

Hypothesis B

Adrenalectomy is known to enhance secretory activity of CRF neurons as reflected in portal blood measurements and the same holds true for stressors (Antoni, 1986). Under both conditions, hypersecretion can be reduced by glucocorticoids. Therefore, the observed effects may not be directly related to corticosterone per se, but may relate to an intrinsic coupling between secretory activity on the one hand and synthesis of CRF and AVP on the other. The observation that dexamethasone reduces stress-induced CRF mRNA may point in the same direction (see above).

Hypothesis C

Several observations indicate that gene transcription can be either enhanced or suppressed by glucocorticoids depending on the actual "bouquet" of transcription factors and therefore may be dependent on the physiological "state" of the cell. For example, Swanson and Simmons (1989) demonstrated that glucocorticoids negatively regulate CRF mRNA in parvocellular neurons of the PVN, but stimulate CRF mRNA in oxytocinergic magnocellular neurons. It is thought that glucocorticoid receptor complexes can act directly upon neuropeptide genes or that the effects are secondary to some other event such as activation of a transcription factor. In CRF parvocellular neurons, several transcription factors have been found, including Brain-2 and *fos*, a proto-oncogene protein, which together with other proteins can form a DNA binding protein complex (Curran *et al*, 1988). In this context, it is worth noting that several stressful stimuli lead to *fos* expression in CRF-positive parvocellular neurons (Ceccatelli *et al*, 1989). In addition, CRF gene transcription can also be regulated by cell surface receptors which are linked to cAMP signal transduction pathways (Seasholtz *et al*, 1988; Copp *et al*, 1989; Dorin *et al*, 1989; Adler *et al*, 1990). From studies with an AVP-producing cell line (GLC-8), Verbeeck *et al* (1991) concluded that corticosteroid receptors, cAMP and protein kinase C can all stimulate AVP gene expression. However, the magnitude and even the direction of the effect of glucocorticoids were found to be dependent on other regulatory factors and the "state" of the cell.

Glucocorticoid-receptor complexes are known to accumulate in the cell nucleus and bind to steroid hormone responsive cis-elements in the control regions of a variety of genes. In many genes this glucocorticoid responsive element (GRE) functions as an enhancer. However, as mentioned above, CRF and AVP as well POMC genes are repressed by glucocorticoids (Davis *et al*, 1986; Gagner *et al*, 1987; Rosendale *et al*, 1987). To explain such paradoxical effects, Drouin and colleagues (1987, 1989) proposed the existence of a negative GRE (NGRE) in addition to the classical GRE. This opens the possibility that depending on the presence of other factors GR complexes bind to either GRE or nGRE on CRF and AVP genes. Recently, Akerblom *et al* (1988) showed that glucocorticoid repression can be mediated by a competition between the GR and the cAMP binding protein (CREB) for binding to overlapping DNA regulatory sequences, representing the GRE and the cAMP responsive element (CRE), respectively. Such a stearic hindrance model may represent a more sophisticated route by which glucocorticoids modulate the expression of certain genes than the simple "switch on" "switch off" mechanisms generally proposed.

Hypothesis D

The much more pronounced increases in AVP than CRF expression in activated CRF parvocellular neurons might relate to the phenomenon that the stability and the translational efficiency of mRNA can be increased by an extension of the mRNA with a poly (A) tail (Palatnik *et al*, 1984; Brawerman 1987). Poly (A) tail extensions of AVP mRNA have been found after water deprivation (for 2-4 hours) and dehydration in parvocellular neurons of the suprachiasmatic nucleus, magnocellular neurons of the supraoptic nucleus and in whole hypothalamic extracts (Carrazana *et al*, 1988; Zingg *et al*, 1988; Carter *et al*, 1990, 1991). Corticosterone receptor complexes may influence the stability of neuropeptide mRNAs, by increasing the poly(A) tail length (Paek *et al*, 1987). Although this may help to explain the relatively large increase in AVP synthesis as compared to that of CRF, it remains to be elucidated whether this phenomenon occurs in parvocellular CRF neurons in rats.

CONCLUSIONS

In conclusion, repeated/chronic stress and adrenalectomy lead to plastic changes in CRF neurons, resulting in an increased production, storage and co-localization of AVP in CRF-positive parvocellular neurons. The mechanism underlying this physiological response of CRF neurons to chronic or repeated stress remains to be elucidated.

ACKNOWLEDGEMENTS

The authors wish to thank Ms. M. Horvathova, Z. Oprsalova, N. Petrikova, I. Szalayova and Mr. R. Binnekade for their technical assistance. This work was supported by funds from a cultural agreement between The Netherlands and Czecho-Slokia.

REFERENCES

Adler, G.K., Smas, C.M., Fiandaca, M., Frim, D.M., and Majzoub, J.A. (1990). Regulated expression of the human corticotropin releasing hormone gene by cyclic AMP. *Molecular and Cellular Endocrinology* **70**, 165-174.

Agnati, L.F., Fuxe, K., Yu, Z.Y., Harfstrand, A., Okret, S., Wikstrom, A.C., Goldstein, M., and Gustafsson, J.A. (1985). Morphometrical analysis of the distribution of corticotrophin-releasing factor, glucocorticoid receptor and phenylethanolamine-N-methyltransferase immunoreactive structures in the paraventricular hypothalamic nucleus of the rat. *Neuroscience Letters* **54**, 147-152.

Akerblom, I.W., Slater, E.P., Beato, M., Baxter, J.C., and Mellon, P.L. (1988). Negative regulation by glucocorticoids through interference with a cAMP responsive enhancer. *Science* **241**, 350-353.

Alonso, G., Siaud, P., and Assenmacher, I. (1988). Immunocytochemical ultrastructural study of hypothalamic neurons containing corticotropin releasing factor in normal and adrenalectomized rats. *Neuroscience* **24**, 553-565.

Antoni, F.A., Palkovits, M., Makara, G.B., Linton, E.A., Lowry, P.J., and Kiss, J.Z. (1983). Immunoreactive corticotropin releasing hormone in the hypothalamo-infundibular tract. *Neuroendocrinology* **36**, 415-423.

Antoni, F.A. (1986). Hypothalamic control of adrenocorticotropin secretion: advances since the discovery of 41-residue corticotropin-releasing factor. *Endocrine Reviews* **7**, 351-378.

Berkenbosch, F., and Tilders, F.J.H. (1988). Effect of axonal transport blockade on corticotropin releasing factor immunoreactivity in the median eminence of intact and adrenalectomized rats: relationship between depletion rate and secretary activity. *Brain Research* **442**, 312-320.

Beyer, H.S., Matta, S.G., and Sharp, B.M. (1988). Regulation of the messenger ribonucleic acid for corticotropin-releasing factor in the paraventricular nucleus and other brain sites of the rat. *Endocrinology* **123**, 2117-2123.

Brawerman, G. (1987). Determinants of RNA stability. *Cell* **48**, 5-6.

Carrazana, E.J., Pasieka, K.B., and Majzoub, J.A. (1988). The vasopressin mRNA poly(A) tract is unusually long and increases during stimulation of vasopressin gene expression *in vivo*. *Molecular and Cellular Biology* **8**, 2267-2274.

Carter, D.A., and Murphy, D. (1990). Vasopressin mRNA in parvocellular neurons of the rat suprachiasmatic nucleus exhibits increased poly(A) tail length following water deprivation. *Neuroscience Letters* **109**, 180-185.

Carter, D.A., and Murphy, D. (1991). Rapid changes in poly(A) tail length of vasopressin and oxytocin mRNAs form a common early component of neurohypophyseal peptide gene activation following physiological stimulation. *Neuroendocrinology* **53**, 1-6.

Ceccatelli, S., Villar, M.J., Goldstein, M., and Hokfelt, T. (1989). Expression of c-fos immunoreactivity in transmitter-characterized neurons after stress. *Proceedings of the National Academy of Sciences (USA)* **86**, 9569-9573.

Copp, R.P., and Samuels, H.H. (1989). Identification of an adenosine 3',5'-monophosphate (cAMP)-responsive region in the rat growth hormone gene: Evidence for independent and synergistic effects of cAMP and thyroid hormone on gene expression. *Molecular Endocrinology* **3**, 790-796.

Curran, T., and Franza, B.R. (1988). *Fos* and *jun*: The AP-1 connection. *Cell* **55**, 395-397.

Davis, L.G., Arentzen, R., Reid, J.M., Manning, R.W., Wolfson, B., Lawrence, K.L., and Baldino, F. (1986). Glucocorticoid sensitivity of vasopressin mRNA levels in the paraventricular nucleus of the rat. *Proceedings of the National Academy of Sciences* (USA) **83**, 1145-1149.

Dorin, R.I., Takahashi, H., Nakai, Y., Fukata, J., Naitoh, Y., and Imura, H. (1989). Regulation of human corticotropin-releasing hormone gene expression by 3',5'-cyclic adenosine monophosphate in a transformed mouse corticotroph cell line. *Molecular Endocrinology* **3**, 1537-1545.

Drouin, J., Charron, J., Gagner, J-P., Jeanrotte, L., Nemer, K.M., Plante, R.K., and Wrange, O. (1987). Pro-opiomelanocortin gene: a model for negative regulation of transcription by glucocorticoids. *Journal of Cellular Biochemistry* **35**, 293-304.

Drouin, J., Trifiro, M.A., Plante, R.K., Nemer, M., Eriksson, P., and Wrange, O. (1989). Glucocorticoid receptor binding to a specific DNA sequence is required for hormone-dependent repression of pro-opiomelanocortin gene transcription. *Molecular and Cellular Biology* **9**, 5305-5314.

Gagner, J-P., and Drouin, J. (1987). Tissue-specific regulation of pituitary pro-opiomelanocortin gene transcription by corticotropin-releasing hormone, 3',5'-cyclic adenosine monophosphate and glucocorticoids. *Molecular Endocrinology* **1**, 677-682.

Gillies, G.E., Linton, E.A., and Lowry, P.J. (1982). Corticotropin releasing activity of the new CRF is potentiated several times by vasopressin. *Nature* **299**, 355-357.

Goeij de, D.C.E., Kvetnansky, R., Whitnall, M.H., Jezova, D., Berkenbosch, F., and

Tilders, F.J.H. (1991). Repeated stress-induced activation of corticotropin-releasing factor neurons enhances vasopressin stores and colocalization with corticotropin-releasing factor in the median eminence of rats. *Neuroendocrinology* 53, 150-159.

Harbuz, M.S., and Lightman, S.L. (1989). Responses of hypothalamic and pituitary mRNA to physical and psychological stress in the rat. *Journal of Endocrinology* 122, 705-711.

Herman, J.P., Schafer, M.K.H., Sladek, C.D., Day, R., Young, E.A., Akil, H., and Watson, S.J., (1989). Chronic electroconvulsive shock treatment elicits up-regulation of CRF and AVP mRNA in select populations of neuroendocrine neurons. *Brain Research* 501, 235-246.

Hisano, S., Tsuruo, Y., Katoh, S., Daikoku, S., Yanaihara, N., and Shibasaki, T. (1987). Intragranular colocalization of arginine vasopressin and methionine-enkephalin-octapeptide in CRF-axons in the rat median eminence. *Cell and Tissue Research* 249, 497-507.

Imaki, T., Nahan, J.L., Rivier, C., Sawchenko, P.E., and Vale, W. (1991). Differential regulation of corticotropin-releasing factor mRNA in rat brain regions by glucocorticoids and stress. *Journal of Neuroscience* 11, 585-599.

Jingami, H., Matsukura, S., Numa, S., and Imura, H. (1985). Effects of adrenalectomy and dexamethasone administration on the level of prepro-corticotropin-releasing factor messenger ribonucleic acid (mRNA) in the hypothalamus and adrenocorticotropin/β-lipotropin precursor mRNA in the pituitary in rats. *Endocrinology* 117, 1314-1320.

Kiss, J.Z., Mezey, E., and Skirboll, L. (1984). Corticotropin-releasing factor-immunoreactive neurons of the paraventricular nucleus become vasopressin positive after adrenalectomy. *Proceedings of the National Academy of Sciences* (USA) 81, 1854-1858.

Kovacs, K., Kiss, K.Z., and Makara, G.B. (1986). Glucocorticoid implants around the hypothalamic paraventricular nucleus prevent the increase of corticotropin-releasing factor and arginine vasopressin immunostaining induced by adrenalectomy. *Neuroendocrinology* 44, 229-234.

Kovacs, K.J., and Mezey, E. (1987). Dexamethasone inhibits corticotropin-releasing factor gene expression in the rat paraventricular nucleus. *Neuroendocrinology* 46, 365-368.

Lechan, R.M., Nestler, J.L., Jacobson, S., and Reichlin, S. (1980). The hypothalamic tuberoinfundibular system of the rat as demonstrated by horseradish peroxidase (HRP) microiontophoresis. *Brain Research* 195, 13-27.

Lightman, S.L., and Young, W.S. (1988). Corticotropin-releasing factor, vasopressin and pro-opiomelanocortin mRNA responses to stress and opiates in the rat. *Journal of Physiology* 403, 511-523.

Lightman, S.L., and Young, W.S. (1989). Influence of steroids on the hypothalamic corticotropin-releasing factor and preproenkephalin mRNA responses to stress. *Proceedings of the National Academy of Sciences* (USA) 86, 4336-4310.

Linton, E.A., Tilders, F.J.H., Hodgkinson, S., Berkenbosch, F., Vermes, I., and Lowry, P.J. (1985). Stress-induced secretion of adrenocorticotropin in rats is inhibited by administration of antisera to ovine corticotropin releasing factor and vasopressin. *Endocrinology* 116, 966-970.

Liposits, Z., and Paull, W.K. (1985). Alterations of the paraventriculo-infundibular corticotropin releasing factor (CRF) immunoreactive neuronal system in long term adrenalectomized rats. *Peptides* 6, 1021-1036.

Merchenthaler, I., Vigh, S., Petrusz, P., and Schally, A.V. (1983). The paraventriculo-infundibular corticotropin releasing factor (CRF) pathway as revealed by immunocytochemistry in long-term hypophysectomized or adrenalectomized rats. *Regulatory Peptides* 5, 295-305.

Paek, I., and Axel, R. (1987). Glucocorticoids enhance stability of human growth hormone mRNA. *Molecular and Cellular Biology* 7, 1496-1507.

Palatnik, C.M., Wilkins, C., and Jacobson, A. (1984). Translational control during early dictyotelium development: Possible involvement of poly (A) sequences. *Cell* 36, 1017-1025.

Rivier, C., and Vale, W. (1983). Interaction of corticotropin-releasing factor and arginine vasopressin on adrenocorticotropin secretion *in vivo*. *Endocrinology* 113, 939-942.

Rivier, C.L., and Plotsky, P.M. (1986). Mediation by corticotrophin releasing factor (CRF) of adenohypophysial hormone secretion. *Annual Review of Physiology* 48, 475-494.

Rosendale, B.E., Jarrett, D.B., and Robinson, A.G. (1987). Identification of a corticotropin-releasing factor-binding protein in the plasma membrane of AtT20 mouse pituitary tumor cells and its regulation by dexamethasone. *Endocrinology* 120, 2357-2366.

Roth, K.A., Weber, E., and Barchas, J.D. (1982). Immunoreactive corticotropin releasing factor (CRF) and vasopressin are co-localized in a subpopulation of the immunoreactive vasopressin cells in the paraventricular nucleus of the hypothalamus. *Life Sciences* 31, 1857-1860.

Sapolsky, R.S., Krey, L.L.C., and McEwen, B.S. (1984). Stress down-regulates corticosterone receptors in a site-specific manner in the brain. *Endocrinology* 114, 287-292.

Sawchenko, P.E., Swanson, L.W., and Vale, W.W. (1984). Co-expression of corticotropin-releasing factor and vasopressin immunoreactivity in parvocellular neurosecretory neurons of the adrenalectomized rat. *Proceedings of the National Academy of Sciences* (USA) 81, 1883-1887.

Sawchenko, P.E. (1987). Evidence for a local site of action for glucocorticoids in inhibiting CRF and vasopressin expression in the paraventricular nucleus. *Brain Research* 403, 213-224.

Schafer, M.K.H., Herman, J.P., Young, E., Thompson, R., Douglas, J., Sherman, T.G., Akil, H., and Watson, S.J. (1987). Gene expression of neuropeptides related to CRF after adrenalectomy. *Neuroscience Abstracts* 13, 583.

Seasholtz, A.F., Thompson, R.C., and Douglas, J.O. (1988). Identification of a cyclic adenosine monophosphate-responsive element in the rat corticotropin-releasing hormone gene. *Molecular Endocrinology* 2, 1311-1319.

Silverman, A.J., and Zimmerman, E.A. (1982). Adrenalectomy increases sprouting in a peptidergic neurosecretory system. *Neuroscience* 7, 2705-2714.

Silverman, A.J., Hou-Yu, A., and Kelley, D.D. (1989). Modification of hypothalamic neurons by behavioral stress. In: Y. Tache, J.E. Morley and M.R. Brown (Eds.), "Neuropeptides and Stress," pp. 23-38. Berlin: Springer Verlag.

Stillman, M.A., Recht, L.D., Rosario, S.L., Seif, S.M., Robinson, A.G., and Zimmerman, E.A. (1977). The effects of adrenalectomy and glucocorticoid replacement on vasopressin and vasopressin-neurophysin in the zona externa of the median eminence in the rat. *Endocrinology* **101**, 42-49.

Suda, T., Tozawa, F., Yamada, M., Ushiyama, T., Tomori, N., Sumitomo, T., Nakagami, Y., and Demura, H. (1988). Insulin-induced hypoglycemia increases corticotropin-releasing factor messenger ribonucleic acid levels in rat hypothalamus. *Endocrinology* **123**, 1371-1375.

Swanson, L.W., and Kuypers, H.G.J.M. (1980). The paraventricular nucleus of the hypothalamus: cytoarchitectonic subdivisions and organization of projections to the pituitary, dorsal vagal complex, and spinal cord as demonstrated by retrograde fluorescence double-labeling methods. *Journal of Comparative Neurology* **194**, 555-570.

Swanson, L.W., Sawchenko, P.E., Rivier, J., and Vale, W.W. (1983). Organization of ovine corticotropin-releasing factor immunoreactive cells and fibers in the rat brain: An immunohistochemical study. *Neuroendocrinology* **36**, 165-186.

Swanson, L.W., and Simmons, D.M. (1989). Differential steroid hormone and neural influences on peptide mRNA levels in CRH cells of the paraventricular nucleus: A hybridization histochemical study in the rat. *Journal of Comparative Neurology* **285**, 413-435.

Tilders, F.J.H., Berkenbosch, F., Vermes, I., Linton, E.A., and Smelik, P.G. (1985). Role of epinephrine and vasopressin in the control of the pituitary-adrenal response to stress. *Federation Proceedings* **44**, 155-160.

Tramu, G., Croix, C., and Pillez, A. (1983). Ability of the CRF immunoreactive neurons of the paraventricular nucleus to produce a vasopressin-like material. *Neuroendocrinology* **37**, 467-469.

Van Oers, J.W.A.M., Tilders, F.J.H., and Berkenbosch, F. (1989). Characterization and biological activity of a rat monoclonal antibody to rat/human corticotropin-releasing factor. *Endocrinology* **124**, 1239-1246.

Verbeeck, M.A.E., Sutano, W., and Burbach, J.P.H. (1991). Regulation of vasopressin mRNA levels in the small cell lung carcinoma cell line GLC-8: Interaction between glucocorticoids and second messengers. *Molecular Endocrinology*, in press.

Whitnall, M.H., Key, S., and Gainer, H. (1987). Vasopressin containing and vasopressin-deficient subpopulations of corticotropin-releasing factor axons are differentially affected by adrenalectomy. *Endocrinology* **120**, 2180-2182.

Whitnall, M.H. (1988). Distributions of pro-vasopressin expressing and pro-vasopressin deficient CRF neurons in the paraventricular hypothalamic nucleus of colchicine-treated normal and adrenalectomized rats. *Journal of Comparative Neurology* **275**, 13-28.

Wolfson, B., Manning, R.W., Davis, L.G., Arentzen, R., and Baldino, F. (1985). Co-localization of corticotropin-releasing factor and vasopressin mRNA in neurones after adrenalectomy. *Nature* **315**, 59-61.

Young, W.S., Mezey, E., and Siegel, R.E. (1986a). Quantitative *in situ* hybridization reveals increased levels of corticotropin-releasing factor mRNA after adrenalectomy in rats. *Neuroscience Letters* **70**, 198-203.

Young, S.W., Mezey, E., and Siegel, R.E. (1986b). Vasopressin and oxytocin mRNAs in adrenalectomized and Brattleboro rats: analysis by quantitative *in situ* hybridization histochemistry. *Molecular Brain Research* **1**, 231-241.

Zimmerman, E.A., Stillman, M.A., Recht, L.D., Antunes, J.L., Carmel, P.W., and Goldsmith, P.C. (1977). Vasopressin and corticotropin-releasing factor: An axonal pathway to portal capillaries in the zona externa of the median eminence containing vasopressin and its interaction with adrenal corticoids. *Annals of the New York Academy of Sciences* **297**, 405-419.

Zingg, H.H., Lefebvre, L., and Almazan, G. (1988). Regulation of poly(a) tail size of vasopressin mRNA. *Journal of Biological Chemistry* **263**, 11041-11043.

PART FOUR

MOLECULAR GENETICS OF NEUROTRANSMITTER ENZYMES

Stress: Neuroendocrine and Molecular Approaches
Edited by R. Kvetnansky, R. McCarty and J. Axelrod

1992 Gordon and Breach Science
Publishers S.A., New York, USA.

GENES OF HUMAN CATECHOLAMINE SYNTHESIZING ENZYMES AND THEIR REGULATION DURING STRESS

T. Nagatsu

Department of Biochemistry
Nagoya University School of Medicine
Nagoya 466 Japan

INTRODUCTION

It has been well established that catecholamine-synthesizing enzymes, especially tyrosine hydroxylase (TH), dopamine-ß-hydroxylase (DBH) and phenylethanolamine N-methyltransferase (PNMT), are induced by stressful stimulation (Kvetnansky *et al*, 1970). Thus the regulation of genetic expression of catecholamine-synthesizing enzymes by stress should be important in understanding various molecular mechanisms of stress and adaptation.

We cloned full-length complementary DNAs (cDNAs) and genomic DNAs of the human catecholamine-synthesizing enzymes, TH, aromatic L-amino acid decarboxylase (AADC), DBH and PNMT, and determined the nucleotide sequences and the deduced amino acid sequences. We discovered multiple messenger RNAs (mRNAs) of human TH, human DBH, and human PNMT (Nagatsu, 1991).

Four types (types 1,2,3, and 4) of human TH mRNAs are produced by alternative mRNA splicing mechanisms from a single gene. We isolated two different cDNAs (types A and B) for human DBH and the genomic DNA, and showed that the two mRNAs are generated through alternative polyadenylation from a single gene. We observed the presence of a minor human PNMT mRNA (type B) in addition to the major mRNA (type A). Type B mRNA of human PNMT carries an approximately 700-bp long untranslated region in the 5'-terminus, suggesting that the two types of human PNMT mRNA are produced from a single gene through the use of two alternative promoters.

The 5'-flanking regions of the genes of human TH, human DBH, and human PNMT contain possible transcription regulatory elements such as a cyclic AMP response element (TH, DBH, and PNMT), a glucocorticoid response element (DBH and PNMT), and an *Sp* 1 binding site (TH and PNMT). Thus, there are common mechanisms of gene expression of the various catecholamine-synthesizing enzymes. The information on the structures of genes of human catecholamine-synthesizing enzymes should be important in understanding the molecular mechanisms of stress and adaptation.

Human Tyrosine Hydroxylase (TH) Gene

Grima *et al* (1987) and we (Kaneda *et al*, 1987; Kobayashi *et al*, 1987) found four types of human TH (types 1-4) by cDNA cloning. Nucleotide sequences analysis of full-length cDNA of types 1 and 2 (Grima *et al*, 1987), type 3 (Kobayashi *et al*, 1987), and type 4 (Kaneda *et al*, 1987), was completed to deduce the amino acid sequences. These mRNAs are constant for the major part but are distinguishable from one another as to the insertion/deletion of 12-and 81-bp sequences, respectively, between the 90th and the 91th nucleotides of type 1. In type 4 mRNA, a 93-bp sequence composed of the 12-and 81- bp sequences is inserted into type 1 TH mRNA. Since this insertion does not alter the reading frame of the protein-coding region, type 4 cDNA codes the longest TH molecules. Southern blot analysis of human genomic DNA suggested that the human TH gene exists as a single gene per haploid DNA, indicating that these different mRNAs are produced through alternative mRNA splicing from a single primary transcript (Kaneda *et al*, 1987).

We (Ichikawa *et al*, 1990) detected two types of TH mRNA corresponding to type 1 and type 2 in adrenal gland and brain of two species of monkeys (*Macaca trus* and *Macaca fuscata*), but only a single form of TH mRNA in *Suncus murinus*, a species of insectivore, and laboratory rats. In contrast to human TH and monkey TH, TH mRNA of rat (Grima *et al*, 1985), mouse (Ichikawa *et al*, 1991), quail (Fauquet *et al*, 1988), cow (D'Mello *et al*, 1988), and *Drosophila* (Neckameyer and Quinn, 1989) were found to be single and similar to human type 1 TH mRNA. We suggest, therefore, that the multiplicity of TH mRNA is specific to primates. Comparison of the amino acid sequences of these forms of TH demonstrate a strong evolutionary conservation of enzyme structure (Nagatsu, 1989; Nagatsu and Ichinose, 1990).

O'Malley et al (1987), Le Bourdelles et al (1988), and we (Kobayashi et al, 1988) isolated genomic clones coding the human TH gene and determined the nucleotide sequence. The nucleotide sequence of the human TH gene revealed that it is composed of 14 exons interrupted by 13 introns spanning approximately 8.5 kb (Kobayashi et al, 1988). The nucleotide sequence of the coding region was the same as that of the type 4 cDNA. The exon/intron organization of the human TH gene confirmed our prediction that the inserted 81-bp fragment is encoded by an independent exon (exon 2), but the inserted 12-bp fragment is encoded by part of an exon (exon 1). The 12-bp insertion sequence is derived from the 3'-terminal portion of exon 1 and the 81-bp insertion sequence is encoded by exon 2. The N-terminal region is encoded by the 5' portion of exon 1, the remaining region from exon 3 to exon 14 is common to all four kinds of mRNA. Since the transcription of this gene seems to start at a single position, we conclude that the four types of human TH mRNA are produced through alternative splicing from the same primary transcript.

We (Kobayashi et al, 1988) have proposed the potential secondary structure of the primary transcript, which may be involved in the inclusion/exclusion of exon 2. Computer-assisted analysis of the secondary structure of the primary transcript led to the prediction of four stable hairpin loops in introns 1 and 2; one hairpin loop in intron 1 (structure A); two hairpin loops in intron 2 (structure B and C); and one hairpin loop between introns 1 and 2 (structure D). Hairpin loops of structures A, B and C may facilitate the inclusion of exon 2 into mRNA by juxtaposing exons 1 and 2 as well as exons 2 and 3. When the loop of structure D is formed, the hairpin loop of structure A should be destroyed because the sequence in intron 1 is common to both structure A and structure D. The hairpin loop of structure D may be involved in joining exon 1 directly to exon 3, preventing the inclusion of exon 2 into mRNA by physically separating exon 2 from exons 1 and 3 through 14-nucleotide perfect-matched stem structures. We assume the presence of trans-acting factors that stabilize the hairpin structure, and presume that they discriminate between structures A and D. When the hairpin loop of structure A is formed, structure D should be destroyed for the same reason as described above.

We examined the expression pattern of multiple types of human TH mRNA in human brain and adrenal medulla by the primer extension method (Kaneda et al, 1990). We investigated the effect of aging on the expression pattern of TH isozymes, and the results suggest that TH type 1 and 2 may be susceptible to brain aging. In adult adrenal medulla, all types of TH mRNA were detected. Relative rations of types 1:2:3:4: were about 0.9:1.0:0.1:0.3. On the other hand, in brain (substantia nigra), type 3 was undetectable and the relative ratios of each type were about 0.5:1.0:0:0.1 in old brains. In both tissues, types 1 and 2 were major species and types 3 and 4 were minor species. Type 3 was almost undetectable in brain and adrenal. In contrast, type 3 was expressed much more than type 4 in human pheochromocytoma, in which the ratio was about 0.9:1.0:0.3:0.1.

We (Kaneda et al, 1991) have succeeded in producing transgenic mice carrying multiple copies of the human TH gene. The transgenes were transcribed correctly and expressed specifically in the brain and adrenal gland. Human TH mRNA levels in the brain were about 50-fold higher than the endogenous mouse TH mRNA. In situ hybridization demonstrated an enormous region-specific expression of the transgenes in the substantia nigra and ventral tegmental area. TH immunoreactivity in these brain regions, though not comparable to the increment of mRNA, was increased significantly by about 3-fold. This observation was also supported by Western blot analysis and TH activity measurements. However, catecholamine levels in transgenic mice were not significantly different from those in non-transgenic mice. These results suggest unknown regulatory mechanisms at translation and /or post-translation for human TH gene expression and for catecholamine levels in transgenic mice.

We (Kabayashi et al, 1988) compared 5'-flanking regions of the human TH gene with the rat TH gene (Lewis et al, 1987). They show about 70% homology and there are many conserved blocks. One conserved sequence is homologous to the consensus sequence of the binding site of the transcription factor Sp 1 which is known to activate transcription initiation of many mammalian genes. Another sequence, TGACGTCA, is homologous to the conserved sequence of the cyclic AMP-response element for transcription activation of various human genes. Combi *et al*, (1989) also reported the presence of *AP* 2 and POU/OCT in 5' flanking regions of rat TH.

Human Dopamine ß-Hydroxylase (DBH) Gene

We (Kobayashi *et al*, 1989) isolated two different cDNAs (types A and B) and genomic DNA of human DBH. We showed that the two mRNAs are generated through alternative polyadenylation from a single gene. Human DBH type A cDNA (clones DBH-1 and DBH-3) contained a 5' -untranslated region of 32 bp, a protein coding region of 1,809 bp (603 amino acids) including the signal peptide sequence (Met 1-Gly 25), and a 3'-untranslated region of 884 bp followed by a poly (A) tail of 21 bp. The protein coding region's sequence corresponded to that reported by Lamouroux *et al* (1987). Human DBH type B cDNA (clones DBH-2 and DBH-4) differs from type A in its 3'-untranslated regions in the absence of a 300 bp sequence (nucleotide 2,394 to 2,693) in type A at the polyadenylation site. The two clones, DBH-3 and DBH-4, corresponded to DBH-1 (type A) and DBH-2 (type B), but there were sequence differences at six nucleotides between the corresponding cDNAs. The difference at nucleotide 910 caused an amino acid change between Ala (DBH-1 and DBH-3 termed DBH/A) and Ser (DBH-2 and DBH-4 termed DBH/S).

We (Ishii *et al*, 1991) succeeded in expressing both DBH/A and DBH/S type A cDNA in COS cells. Both of the expressed proteins showed enzyme activities and immunoreactivities. The two proteins had similar kinetic constants, but had different homospecific activities (activities per enzyme protein); the homospecific activity of human DBH/S was low, approximately one thirteenth of human DBH/A. Human serum DBH that is derived from sympathetic nerve terminals is varied among individuals and is genetically determined. The relationship between the two forms of human DBH and these clinical variants of DBH of low activity remains to be examined further.

We (Kobayashi *et al*, 1989) sequenced human DBH genomic DNAs. The human DBH gene spans approximately 23 kb and consists of 12 exons interrupted by 11 introns. Exon 12 encoded the 3'-terminal region of 1,013 bp of type A, including the 300 bp sequence. Northern blot hybridization and S1 nuclease mapping experiments supported the conclusion that alternative use of two polyadenylation sites from a single DBH gene generates two different mRNA types, types A and B. The ratio of type A to type B mRNAs in human pheochromocytoma was approximately 1.0 to 0.2. The functional significance of the production of multiple mRNAs having different 3'-untranslated regions through alternative polyadenylation is unknown. The 3'-untranslated region may be involved in mRNA stability and translational efficiency.

We determined the nucleotide sequence of the promoter region. We found possible transcription regulatory elements, TATA, CCAAT, CACCC, GC boxes, a cyclic AMP response element (CRE), an AP-2 element, and a glucocorticoid response element (GRE) near the transcriptional initiation site of the human DBH gene.

Human Phenylethanolamine N-Methyltransferase (PNMT) Gene

We (Kaneda et al, 1988) reported the complete nucleotide sequence of a full-length human PNMT cDNA and the deduced amino acid sequence of the enzyme. Human PNMT consists of 282 amino acid residues with a predicted molecular weight of 30,833, including initial methionine.

We (Sasaoka et at, 1989) also isolated genomic DNA for human PNMT from a human genomic library using cloned human PNMT as a probe, and determined the nucleotide sequence. PNMT is encoded in a single gene which consists of three exons. We observed the presence of a minor PNMT mRNA (type B), besides the major mRNA (type A) reported previously by us (Kaneda et al, 1988) by Northern blot hybridization. The type B human PNMT mRNA carries an approximately 700 nucleotide-long untranslated region in the 5' terminus. This suggests that two types of mRNA are produced from a single gene through the use of two alternative promoters. A TATA-like sequence is located 30 bp upstream from the cap site of type A mRNA. Upstream from the cap site, there are three sequences resembling Sp 1 binding sites and sequences of glucocorticoid response elements (GRE), with the latter also found in the first intron.

cDNA and genomic DNA of PNMT were cloned from humans (Kobayashi et al, 1988; Baetge et al, 1988), as well as from cows (Baetge et al, 1988; Batter et al, 1988). Potential binding sites for glucocorticoid response element (GRE) and Sp 1 were found in the 5' flanking region of the gene among various species.

CONCLUSIONS

Cloning and nucleotide sequencing of cDNAs and genomic DNAs of TH, DBH, and PNMT provide useful information on gene expression of catecholamine-synthesizing enzymes during stressful stimulation. There are several common putative transcription-regulatory sequences in the 5' flanking region in the genes for the catecholamine-synthesizing enzymes (Table 1).

For example, *Sp* 1 is common to TH and PNMT, while a cyclic AMP-response element and AP 2 are common to TH and DBH. On the other hand, a glucocorticoid response element is common to DBH and PNMT. This could explain simultaneous induction of TH, DBH, and PNMT following exposure to stress.

It should be noted that human and monkey TH molecules have multiple forms although other species of animals have only one type of TH corresponding to human TH type. It is possible in humans exposed to stress that only one type of TH isoenzyme might be induced in response, and that, if the induction was different among the human TH isoenzymes, this might be related to the different response to stress of humans and monkeys from other animal species.

TABLE 1. Putative transcription-regulatory sequences in the genes for catecholamine-synthesizing enzymes.

Enzyme	Transcription-regulatory sequences
Tyrosine hydroxylase (TH)	cyclic AMP response element
	SP 1
	AP 1
	AP 2
	POU/OCT
Dopamine ß-hydroxylase (DBH)	cyclic AMP-response element
	glucocorticoid response element
	AP 2
Phenylethanolamine/ N-methyltransferase (PNMT)	glucocorticoid response element
	SP 1

REFERENCES

Baetge, E.E., Behringer, R.R., Messing, A., Brinster, R.L., and Palmiter, R.D. (1988). Transgenic mice express the human phenylethanolamine N-methyltransferase gene in adrenal medulla and retina. *Proceedings of the National Academy of Sciences* (USA) **85**, 3648-3652.

Baetge, E.E., Suh, Y.H., and Joh, T.H. (1986). Complete nucleotide and deduced amino acid sequence of bovine phenylethanolamine N-methyltransferase: partial amino acid homology with rat tyrosine hydroxylase. *Proceedings of the National Academy of Sciences* (USA) **83**, 5454-5458.

Batter, D.K., D'Mello, S.R., Turzai, L.M., Hughes, H.B. III., Gioio, A.E., and Kaplan, B.B. (1988). The complete nucleotide sequence and structure of the gene encoding bovine phenylethanolamine N-methyltransferase. *Journal of Neuroscience Research* **19**, 367-376.

Combi, F., Fung, B., and Chikaraish, D. (1989). 5' Flanking DNA sequences direct cell-specific expression of rat tyrosine hydroxylase. *Journal of Neurochemistry* **53**, 1656-1659.

D'Mello, S.R., Weisberg, S.P., Stachowiak, M.K., Turzai, L.M., Gioio, A.E., and Kaplan, B.B. (1988). Isolation and nucleotide sequence of a cDNA clone encoding bovine adrenal tyrosine hydroxylase: comparative analysis of tyrosine hydroxylase gene products. *Journal of Neuroscience Research* **19**, 440-449.

Fauguet, M., Grima, B., Lamouroux, A., and Mallet, J. (1988). Cloning of quail tyrosine hydroxylase: amino acid homology with other hydroxylases discloses functional domains. *Journal of Neurochemistry* **50**, 142-148.

Grima, B., Lamouroux, A., Blanot, F., Biguet, N.F., and Mallet, J. (1985). Complete coding sequence of rat tyrosine hydroxylase mRNA. *Proceedings of the National Academy of Sciences* (USA) **82**, 617-621.

Grima, B., Lamouroux, A., Boni, C., Jullien, J.-F., Javog-Agid, F., and Mallet, J. (1987). A single human gene encoding multiple tyrosine hydroxylases with different predicted functional characteristics. *Nature* **326**, 707-711.

Ichikawa, S., Ichinose, H., and Nagatsu, T. (1990). Multiple mRNAs of monkey tyrosine hydroxylase. *Biochemical and Biophysical Research Communications* **173**, 1331-1336.

Ichikawa, S., Sasaoka, T., and Nagatsu, T. (1991). Primary structure of mouse tyrosine hydroxylase deduced from its cDNA. *Biochemical and Biophysical Research Communications* **176**, 1610-1616.

Ishii, A., Kobayashi, K., Kiuchi, K., and Nagatsu, T. (1991). Expression of two forms of human dopamine-β-hydroxylase in COS cells. *Neuroscience Letters* **125**, 25-28.

Kaneda, N., Kobayashi, K., Ichinose, H., Kishi, F., Nakasawa, A., Kurosawa, Y., Fujita, K., and Nagatsu, T. (1987). Isolation of novel cDNA clone for human tyrosine hydroxylase: alternative RNA splicing produces four kinds of mRNA from a single gene. *Biochemical and Biophysical Research Communications* **146**, 971-975.

Kaneda, N., Ichinose, H., Kobayashi, K., Oka, K., Kishi, F., Nakazawa, A., Kurosawa, Y., Fujita, K., and Nagatsu, T. (1988). Molecular cloning of cDNA and chromosomal assignment of the gene for human phenylethanolamine N-methyltransferase, the enzyme for epinephrine biosynthesis. *Journal of Biological Chemistry* **263**, 7672-7677.

Kaneda, N., Kobayashi, K., Ichinose, H., Sasaoka, T., Ishii., A., Kiuchi, K., Kurosawa, Y., Fujita, K., and Nagatsu, T. (1990). Molecular biological approaches to catecholamine neurotransmitters and brain aging. In: T. Nagatsu and O. Hayaishi (Eds.), "Aging of the Brain: Cellular and Molecular Aspects of Brain Aging and Alzheimer's Disease," pp. 53-66. Tokyo: Japan Scientific Societies Press.

Kaneda, N., Sasaoka, T., Kobayashi, K., Kiuchi, K., Nagatsu, I., Kurosawa, Y., Fujita, K., Yokoyama, M., Nomura, T., Katsuki, M., and Nagatsu, T. (1991). Tissue-specific and high-level expression of human tyrosine hydroxylase gene in transgenic mice. *Neuron* 6, 583-594.

Kobayashi, K., Kaneda, N., Ichinose, H., Kishi, F., Nakazawa, A., Kurosawa, Y., Fujita, K., and Nagatsu, T. (1987). Isolation of a full-length cDNA clone encoding human tyrosine hydroxylase type 3. *Nucleic Acid Research* 15, 6733.

Kobayashi, K., Kaneda, N., Ichinose, H., Kishi, F., Nakazawa, A., Kurosawa, Y., Fujita, K., and Nagatsu, T. (1988). Structures of the human tyrosine hydroxylase gene: alternative splicing from a single gene accounts for generation of four mRNA types. *Journal of Biochemistry* 103, 907-912.

Kobayashi, K., Kurosawa, Y., Fujita, K., and Nagatsu, T. (1989). Human dopamine ß-hydroxylase gene: two mRNA having different 3'-terminal regions are produced through alternative polyadenylation. *Nucleic Acid Research* 17, 1089-1102

Kvetnansky, R., Weise, V.K., and Kopin, I.J. (1970). Elevation of adrenal tyrosine hydroxylase and phenylethanolamine-N-methyltransferase by repeated immobilization of rats. *Endocrinology* 87, 744-749.

Lamouroux, A., Vigny, N., Faucon Biguet, N., Darmon, M.C., Franck, R., Henry, J. P., and Mallet, J. (1987). The primary structure of human dopamine-ß-hydroxylase: insights into the relationship between the soluble and the membrane-bound forms of the enzyme. *EMBO Journal* 6, 3931-3937.

Le Bourdelles, B., Boularand, S., Boni, P., Horellou, P., Dumas, S., Grima, B., and Mallet, J. (1988). Analysis of the 5' region of the human tyrosine hydroxylase gene: combined patterns of exon splicing generate multiple regulated tyrosine hydroxylase iso-forms. *Journal of Neurochemistry* 50, 988-991.

Lewis, E.J., Harrington, C.A., and Chikaraishi, D.M. (1987). Transcriptional regulation of the tyrosine hydroxylase gene by glucocorticoids and cyclic AMP. *Proceedings of the National Academy of Sciences* (USA) 84, 3550-3554.

Nagatsu, T. (1989). The human tyrosine hydroxylase gene. *Cellular and Molecular Neurobiology* 9, 313-321.

Nagatsu, T. (1991). Genes for catecholamine-synthesizing enzymes. *Neuroscience Research*, in press.

Neckameyer, W.S., and Quinn, W.G. (1989). Isolation and characterization of the gene for *Drosophila* tyrosine hydroxylase. *Neuron* 2, 1167-1175.

O'Malley, K.L., Anhalt, M.J., Martin, B.M., Kalsoe, J.R., Winfield, S.L., and Ginns, E.I. (1987). Isolation and characterization of the human tyrosine hydroxylase gene: identification of 5' alternative splice sites responsible for multiple mRNA. *Biochemistry* 26, 6910-6914.

Sasaoka, T., Kaneda, N., Kurosawa, Y., Fujita, K., and Nagatsu, T. (1989). Structure of human phenylethanolamine N-methyltransferase gene: existence of two types of mRNA with different transcription initiation sites. *Neurochemistry International* 15, 555-565.

Stress: Neuroendocrine and Molecular Approaches
Edited by R. Kvetnansky, R. McCarty and J. Axelrod

1992 Gordon and Breach Science
Publishers S.A., New York, USA.
Photocopying permitted by license only.

GENE EXPRESSION AND REGULATION OF MONOAMINE ENZYMES DURING STRESS

T. C. Wessel, K.-T. Kim, K. S. Kim, J. M. Carroll, H. Baker and T. H. Joh

Laboratory of Molecular Neurobiology, The Burke Medical Research Institute, Cornell University Medical College, White Plains, New York

The monoamine neurotransmitters, the catecholamines and serotonin, play an important role in normal brain function and during stress but are also involved in a variety of neurologic and psychiatric diseases that affect tens of millions of people worldwide. For many years, neurobiological and behavioral research has focused on the factors that control monoamine neurotransmitter synthesis, transport and release as well as pre- and postsynaptic receptor activation. Adaptive changes in catecholamine and serotonin tissue levels are achieved primarily through the regulation of biosynthetic processes. The biosynthesis of the catecholamine neurotransmitters, dopamine (DA), norepinephrine (NE) and epinephrine (EPI), is catalyzed by four enzymes: tyrosine hydroxylase (TH), aromatic L-amino acid decarboxylase (AADC), dopamine ß-hydroxylase (DBH) and phenylethanolamine N-methyltransferase (PNMT). Serotonin biosynthesis requires two enzymes, tryptophan hydroxylase (TPH) and AADC. From the perspective of the molecular neurobiologist, the expression and regulation of these neurotransmitter enzymes have several unique characteristics.

The catecholamines DA, NE and EPI are synthesized, stored and released by their respective phenotype-specific neurons, namely dopaminergic, noradrenergic and adrenergic neurons. Accordingly, the genes encoding TH and AADC are active in virtually all catecholamine neurons. DBH is not expressed in DA neurons, but is present in the noradrenergic and adrenergic neurons of the pons and medulla oblongata as well as chromaffin cells outside the neuraxis. PNMT is found in two small groups of adrenergic neurons in the brainstem and of course in the enormous population of EPI-containing cells of the adrenal medulla. This indicates that certain nuclear mechanisms must exist to produce the tissue-specific expression of these biosynthetic enzymes.

Another intriguing feature of monoamine cells is that the levels of enzyme activity and of protein content for individual enzymes in neuronal and chromaffin cells are not static but highly regulated and change to different degrees in various catecholaminergic and serotonergic systems. For instance, enzyme activities in sympathetic neurons, the adrenal medulla and the noradrenergic neurons of the central nervous system can be increased by cold stress or immobilization stress (Kvetnansky, 1973) or by the administration of drugs, particularly reserpine (Mueller *et al*, 1969; Joh *et al*, 1973; Reis *et al*, 1975). Recently, it has been shown that the increase in TH activity following reserpine is mostly due to elevated transcriptional activity of the TH gene resulting in the production of newly synthesized enzyme protein (Tank *et al*, 1985; Richard *et al*, 1988; Faucon-Biguet *et al*, 1986). But there are other, physiologically more relevant factors that increase the steady state levels of enzyme mRNAs as well: glucocorticoids increase TH, DBH and PNMT mRNAs in the adrenal medulla of hypophysectomized rats (Stachowiak *et al*, 1988; Jiang *et al*, 1989) and in chromaffin cell cultures (Baetge *et al*, 1981; Lewis *et al*, 1983, 1987). Increased levels of cAMP are thought to cause the elevation of these enzyme mRNAs in cultured cells. It is now known that there are several cis-acting elements, including CRE, AP-1, SP-1 and GRE, in the regulatory regions of these enzymes (Harrington *et al*, 1987; Cambi *et al*, 1989; Coker *et al*, 1988; Rose *et al*, 1990). The function of these elements is the focus of intense research and certainly contributes to the changes in the steady state levels of the enzyme mRNAs.

Interestingly, there is increasing evidence that catecholamine enzymes are expressed and regulated in a parallel fashion *in vivo* (Reis *et al*, 1975; Wessel *et al*, 1991). When TH activity increases, DBH and PNMT activities also increase. Furthermore, in those cell culture systems that exhibit TH activity, DDC and DBH are invariably co-expressed and simultaneously upregulated. However, PNMT gene expression has never been detected in a continuous cell line, although up-regulation of this gene has been demonstrated convincingly in EPI-producing cells of the bovine adrenal medulla in primary culture (Ross et al, 1990).

Another fascinating observation concerns the homologies that exist among TH and TPH. TH and TPH are the rate-limiting enzymes in the catecholamine and serotonin biosynthetic pathways, respectively. TH and TPH as well as phenylalanine hydroxylase (PH) are members of the aromatic L-amino acid hydroxylase gene family (Darmon et al, 1986; Grenett et al, 1987). The primary structure of these enzymes contains a high degree of homology in their nucleotide sequences as well as their amino acid sequences. Thus, the two neuronal enzymes, TH and TPH, are structurally similar. TPH catalyzes the conversion of tryptophan to 5-hydroxytryptophan, which is the first step in serotonin (5-HT) biosynthesis. Interestingly, TPH protein isolated from pineal gland versus the dorsal raphe nucleus differs in several biochemical properties, such as molecular weight, pI and specific activity; furthermore, there is a striking difference in the relative abundance of TPH mRNA per cell in these two structures (Grenett et al, 1987; Dumas et al, 1989; Kim et al, 1991).

In order to understand fully the characteristic properties of these enzymes and their function in brain, it is imperative to elucidate the mechanisms underlying gene expression and regulation of these neurotransmitter biosynthetic enzymes. We have been investigating the biochemistry and molecular biology of the monoamine synthesizing enzymes for many years, and this report details some of our more recent results.

Parallel Regulation of the Catecholamine Biosynthetic Enzymes

Reserpine administered daily for 4 days (2 mg/kg) or given as one acute injection of 5 to 10 mg/kg produces a rise in the activity and amount of TH in the superior cervical ganglia and adrenal gland (Kvetnansky, 1973; Mueller et al, 1969; Joh et al, 1973). This treatment also will increase TH and DBH in central noradrenergic neurons of the locus coeruleus. As these increases depend on the integrity of preganglionic innervation, this phenomenon has been referred to as transsynaptic induction (Kvetnansky, 1973). Previous studies have demonstrated that the increase in TH activity following reserpine was observed after accumulation of TH mRNA in responsive tissues (Faucon-Biguet *et al*, 1986; Wessel *et al*, 1991). Similar results have been obtained very recently for DBH (McMahon *et al*, 1990). However, the reaction of AADC to the administration of reserpine seems to differ from the rest of the enzymes (Joh *et al*, 1973; Reis *et al*, 1975), since the activity and quantity of AADC protein do not change significantly. Since enzyme activity and protein levels are determined in dissected tissue homogenates that contain the enzyme in the relatively small volume of the catecholaminergic cell bodies and their local processes, minor changes in enzyme activity and in the number of active enzyme molecules may be difficult to detect. This is particularly pertinent with respect to AADC activity, since AADC expression is not limited to catecholaminergic or serotonergic neurons of the central nervous system (Jaeger *et al*, 1983). The neurotransmitter substances in this third group of AADC-immunopositive neurons are unknown. For this reason it has been very useful to apply immunohistochemical and *in situ* hybridization techniques to gain a semiquantitative measure of enzyme protein and enzyme mRNA in individual cells or cell clusters.

More importantly, these techniques enabled us to examine the temporal relationship between mRNA accumulation and increases in protein content for all of the catecholamine synthesizing enzymes. Thus, we have sought to verify that *de novo* TH and DBH protein production is preceded by specific mRNA accumulation following the administration of reserpine and that non-specific neuronal cell markers are not affected. Our second objective was to determine whether AADC and PNMT could be regulated in a similar parallel fashion along with TH and DBH.

For *in situ* hybridization experiments, adult Sprague-Dawley rats were anesthetized with pentobarbital and rapidly perfused transcardially with 0.9% sodium chloride, containing 0.5% nitrite and 1,000 U heparin. This was followed by rapid perfusion with ice-cold 4% formaldehyde in 0.1 M sodium phosphate buffer. Brains were immediately removed, cut into blocks, and submerged in the ice-cold fixative where they remained for one hour. The blocks were then rinsed twice with phosphate buffer and cryoprotected by storing the tissue in 30% sucrose overnight at 4°C. Tissue sections of 40μm thickness were cut on a sliding microtome and stored in 20 ml glass vials filled with 2x SCC (0.3 M sodium chloride/0.03 M sodium citrate) with 10mM DTT (dithiothreitol) at 4°C. This storage solution was then replaced with pre-hybridization buffer containing 50%

formaldehyde, 10% dextran sulfate, 2x SSC, 1x Denhardt's solution, 50mM DTT, and 0.5 mg/ml sonicated and denatured salmon sperm DNA. Prehybridization was carried out for 60 minutes at 48°C. Labeled denatured probe (1x10^7 cpm) was then added to the vial in 100 μl of the same pre-hybridization buffer. Hybridization was performed overnight at 48°C. After extensive washes in decreasing concentrations of SSC (lowest 0.1x), tissue sections were mounted onto gelatin-subbed slides, which were dehydrated through graded ethanols (70, 90, 100%) and dipped in Kodak NTB-2 after drying. After storage in light-tight boxes at 4°C for 7 to 21 days, the slides were developed in Kodak D-19, counterstained with cresyl violet and cover-slipped with Permount.

One great advantage of this particular technique is that all tissue sections can be hybridized with a given radio-labeled probe under exactly the same conditions as all floating sections are exposed to the same hybridization buffer, probe concentrations and washing conditions. Treatment or control animal sections can be easily distinguished by creating a small puncture hole in the tissue block before cutting on the sliding microtome.

Our results indicate that reserpine will trigger an increased transcription of all the catecholamine synthesizing enzymes, although this response is neither quantitatively nor temporally uniform for the different enzymes involved. As depicted in Figure 1, the grain density observed above the region of the noradrenergic cells of the locus coeruleus is much greater in the reserpinized animal for TH and DBH mRNA when compared to a control animal. AADC also shows some increase although it is more modest than those observed for the other two enzyme mRNA species.

Transcriptional Regulation: Putative Enhancer-like Sequences

Over the past several years, the complementary DNAs for all of these enzymes have been cloned and sequenced (Table 1). These cDNAs have been the essential tools utilized by many laboratories to isolate and characterize the genomic structure of several monoamine synthesizing genes in different animal species. Several putative enhancer-like sequence motifs have been identified which are thought to respond to various factors, such as hormones and second messengers. The rate-limiting enzyme in the catecholamine biosynthetic pathway is TH which has been the interest of many

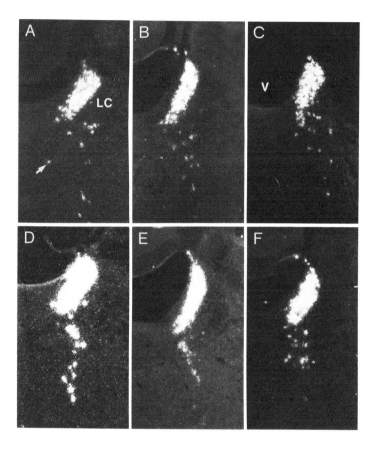

FIGURE 1. *In situ* hybridization study demonstrating mRNA increases for TH, AADC and DBH in the rat locus coeruleus (LC) 24 hours after a single injection of reserpine (10 mg/kg). The most dramatic increase is observed in TH: in panel A, individual cells are clearly visible in the LC and in the more ventrally located neurons (arrow) of the saline-injected rat; in B, the silver grain density overlying the LC cells is dramatically increased following reserpine treatment. The responses for AADC (panels C and D) and DBH (E and F) are more modest but discernable, as the boundaries between individual cells vanish in the reserpine-treated animal. Tissue sections of each group were hybridized to their respective [^{35}S]-labeled cDNA probes together as floating sections in a single vial to assure uniform conditions. (v = fourth ventricle).

TABLE 1.   Monoamine neurotransmitter enzyme cDNAs.

Enzymes	Species	mRNAs	References	
Tyrosine Hydroxylase	rat	2.1 kb	Lamouroux et al	1982
(TH)	rat		Lewis et al	1983
	rat		Grima et al	1985
	human	3 diff.	Grima et al	1987
	human	4 diff.	Kaneda et al	1987
	human	4 diff.	Kobayashi et al	1988
	bovine	2.0 kb	D'Mello et al	1988
Aromatic L-amino acid	bovine	2.3 kb	Albert et al	1987
decarboxylase (AADC)	bovine		Kang and Joh	1990
	human	2.1 kb	Ichinose et al	1989
	rat	2.0 kb	Tanaka et al	1989
Dopamine	human	3.0 kb	Lamouroux et al	1987
ß-hydroxylase	human	2 mRNAs	Kobayashi et al	1989
(DBH)	bovine	2.9 kb	Hwang et al	1988
	bovine	4.5 kb	Hwang et al	1989
	bovine	membrane	Taljanidisz et al	1989
	bovine	2.2 kb	Wu et al	1990
	bovine	2.4 kb	Lewis et al	1990
	rat	2.3 kb	Kim et al	1990
	rat	2.4 kb	McMahon et al	1990
Phenlethanolamine	bovine	1.1 kb	Baetge et al	1983
N-methyltransferase	bovine		Baetge et al	1986
(PNMT)	bovine		Batter et al	1987
	rat		Mezey	1989
	rat	(gene)*	Ross et al	1990
Tryptophan	rabbit (pineal)		Darmon et al	1986
Hydroxylase	rat (pineal)		Grenett et al	1987
(TPH)	rat (pineal)		Darmon et al	1988
	rat (CNS)		Kim et al	1991

*5' end sequences identified.

investigators.   Recently, with the isolation of genomic sequences, attention has focused on the structure and function of the upstream regulatory elements in the TH gene. We have concentrated our efforts on the analysis of those transcriptional components underlying the tissue-specific and regulated expression of TH. In the upstream region of the rat TH gene, a putative cAMP response element (CRE) is present at -45 bp (see Figure 2), and AP-1 and AP-2 consensus sequences are found at -206 and -221 bp, respectively. Fusion constructs containing 503 and 151 bp of 5' flanking sequences from the rat TH gene linked to bacterial CAT were assayed by transient transfection of mammalian cell lines. Both TH (503/+25)CAT and TH(-151/+25)CAT contained sequences sufficient to drive CAT transcription in several cell types including CV1 and Ltk⁻ fibroblasts, C6 glioma and BE(2)C, a human neuroblastoma line. CAT activity was detected in TH-expressing and non-expressing cell types which may indicate that elements determining cell type specific expression reside outside this region. Though activity was measurable in all cell types tested, variations in basal expression were observed. In CV-1 and Ltk⁻ fibroblast lines, basal levels of THCAT expression were typically lower than C6 or BE(2)C cells transfected with the same precipitate and were similar to the promoterless pOCAT controls. The C6, HeLa and BE(2)C cells supported basal expression greater than 3 times that of pOCAT.

Since TH transcription is increased in response to activation of second messenger systems (Lewis et al, 1986, 1987; Harrington et al, 1987; Evinger et al, 1987) and an element similar in sequence to cyclic AMP regulated elements (CRE) was observed at -45 bp in the TH gene, its function was tested in transient assays by treatment of the cells with forskolin. Following transfection with either of the THCAT constructs, CAT activity in all cell lines tested was responsive to elevation of intracellular cyclic AMP by forskolin and activation of protein kinase C by TPA (12-O-tetradecanoylphorbol-13-acetate). The forskolin/TPA-induced levels in CV-1 or Ltk⁻ cells were high and comparable to C6 or BE(2)C and at least 10 fold higher than pOCAT in all cell lines. Depending on the cell type and the basal expression, the relative induction ranged from 2.5 to greater than 10 fold over untreated controls. Increased CAT activity in treated cultures was measurable within 4 hours of treatment.

Activation of protein kinase C has also been shown to increase steady state levels of TH mRNA (Evinger et al, 1987; Stachowiak et al, 1989). Treatment of BE(2)C cells with the phorbol ester, TPA, at concentrations from 50 to 500 nM resulted in small but consistent increases in TH-driven CAT expression. These inductions were on the order of 1.5 to 2-fold and of lesser magnitude than the forskolin effect. In the BE(2)C cells, the effect of TPA plus forskolin was additive. Whether the activation of these two second messenger systems

a.

-503
Bam H1
GATCCAGCCCAGTGCCAGCACATATACCGACTGGGGCAGTGAAT

AGATAGTACACTTTGTTACATGGGCTGGGGGGAACATGGCCCAT

GTCCTGGAGGGGACTTTATGACAGACATCCAAAAATCCAGTGAGA

GGGCTTCTAGATTTGTCTCCAAAGGTTATAGTTCTAACATGAGCCC

TTAGGAAATCAGCATGGTTCTCCCTGTGTGCCCTGGTTTGGTTAGA

GAGCTCTAGCGGTCTCCTGTCCCACAGAATACCAGCCAGCCCCTG
                        AP-2   FSE/AP-1
CCCTACGTCGT **GCCTCGGG** CTGAGGG **TGATTCA** GAGGCAGGT
                        POU/OCT   Stu1
GCCTGTGACAGTGG **ATGCAATT** AGATCTAATGGGACGGAGGCCT

TTCTCGTCGCCCTCGCTCCATGCCCACCCCCGCCTCCCTCAGGCA

CAGCAGGCGTGGAGAGGATGCGCAGGAGGTAGGAGGTGGGGGAC
                    CRE
CCAGAGGGGCTT **TGACGTCAG** CCTGGCC **TTTAAA** GAGGGCGCC

TGCCTGGCGAGGGC TGTGGAGACAGAACTCGGGACCACC^Alu I
            +1   +27

b.

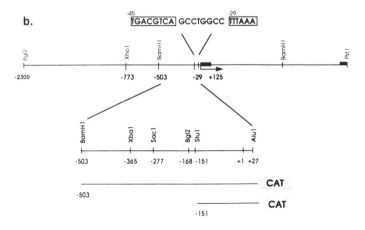

FIGURE 2. A. 1.5 kb BamHI fragment strongly hybridized to a probe from the 5' end of the full-length cDNA and was sequenced by dideoxy chain termination. The region from the 5' BamHI to the AluI site is shown here. Sequences with similarity to previously identified regulatory elements are shown in bold. B. THCAT fusion gene constructs: the BamHI/AluI fragment was inserted in front of the coding sequence of bacterial CAT and used in transient transfections. The truncated form, TH(-151/+25)CAT, was prepared by digestion with StuI. The sites of insertion were confirmed by restriction digests and sequence analysis.

converges on the same or different elements in the TH gene needs to be investigated further by fine mapping of the region.

The element or group of elements responsible for this induction appear to be within 151 bp of the transcription start site. In BE(2)C transfectants forskolin and TPA induced CAT expression. This suggests that the essential promoter elements were present in the shorter construct and cyclic AMP induction did not require sequences upstream of -151 bp.

Treatment of the BE(2)C transfectants with dexamethasone failed to enhance TH(-503/+25)CAT expression. In the same set of transfected cultures forskolin and TPA induced expression. The element(s) conferring steroid responsivity must lie outside this 503 bp of the 5' upstream region.

In summary, activators of cAMP or the protein kinase C signal transduction pathways, i.e. forskolin or the phorbol ester TPA, induce CAT expression in transfected cells that contain constructs of either - 151 bp or -503 bp length (Carroll *et al*, 1991). This suggests that those elements responsive to cAMP and protein kinase C stimulation lie within -151 bp of the transcription initiation site. The results from these experiments suggest that elements conferring cell type-specific expression may reside at least in part beyond the -503 bp region.

Tryptophan Hydroxylase in the Central Nervous System

Tryptophan hydroxylase catalyzes the first and rate-limiting step in the serotonin biosynthetic pathway. In the endocrine pineal gland (PG), TPH is the first enzyme in the melatonin biosynthetic pathway. It has been known that TPH in the pineal gland and in the dorsal raphe nucleus (DRN), which represents the largest population of serotonergic neurons in the CNS, differs in its biochemical properties (Kim *et al*, 1991; Jaquier *et al*, 1969). Moreover, by northern blot analysis, TPH cDNAs derived from both rabbit and rat PG fail to hybridize to poly A^+ RNA from the DRN, although they hybridize strongly to poly A^+ RNA from the PG (Dumas *et al*, 1989; Kim *et al*, 1991). Furthermore, by *in situ* hybridization histochemistry, TPH cDNA probes derived from the PG reacted very weakly with rat DRN (Dumas *et al*, 1989). These observations imply that the steady state level of TPH mRNA in the pineal gland is much higher than that in the DRN. In striking contrast, the enzymatic activity of TPH in the DRN is much higher than in the pineal gland. The findings suggested two possibilities: (a)

TPH from rat DRN and PG may be encoded by different mRNA transcripts; or (b) the steady state level of TPH mRNA in the serotonergic cells of DRN is extremely low, but the translational rate of the TPH mRNA in the DRN is far greater than in the pineal gland (Kim *et al*, 1991). Thus, our study was designed to determine whether TPH mRNA from the DRN is (i) identical to the gene product in the PG; (ii) a product of alternative slicing; or (iii) the product of a separate gene with a possible partial sequence identity.

Using the polymerase chain reaction, various fragments of pineal TPH cDNA were amplified. Poly A$^+$RNA prepared from rat DRN and oligonucleotide primers based on the published coding sequence of rat pineal TPH cDNA were used, and the resulting reaction products, containing the presumptive coding region of DRN-TPH, were cloned, sequenced, and compared to their counterparts from the pineal gland. The nucleotide sequence of our DRN-TPH cDNA (Kim *et al*, 1991) was identical to that of the PG-TPH cDNA sequence previously reported (Grenett *et al*, 1987; Dumas *et al*, 1989). However, our attempt to identify TPH message in the DRN by northern blot analysis using both PG-TPH and DRN-TPH cDNAs failed, although TPH message from the pineal gland was clearly detected with both cDNA probes.

In addition, *in situ* hybridization of TPH mRNA in the rat DRN using a cloned TPH cDNA fragment was carried out to examine whether the steady-state level of TPH mRNA in the DRN is detectable. A 0.8-kb cDNA fragment (corresponding to bases 475-1336 of the cDNA sequence) was labeled with [^{35}S]-dCTP by random-priming, to a specific activity of approximately 1×10^9 cpm/μg. Sections were cut through the midbrain region of adult Sprague-Dawley rats and subjected to the same floating section hybridization technique outlined above.

In contrast to the previously reported results, *in situ* hybridization in adjacent tissue sections revealed a distribution pattern of TPH message that overlapped with the area of AADC mRNA containing cell bodies in the DRN and in the median raphe (Figure 3). We observed that the number of AADC-positive cell bodies was slightly larger than that of TPH-positive neurons both in immunohistochemical and *in situ* hybridization experiments. The most likely explanation for this finding is that the difference in numbers is made up by TH/AADC-positive dopaminergic neurons in this area, as demonstrated by the localization of TH-positive cells in the midline of

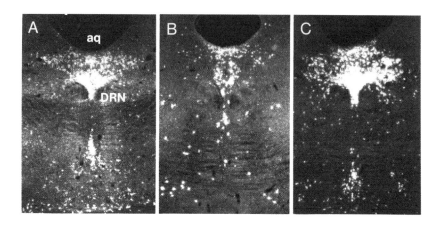

FIGURE 3. *In situ* hybridization with random-primed rat cDNA probes to (A) TPH, (B) AADC and (C) TH in adjacent sections of the rat midbrain. The region of the dorsal raphe nucleus (DRN) is visualized as a triangular cluster of neurons ventral to the aqueduct (aq). The exposure time for TPH was four weeks because of the lower abundance of TPH mRNA in this and other serotonergic groups in the central nervous system. This is in contrast to the strong hybridization signal seen after just six days of photoemulsion exposure for AADC (panel B) and TH (panel C).

the DRN. No measurable changes in TPH, TH or AADC mRNA levels were detected 24 hours after reserpine treatment in the region of the DRN. This is analogous to the lack of response observed in the substantia nigra pars compacta after reserpine injection. Further characterization of the regulatory elements in the upstream region of the TPH gene will be necessary to determine what second messenger systems are involved in the regulation of this gene.

In conclusion, we hope to gain a more complete understanding of the regulation of these important genes in the nervous system by using a combination of neurobiological and molecular techniques. Ultimately, a direct link should be established between receptor activation, second messenger signals, expression and modification of nuclear proteins that lead to the rapid induction of catecholamine genes during stress.

REFERENCES

Baetge, E.E., Kaplan, B.B., Reis, D.J., and Joh, T.H. (1981). Translation of tyrosine hydroxylase from poly(A)mRNA in pheochromocytoma cells (PC12) is enhanced by dexamethasone. *Proceedings of the National Academy of Sciences* (USA) **78**, 1269-1273.

Baetge, E.E., Moon, H.M., Kaplan, B.B., Park, D.H., Reis, J., and Joh, T.H. (1983). Identification of clones containing DNA complementary to phenylethanolamine N-methyltransferase mRNA. *Neurochemistry International* **5**, 611-617.

Baetge, E.E., Suh, Y-H., and Joh, T.H. (1986). Complete nucleotide and deduced amino acid sequence of bovine phenylethanolamine N-methyltransferase: partial homology with rat tyrosine hydroxylase. *Proceedings of the National Academy of Sciences* (USA) **83**, 5454-5458.

Batter, D.K., D'Melio, S.R., Turzai, L.T., Hughes, H.B. III, Gioio, A.E., and Kaplan, B.B. (1988). The complete nucleotide sequence and structure of the gene encoding bovine phenylethanolamine N-methyltransferase. *Journal of Neurochemistry* **19**, 367-376.

Cambi, F., Fung, B., and Chikaraishi, D.M. (1989). 5' Flanking DNA sequences direct cell-specific expression of rat tyrosine hydroxylase. *Journal of Neurochemistry* **53**, 1656-1659.

Carroll, J., Kim, K.S., Goodman, H.M., and Joh, T.H. (1991). Effects of second messenger system activation of functional expression of tyrosine hydroxylase fusion gene constructs in neuronal and non-neuronal cells. *Journal of Molecular Neuroscience*, in press.

Coker III, G.T., Vinnedge, L., and O'Malley, K. (1988). Characterization of rat and human tyrosine hydroxylase genes: functional expression of both promoters in neuronal and non-neuronal cell types. *Biochemical and Biophysical Research Communications* **157**, 1341-1347.

Darmon, M.C., Grima, B., Cash, C.D., Maitre, M., and Mallet, J. (1986). Isolation of a rat pineal gland cDNA clone homologous to tyrosine and phenylalanine hydroxylases. *FEBS Letters* **206**, 43-46.

Darmon, M.C., Guibert, B., Leviel, V., Ehlet, M., and Mallet, J. (1988). Sequence of two mRNAs encoding active rat tryptophan hydroxylase. *Journal of Neurochemistry* **51**, 312-316.

D'Mello, S.R., Weisberg, E.P., Stachowiak, M.K., Turzai, L.M., Gioio, A.E., and Kaplan, B.B. (1988). Isolation and nucleotide sequence of a cDNA clone encoding bovine adrenal tyrosine hydroxylase: comparative analysis of TH gene products. *Journal of Neuroscience Research* **19**, 440-449.

Dumas, S., Darmon, M.C., Delort, J., and Mallet, J. (1989). Differential control of tryptophan hydroxylase expression in raphe and in pineal gland: evidence for a role of translation efficiency. *Journal of Neuroscience Research* **24**, 537-547.

Evinger, M., Carroll, J., Hyman, S., Goodman, H., and Joh, T. (1987). Differential regulation of the mRNAs encoding for the catecholamine enzymes tyrosine hydroxylase and phenylethanolamine N-methyltransferase in bovine chromaffin cells. *Neuroscience Abstracts* **13**, 1086.

Grenett, H.E., Ledley, F.D., Reed, L.L., and Woo, S.L.C. (1987). Full-length cDNA for rabbit tryptophan hydroxylase: functional domains and evolution of aromatic amino acid hydroxylases. *Proceedings of the National Academy of Sciences* (USA) **84**, 5530-5534.

Faucon-Biguet, N., Buda, M., Lamouroux, A., Samolyk, D., and Mallet, J. (1986). Time course of the changes of TH mRNA in rat brain and adrenal medulla after a single injection of reserpine. *EMBO Journal* **5**, 287-291.

Grima, B., Lamouroux, A., Blanot, F., Faucon-Biguet, N., and Mallet, J. (1985). Complete coding sequence of rat tyrosine hydroxylase mRNA. *Proceedings of the National Academy of Sciences* (USA) **82**, 617-621.

Grima, B., Lamouroux, A., Boni, C., Julien, J-F, Javoy-Agid, F., and Mallet, J. (1987). A single human gene encoding multiple tyrosine hydroxylases with different predicted functional characteristics. *Nature* **326**, 707-710.

Harrington, C.A., Lewis, E.J., Krzemien, D., and Chikaraishi, D.M. (1987). Identification and cell type specificity of the tyrosine hydroxylase gene promotor. *Nucleic Acid Research* **15**, 2363-2384.

Hwang, O., Smith, J., and Joh, T.H. (1988). Characterization and sequence analysis of bovine dopamine ß-hydroxylase cDNA. *Neuroscience Abstracts* **14**, 26.

Hwang, O., Kim, K.S., and Joh, T.H. (1989). Sequence analysis of bovine dopamine β-hydroxylase cDNA. *Neuroscience Abstracts* **15**, 818.

Jaeger, C.B., Teitelman, G., Joh, T.H., Albert, V.R., Park, D.H., and Reis, D.J. (1983). Some neurons of rat central nervous system contain aromatic L-amino acid decarboxylase but not monoamines. *Science* **219**, 1233-1235.

Jaquier, E., Robinson, D.S., Lovenberg, W., and Sjoerdsma, A. (1969). Further studies on tryptophan hydroxylase in rat brainstem and beef pineal. *Biochemical Pharmacology* **18**, 1071-1081.

Jiang, W., Uht, R., and Bohn, M.C. (1989). Regulation of phenylethanolamine N-methyltransferase (PNMT) mRNA in the rat adrenal medulla by corticosterone. *International Journal of Developmental Neuroscience* **7**, 513-520.

Joh, T.H., Geghman, C., and Reis, D.J. (1973). Immunochemical demonstration of increased accumulation of tyrosine hydroxylase protein in sympathetic ganglia and adrenal medulla elicited by reserpine. *Proceedings of the National Academy of Sciences* (USA) **70**, 2767-2771.

Kaneda, N., Kobayashi, K., Ichinose, H., Kishi, F., Nakazawa, A., Kurosawa Y., Fujita, K., and Nagatsu, T. (1987). Isolation of a novel cDNA clone for human tyrosine hydroxylase: Alternative RNA splicing produces four kinds of mRNA from a single gene. *Biochemical and Biophysical Research Communications* **146**, 971-975.

Kim, K.S., Wessel, T.C., Stone, D.M., Carver, C.H., Joh, T.H., and Park, D.H. (1991). Molecular cloning and characterization of cDNA encoding tryptophan hydroxylase from rat central serotonergic neurons. *Molecular Brain Research* **9**, 277-283.

Kim, K.T., Wessel, T., Kim, K.S., Carver, C., and Joh, T.H. (1990). Characterization of two cDNA forms of bovine and rat dopamine β-hydroxylase. *Neuroscience Abstracts* **16**, 69.

Kobayashi, K., Kaneda, N., Ichinose, H., Kishi, F., Nakazawa, A., Kurosawa, Y., Fujita, K., and Nagatsu, T. (1988). Structure of the human tyrosine hydroxylase gene:

Alternative splicing from a single gene accounts for generation of four mRNA types. *Biochemistry Journal* **103**, 907-912.

Kobayashi, K., Kurosawa, Y., Fujita, K., and Nagatsu, T. (1989). Human dopamine β-hydroxylase gene: two mRNA types having different 3'-terminal regions are produced through alternative polyadenylation. *Nucleic Acid Research* **17**, 1089-1102.

Kvetnansky, R. (1973). Transsynaptic and humoral regulation of adrenal catecholamine synthesis in stress. In: E. Usdin and S. Snyder, "Frontiers in Catecholamine Research," pp. 223--229. New York: Pergamon Press.

Lamouroux, A., Faucon-Biguet, N., Samolyk, D., Privat, A., Salomon, J.C., Pujol, J-F., and Mallet, J. (1982). Identification of cDNA clones coding for rat tyrosine hydroxylase antigen. *Proceedings of the National Academy of Sciences* (USA) **79**, 3881-3885.

Lamouroux, A., Vigny, A., Faucon-Biguet, N., Carmon, M.C., Frank, R., Henry, J-P., and Mallet, J. (1987). The primary structure of human dopamine β-hydroxylase: insights into the relationship between the soluble and the membrane-bound forms of the enzyme. *EMBO Journal* **6**, 3931-3937.

Lewis, E.J., Tank, A.W., Weiner, N., and Chikaraishi, D.M. (1983). Regulation of tyrosine hydroxylase mRNA by glucocorticoid and cAMP in a rat pheochromocytoma cell line. *Journal of Biological Chemistry* **258**, 14632-14637.

Lewis, E.J., and Chikaraishi, D.M. (1986). Regulated expression of the tyrosine hydroxylase gene by epidermal growth factor. *Molecular and Cell Biology* **7**, 3332-3336.

Lewis, E.J., Harrington, C.A., and Chikaraishi, D.M. (1987). Transcriptional regulation of the tyrosine hydroxylase gene by glucocorticoid and cyclic AMP. *Proceedings of the National Academy of Sciences* (USA) **84**, 3550-3554.

Lewis, E.J., Allison, S., Fader, D., Claflin, V., and Baizer, L. (1990). Bovine dopamine β-hydroxylase cDNA. *Journal of Biological Chemistry* **265**, 1021-1028.

McMahon, A., Geertman, R., and Sabban, E.L. (1990). Rat dopamine β-hydroxylase: molecular cloning and characterization of the cDNA and regulation of the mRNA by reserpine. *Journal of Neuroscience Research* **25**, 395-404.

Mezey, E. (1989). Cloning of the rat adrenal medullary phenylethanolamine N-methyltransferase. *Nucleic Acid Research* **17**, 2125.

Mueller, R.A., Thoenen, H., and Axelrod, J. (1969). Increase in tyrosine hydroxylase activity after reserpine administration. *Journal of Pharmacology and Experimental Therapeutics* **169**, 74-79.

Reis, D.J., Joh, T.H., and Ross, R.A. (1975). Effects of reserpine on activities and amounts of tyrosine hydroxylase and dopamine β-hydroxylase in catecholamine neuronal system in rat brain. *Journal of Pharmacology and Experimental Therapeutics* **193**, 775-784.

Richard. F., Faucon-Biguet, N., Labatut, L.D., Mallet, J., and Buda, M. (1988). Modulation of tyrosine hydroxylase gene expression in rat brain and adrenals by exposure to cold. *Journal of Neuroscience Research* **20**, 32-37.

Ross, M.E., Evinger, M.J., Hyman, S.E., Carroll, J.M., Mucke, L., Comb, M., Reis, D.J., Joh, T.H., and Goodman, H.M. (1990). Identification of a functional glucocorticoid response element in the phenylethanolamine N-methyltransferase promotor using fusion genes introduced into chromaffin cells in primary culture. *Journal of Neuroscience* **10**, 520-530.

Stachowiak, M.K., Rigual, R.J., Lee, P.H.K., Viveros, O.H., and Hong, J.S. (1988).
    Regulation of tyrosine hydroxylase and phenylethanolamine N-
    methyltransferase mRNA levels in the sympathoadrenal system by the
    pituitary-adrenocortical axis. *Molecular Brain Research* **3**, 275-286.

Stachowiak, M.K., Hong, J.S., and Viveros, O.H. (1990). Coordinate and differential
    regulation of phenylethanolamine N-methyltransferase, tyrosine hydroxylase
    and proenkephalin mRNAs by neural and hormonal mechanisms in cultured
    adrenal medullary cells. *Brain Research* **510**, 277-288.

Taljanidisz, J., Stewart, L., Smith, A.J., and Klinman, J.P. (1989). Structure of bovine
    adrenal dopamine ß-monooxygenase, as deduced from cDNA and protein
    sequencing: evidence that the membrane bound form of the enzyme is
    anchored by an uncleaved signal peptide. *Biochemistry* **28**, 10054-10061.

Tank, A.W., Lewis, E.J., Chikaraishi, D.M., and Weiner, N. (1985). Elevation of RNA
    coding for tyrosine hydroxylase in rat adrenal gland by reserpine treatment
    and exposure to cold. *Journal of Neurochemistry* **45**, 1030-1033.

Wessel, T., and Joh, T.H. (1991). Parallel regulation of catecholamine synthesizing
    enzymes during the induction by the reserpine administration. Manuscript in
    preparation. (See also Neuroscience Abstracts **16**, 800).

Wu, H-J., Parmer, R.J., Koop, A.H., Rozansky, D.J., and O'Connor, D.T. (1990).
    Molecular cloning, structure, and expression of dopamine ß-hydroxylase from
    bovine adrenal medulla. *Journal of Neurochemistry* **55**, 97-105.

Stress: Neuroendocrine and Molecular Approaches
Edited by R. Kvetnansky, R. McCarty and J. Axelrod

1992 Gordon and Breach Science
Publishers S.A., New York, USA.
Photocopying permitted by license only.

TRANS-SYNAPTIC MODULATION OF RAT ADRENAL TYROSINE HYDROXYLASE GENE EXPRESSION DURING COLD STRESS

L. L. Miner, A. Baruchin and B. B. Kaplan

Molecular Neurobiology and Genetics Program, Western Psychiatric
Institute and Clinic, University of Pittsburgh School of Medicine
Pittsburgh, PA 15213 USA

INTRODUCTION

It is well established that chronic cold exposure enhances the synthesis and release of catecholamines from adrenal chromaffin cells, an adaptive response which is maintained by elevation in the activity of the catecholamine biosynthetic enzymes (for review, see Kaplan *et al*, 1987). The most pronounced alterations occur in tyrosine hydroxylase [TH; tyrosine 3-monooxygenase: L-tyrosine, tetrahydropteridine: oxygen oxidoreductase (3-hydroxylating); EC 1.14.16.2], the initial and rate-limiting enzyme in the catecholamine biosynthetic pathway. Under conditions of chronic cold exposure, alterations in TH activity are accompanied by concomitant increases in the levels of TH mRNA and protein (Stachowiak *et al*, 1985; Tank *et al*, 1985; Richard *et al*, 1988). The results of denervation experiments demonstrate that the cold-induced alterations in TH gene expression are neurally mediated, requiring an intact sympathetic innervation to the gland (Fluharty *et al*, 1985; Stachowiak *et al*, 1986).

In contrast to the effects of several other physiological stressors (e.g., glucoprivation, hypotension, and immobilization), cold exposure does not induce a rapid and robust activation of TH enzyme activity.

Rather, significant increases in TH activity are not observed for approximately 24 hours after the onset of stress with maximal increases of three- to four-fold occurring after 5-7 days (Fluharty *et al*, 1985; Baruchin *et al*, 1990). Therefore, the molecular events that occur during the initial stages of the adaptive response to cold exposure are unclear.

In this chapter, we briefly summarize our most recent findings on the effects of cold stress on adrenal TH gene expression with special emphasis placed upon the molecular alterations which occur during the first few days of cold exposure. Our findings indicate that cold stress results in surprisingly rapid elevations in adrenomedullary TH mRNA levels, and that the gradual increase in TH activity is mediated by multiple control mechanisms that include both pre- and post-translational regulation. In addition, the results of preliminary studies on the molecular mechanism(s) mediating the cold-induced, trans-synaptic modulation of adrenal TH gene transcription are presented.

MATERIAL AND METHODS

Animals and Procedures

Male Sprague-Dawley rats (200-250 gm) were used throughout these studies. Animals were housed in individual wire-mesh cages on a 12 hour light/dark cycle. In the stress paradigm employed, rats with clipped fur were placed in a 5°C chamber for the times specified with food and water provided *ad libitum*. Control animals were maintained at 22°C. In some experiments, animals that were handled and exposed to a running razor served as sham-shaved controls. All animals were sacrificed by decapitation, and adrenomedullary tissue isolated from frozen adrenal glands using a stainless steel punching needle.

RNA and Protein Analyses. Total adrenomedullary RNA was isolated from frozen tissue using a guanidine isothiocyanate/CsCl gradient procedure (Kaplan *et al*, 1979). The relative abundance of TH mRNA was estimated by RNA dot-blot hybridization and northern analysis as previously described (Stachowiak *et al*, 1986; Baruchin *et al*, 1990). In these experiments, a 350-bp cDNA fragment coding for rat PC-12 cell TH mRNA (Lewis *et al*, 1983) or a full-length bovine TH cDNA clone

(D'Mello *et al*, 1988) were used as the hybridization probes. Hybridization signals were detected by autoradiography and signals quantified by densitometric scanning (A600 nm) of the autoradiograms.

Levels of TH protein were estimated in adrenomedullary tissue extracts by immunoblot analysis (Weisberg *et al*, 1989; Baruchin *et al*, 1990). Proteins were size-fractionated on 10% polyacrylamide-SDS slab gels, and were electrophoretically transferred to nitrocellulose or Immobilon-P filters as described previously. TH-immunoreactive species were detected by polyclonal and monoclonal antisera directed against TH purified from rat PC-12 cells, and were generously provided by Drs. A.W. Tank (University of Rochester School of Medicine; Rochester, NY) and G. Kapatos (Sinai Research Institute; Detroit, MI), respectively. Levels of TH immunoreactivity were quantified by densitometric scanning at 600 nm, and relative TH abundance was determined by interpolation of the plot of absorbance peak area versus total protein from control animals. TH activity was measured by the coupled decarboxylase assay of Waymire *et al* (1971) as modified by Kapatos and Zigmond (1979). In these experiments, a saturating concentration of cofactor (3 mM, 6MPH$_4$) and an optimal pH (6.8) were employed to minimize any changes in TH activity caused by short-term activation and inhibition by catecholamines.

Nuclear protein extract preparation. Adrenal medullary tissue was homogenized in hypotonic buffer containing 10 mM Hepes (pH 8.3), 1.5 mM MgCl$_2$, 1 mM dithiothreitol (DTT), 1 mM phenylmethane-sulfonyl fluoride (PMSF), 0.1% leupeptin and aprotinin. Lysates were centrifuged at 1000g for 5 minutes at 4°C, and pellets washed once in the above buffer. Proteins were extracted from the crude nuclear pellet by incubation in 30 mM Hepes (pH 8.3), 25% glycerol, 0.3 mM EDTA, 450 mM NaCl, 12 mM MgCl$_2$,1 mM PMSF, 1 mM DTT, 0.1% leupeptin and aprotinin, and centrifuged at 12,000g for 15 minutes at 4°C. The supernatant was retained and diluted with 2 volumes of the extraction buffer lacking NaCl, and was stored at -20°C until use.

Preparation of DNA probes. A 266 bp fragment of the bovine TH promoter region (PSTI/NAEI digest, -246 to +12 relative to the transcription start site) was isolated and subcloned into the PSTI/SSTII sites of Bluescript SK⁻. For the following analyses, the 266 bp insert was digested with the restriction enzyme SAU96 yielding three DNA fragments; T1, +12 to -57, T2, -58 to -188 and T3, -189 to -246, the latter of which contains a putative AP-1 consensus sequence (TGATTCAGA). A fourth fragment, T4, -269 to -278, was obtained by PCR amplification of the region -330 to -228 in the 1.5 kb bovine TH promoter followed by a HPAII digest. These fragments were radioactively end-labelled using [α-^{32}P]dCTP in a fill-in reaction with Klenow fragment and subsequently gel purified for use in the gel mobility shift assays.

Gel mobility shift assay. DNA-protein binding reactions were conducted at 20°C for 20 minutes in binding buffer (50% glycerol, 60 mM Hepes, pH 8.3, 300 mM NaCl, 25 mM MgCl$_2$, 20 mM Tris-HCl, 5 mM EDTA, 5 mM DTT) with 1 μg poly(dI-dC), 2 μg bovine serum albumin and 0.1-0.5 ng of [^{32}P]labeled target DNA. Samples were size-fractionated on 6% polyacrylamide gels in 25 mM Tris-borate (pH 8.3), 0.5 mM EDTA at 50 mA for 1 hour. Gels were subsequently dried and exposed to Kodak X-Omat AR film overnight with a Cronex intensifying screen.

RESULTS

Time Course of the Effects of Cold Stress on Adrenomedullary TH Gene Expression

To explore more thoroughly the effects of cold exposure on adrenal TH gene expression, the temporal relationships among cold-induced alterations in TH RNA, protein levels, and enzyme activity were investigated. The results of these studies are summarized in Figure 1. In contrast to the gradual increase observed in TH enzyme activity, cold stress resulted in a remarkably rapid elevation of TH mRNA levels, as judged by RNA dot-blot hybridization assay. After 1 hour of cold exposure, levels of total cellular TH mRNA increased approximately 50-80% compared to non-stressed control values with maximal increases of 300-400% occurring within 3-6 hours. Northern analysis of polysomal RNA preparations from control and animals exposed to cold for 1 hour also revealed similar increases in levels of TH mRNA (approximately 70%). These alterations in mRNA levels were followed by a two-fold increase in TH immunoreactivity by 24 hours with maximal increases in protein levels (300-500% of control values) occurring at approximately 72-96 hours. Despite the relatively

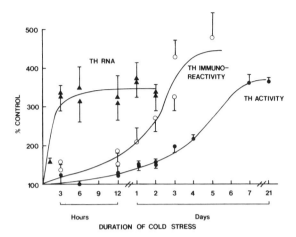

FIGURE 1. Time course of the effects of cold exposure on adrenomedullary levels of TH RNA, protein, and enzyme activity. Animals were exposed to cold (5° C) for the time periods specified, and TH mRNA abundance and immunoreactivity estimated by RNA dot-blot hybridization and immunoblot analysis as described in Material and Methods. Values are given as percent non-stressed controls ($\pm$S.E.M.) and represent the means of single samples pooled from 5-7 animals. (Data adopted from Baruchin *et al*, 1990).

large increases in TH mRNA and protein content, only modest increases in TH activity were detected prior to 72 hours of cold stress, as monitored by a standard *in vitro* activity assay. Taken together, the results of this time course study indicate that cold-induced alterations in adrenal TH activity are mediated by multiple cellular control mechanisms which appear to involve both pre- and post-translational regulation.

Effects of Cold Stress on the Interaction of Adrenomedullary Nuclear Proteins with the 'Promoter' Region of the TH Gene

During the past five years, significant insights have been gained into the molecular mechanisms controlling eukaryote gene expression. It is now well established that regulation of transcriptional activity is effected through the interaction of protein factors or complexes with cis-acting regulatory elements situated in the 5' flanking region of the gene (for review, see Maniatis *et al*, 1987; Jones *et al*, 1988). In this regard, the trans-synaptic modulation of neuronal gene expression often involves the induction of cellular immediate early genes (IEGs). The products of a number of IEGs have been characterized and encode regulatory proteins or transcription factors which could mediate the longer-term responses of neurons to multiple trans-synaptic signals (for review, see Sheng and Greenberg, 1990). To explore the trans-synaptic regulation of adrenal TH gene expression, we have initiated studies designed to define the cis-acting regulatory elements and transcription factors which mediate cold-induced alterations in adrenal TH gene expression.

In our initial experiments, attention was focused on the proximal region of the TH 'promoter', and the 5' flanking region of the bovine TH gene used to construct a series of molecular probes. The strategy employed in the subcloning of these DNA fragments and the location of several putative regulatory elements are illustrated in Figure 2. The results of a previous cross-species analysis of the nucleotide sequence of the TH 'promoter' had revealed strong sequence and positional conservation of seven regions located within 300 bp of the transcription initiation site (D'Mello et al, 1989). These elements (9-21 bp in length) manifest 80-100% sequence similarity, and are located in the same position in the 5' flanking region of the bovine, rat, and human TH genes. Six of the elements identified share significant sequence identity to the consensus sequences established for well characterized regulatory elements to include the cAMP-(CRE) and phorbol ester (TPA)-response elements, and binding sites for the AP-2 and SP-1 transcription factors (Figure 2). The high degree of sequence and positional conservation of these regions, as well as their similarity to known cis-acting regulatory elements, suggest that these regions may act to modulate the transcriptional activity of the TH gene.

To assess the effects of cold stress on the binding of nuclear proteins to the TH promoter, radiolabeled DNA fragments were mixed with crude adrenomedullary nuclear protein extracts prepared from normal and cold-stressed rats, and DNA-protein complexes were resolved from unbound DNA using a polyacrylamide gel mobility shift assay. A typical example of this experiment is shown in Figure 3. Two of the four DNA fragments studied (T1 and T3) showed stress-induced alterations in gel migration patterns. These shifts were observed within 1 hour of cold exposure and were still visible 48 hours after the onset of stress. No differences were observed in the gel mobility patterns of the T4 and T2 fragments at 1 and 3 hours of cold exposure. The cold-

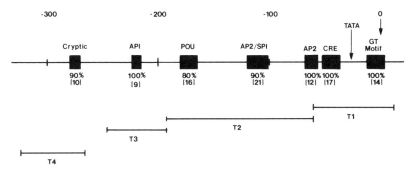

FIGURE 2.   Position of putative cis-acting regulatory elements within the proximal 5' flanking region of the bovine TH gene. Regions displaying high sequence and positional conservation in the bovine, rat, and human TH genes are shown as boxes. The cross-species percentage sequence identity among these elements and length (nt) of the elements (in parenthesis) is shown below the line. Numbers at the top of the figure indicate distance from the transcription initiation site (0). The DNA fragments (T1-T4) employed in the mobility shift analyses are shown at the bottom of the diagram.

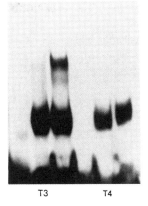

TH Promoter Fragment

FIGURE 3.   The effect of cold stress on the binding of adrenomedullary nuclear proteins to the promoter region of the TH gene.   [^{32}P]end-labelled target DNA fragments were mixed with nuclear protein extracts prepared from control animals and rats exposed to 1 hour of cold. Equal amounts of protein extract (2 µg) were used for each reaction. DNA-protein complexes were separated from unbound DNA by PAGE as described in Materials and Methods. Lanes 1, unreacted DNA fragment; 2, control nuclear protein; 3, nuclear proteins from 1 hour of cold exposure.

induced alteration in the migration pattern of the T3 fragment could be abolished by denervation of the adrenal gland, indicating that the observed effects were neurally mediated (data not shown).

Since the T3 fragment contains sequences similar to those comprising a phorbol ester-inducible promoter element (AP-1; Angel *et al*, 1987; Lee *et al*, 1987; Sassone-Corsi *et al*, 1988) that binds *c-Fos/c-Jun* complexes (Bohmann *et al*, 1987; Rauscher *et al*, 1988), an attempt was made to attenuate the stress-induced alterations in the T3 migration pattern using a synthetic double-stranded oligodeoxyribonucleotide containing the AP-1 consensus sequence. Preincubation of the protein extracts with this 22-mer (5'GATCCGGCTGACTCATCACTAG) eliminated the cold-induced changes in the T3 migration pattern. In contrast, the formation of these gel shifts was unaffected by exposure of the protein extracts to a synthetic oligonucleotide containing a generic binding site for the SP-1 transcription factor (5' GATCGATCGGGCGGGGCGATC). Preincubation of the nuclear protein extracts with antiserum prepared against the *c-Jun*/AP-1 complex (Oncogene, Manhasset, NY) also inhibited formation of the gel mobility shifts induced by cold exposure. The inhibitory effects of the anti-*c-Jun* antibodies were abolished by heat-denaturation of the antiserum. These data suggest that cold-induced alterations in adrenal TH gene expression may be mediated, at least in part, by the interaction of the *c-Fos/c-Jun* complex with the putative AP-1 binding site present in the 5' flanking region of the TH gene.

DISCUSSION

In this communication, we have explored the temporal relationships among cold-induced changes in adrenomedullary TH RNA and protein levels, and enzyme activity. The results of this time course study indicate that stress-induced alterations in TH activity are mediated by multiple regulatory mechanisms. Most notable are the remarkably rapid increases observed in TH mRNA levels and the differences in the induction profiles of TH activity and protein. For example, at no point during the initial 72 hours of cold exposure were alterations in enzyme activity concomitant with those manifest in TH immunoreactivity. These findings suggest that the newly synthesized TH molecules are catalytically "inactive," at least under *in vitro* assay conditions. Although the basis for this phenomenon is unknown, post-translational modification of the enzyme by proteolytic cleavage seems unlikely, as we have not observed size differences in the TH molecules synthesized during cold exposure. In this regard, the sensitivity of the immunoblot procedure employed in this analysis is sufficient to detect size differences in as few as seven amino acids (D'Mello *et al*, 1988). A more likely alternative explanation is the requirement of a covalent modification (e.g., phosphorylation) that mediates the assembly or conversion of the newly synthesized enzyme to a basally active configuration.

It is noteworthy that a similar phenomenon has been observed in rat pheochromocytoma cells (PC-18) treated with 8-bromocyclic AMP or dexamethasone (Tank *et al*, 1986ab). In these studies, five-fold increases in TH mRNA levels were observed as early as 6 hours after drug treatment, whereas significant increases in TH activity were not detected until 24 hours. Importantly, maximal increases in enzyme activity were not achieved until 4 days. Therefore, the pattern of TH induction reported here may not be a unique feature of the cold stress response, but could represent a general characteristic of the adaptive response of chromaffin cells to marked increases in their functional requirements.

The surprisingly rapid induction of TH mRNA levels caused by cold exposure is intriguing, and provides an excellent physiological model system in which to investigate the molecular mechanism(s) which mediate the trans-synaptic modulation of TH gene expression. The results of our preliminary gel shift mobility analysis indicate that cold stress causes rapid alterations in the binding of adrenomedullary nuclear proteins to the 5' flanking region of the TH gene. The results of the denervation experiments indicate that these alterations are neurally mediated, requiring an intact sympathetic innervation to the gland. Interestingly, one of the prominent alterations in nuclear protein binding involves a highly conserved region located 213-268 bp distal to the transcription initiation site. This region contains an element which bears striking similarity to the AP-1 consensus sequence. The cold-induced shifts in the migration pattern of this DNA fragment could in fact be abolished by competition with a synthetic oligonucleotide containing the AP-1 consensus sequence and preincubation of the nuclear protein extracts with antibodies to *c-Jun*. These findings suggest that this putative AP-1 site, first detected by comparative sequence analysis, is functional in the TH gene, and that the trans-synaptic induction of adrenomedullary TH gene expression is mediated, at least in part, by a phorbol ester-responsive element, a target site for the c-Fos/c-Jun transcription complex.

Induction of c-Fos and c-Jun occurs in response to a wide variety of stimuli in numerous cell types. In cultured neuronal cells, neurotransmitters, growth factors, and agents which stimulate voltage-gated calcium influx have been shown to be potent inducers of proto-oncogene expression (see Morgan and Curran, 1989). These gene products may act as transcriptional regulators that couple extracellular signals to target gene expression. Germane to this discussion are the recent findings of Curran and colleagues (Sonnenberg et al, 1989). In this work, evidence is provided to suggest that interaction of the c-Fos/c-Jun heterodimeric complex with the AP-1 site mediates up-regulation of rat proenkephalin gene in hippocampus in response to pentylenetetrazol-induced seizures. Similar to the situation in the proenkephalin gene, the AP-1 site in the rat TH 'promoter' is located in a region that appears essential for both basal and stimulated transcriptional activity, as judged by DNA-mediated transient cell transfection analyses (Cambi et al, 1989; Gizang-Ginsberg and Ziff, 1990). Therefore, these proto-oncogene transcription factors could play an important role in the trans-synaptic modulation of TH gene transcription and the adaptive response of adrenal chromaffin cells to physiological stress.

ACKNOWLEDGMENTS

The authors are grateful to Drs. A.W. Tank (University of Rochester School of Medicine; Rochester, NY) and G. Kapatos (Sinai Research Institute; Detroit, MI) for the generous gift of TH antisera, and to Ms. L. Phillip for assistance in the preparation of the manuscript. We express our appreciation to Dr. E.P. Weisberg for valuable technical assistance provided throughout the course of this investigation. This work was supported by USPHS grant MH29670. L. Miner and A. Baruchin were supported by ADAMHA Training Grant MH18273.

REFERENCES

Angel, P., Imagawa, M., Chiu, R., Stein, B., Imbra, R.J., Rahmsdorf, H.J., Jonat, C., Herrlich, P., and Karin, M. (1987). Phorbol ester-inducible genes contain a common cis element recognized by a TPA-modulated transacting factor. *Cell* **49**, 729-739.

Baruchin, A., Weisberg, E.P., Miner, L.L., Ennis, D., Nisenbaum, L.K., Naylor, E., Stricker, E.M., Zigmond, M.J., and Kaplan, B.B. (1990). Effects of cold exposure on rat adrenal tyrosine hydroxylase: An analysis of RNA, protein, enzyme activity, and cofactor levels. *Journal of Neurochemistry* **54**, 1769-1775.

Bohmann, D., Bos, T.J., Admon, A., Nishimura, T., Vogt, P.K., and Tjian, R. (1987). Human proto-oncogene *c-jun* encodes a DNA binding protein with structural and functional properties of transcription factor AP-1. *Science* **238**, 1386-1392.

Cambi, F., Zung, B., and Chikaraishi, D. (1989). 5' flanking DNA sequences direct cell-specific expression of rat tyrosine hydroxylase. *Journal of Neurochemistry* **53**, 1656-1659.

D'Mello, S.R., Weisberg, E.P., Stachowiak, M.K., Turzai, L.M., Gioio, A.E., and Kaplan, B.B. (1988). Isolation and nucleotide sequence of a cDNA clone encoding bovine adrenal tyrosine hydroxylase: Comparative analysis of tyrosine hydroxylase gene products. *Journal of Neuroscience Research* **19**, 440-449.

D'Mello, S.R., Turzai, L.M., Gioio, A.E., and Kaplan, B.B. (1989). Isolation and structural characterization of the bovine tyrosine hydroxylase gene. *Journal of Neuroscience Research* **23**, 31-40.

Fluharty, S.J., Snyder, G.L., Zigmond, M.J., and Stricker, E.M. (1985). Tyrosine hydroxylase activity and catecholamine biosynthesis in the adrenal medulla of rats during stress. *Journal of Pharmacology and Experimental Therapeutics* **233**, 32-38.

Gizang-Ginsberg, E., and Ziff, E.B. (1990). Nerve growth factor regulates tyrosine hydroxylase transcription through a nucleoprotein complex that contains *c-Fos*. *Genes and Development* **4**, 477-491.

Jones, N.C., Rigby, W.J., and Ziff, E.B. (1988). Trans-acting protein factors and the regulation of eukaryotic transcription: Lessons from studies on DNA tumor viruses. *Genes and Development* **2**, 267-281.

Kapatos, G., and Zigmond, M.J. (1979). Effect of haloperidol on dopamine synthesis and tyrosine hydroxylase in striatal synaptosomes. *Journal of Pharmacology and Experimental Therapeutics* **208**, 468-475.

Kaplan, B.B., Bernstein, S.L., and Gioio, A.E. (1979). An improved method for the rapid isolation of brain ribonucleic acid. *Biochemistry Journal* **183**, 181-184.

Kaplan, B.B., Stachowiak, M.K., Stricker, E.M., and Zigmond, J. (1987). Regulation of adrenal tyrosine hydroxylase gene expression during cold stress. In: S. Kaufman (Ed.), "Amino Acids in Health and Disease: New Perspectives," pp. 71-74. New York: Alan R. Liss, Inc.

Lee, W., Mitchell, P., and Tjian, R. (1987). Purified transcription factor AP-1 interacts with TPA-inducible enhancer elements. *Cell* **49**, 741-752.

Lewis, E.J., Tank, A.W., Weiner, N., and Chikaraishi, D.M. (1983). Regulation of tyrosine hydroxylase mRNA by glucocorticoids and cyclic AMP in a rat pheochromocytoma cell line. *Journal of Biological Chemistry* **258**, 14632-14637.

Maniatis, T., Goodbourn, S., and Fischer, J.A. (1987). Regulation of inducible and tissue-specific gene expression. *Science* **236**, 1237-1244.

Morgan, J.I., and Curran, T. (1989). Stimuli-transcription coupling in neurons: Role of cellular immediate-early genes. *Trends in Neuroscience* **12**, 459-462.

Rauscher F.J., Sambucelli, L.C., Curran, T., Distel, R.J., and Spiegelman, B.M. (1988). A common DNA binding site for *Fos* protein complexes and transcription factor AP-1. *Cell* **52**, 471-480.

Richard, F., Faucon-Biguet, N., Labatut, R., Rollet, D., Mallet, J., and Buda, M. (1988). Modulation of tyrosine hydroxylase gene expression in rat brain and adrenals by exposure to cold. *Journal of Neuroscience Research* **20**, 32-37.

Sassone-Corsi, P., Lamph, W.W., Kamps, M., and Verma, I.M. (1988). *Fos*-associated cellular p39 is related to nuclear transcription factor AP-1. *Cell* **54**, 553-560.

Sheng, M., and Greenberg, M.D. (1990). The regulation and function of *c-fos* and other immediate early genes in the nervous system. *Neuron* **4**, 477-485.

Sonnenberg, J.L., Rauscher, F.J., Morgan, J.I., and Curran, T. (1989). Regulation of proenkephalin by *Fos* and *Jun*. *Science* **246**, 1622-1625.

Stachowiak, M.K., Sebbane, R., Stricker, E.M., Zigmond, M.J., and Kaplan, B.B. (1985). Effect of chronic cold exposure on tyrosine hydroxylase mRNA in rat adrenal gland. *Brain Research* **359**, 356-359.

Tank, A.W., Lewis, E.J., Chikaraishi, D.M., and Weiner, N. (1985). Elevation of RNA coding for tyrosine hydroxylase in rat adrenal gland by reserpine treatment and exposure to cold. *Journal of Neurochemistry* **45**, 1030- 1033.

Tank, A.W., Curella, P., and Ham, L. (1986a). Induction of mRNA for tyrosine hydroxylase by cyclic AMP and glucocorticoids in a rat pheochromocytoma cell line: Evidence for the regulation of tyrosine hydroxylase synthesis by multiple mechanisms in cells exposed to elevated levels of both inducing agents. *Molecular Pharmacology* **30**, 497-503.

Tank, A.W., Ham, L., and Curella, P. (1986b). Induction of tyrosine hydroxylase by cyclic AMP and glucocorticoids in a rat pheochromocytoma cell line: Effect of the inducing agents alone or in combination on the enzyme levels and rate of synthesis of tyrosine hydroxylase. *Molecular Pharmacology* **30**, 486-496.

Waymire, J.C., Bjur, R., and Weiner, N. (1971). Assay of tyrosine hydroxylase by the coupled decarboxylation of dopa formed from 1-^{14}C-L-tyrosine. *Analytical Biochemistry* **43**, 588-600.

Weisberg, E.P., Baruchin, A., Stachowiak, M.K., Stricker, E.M., Zigmond, M.J., and Kaplan, B.B. (1989). Isolation of a rat adrenal cDNA clone encoding phenylethanolamine N-methyltransferase and cold-induced alterations in adrenal PNMT mRNA and protein. *Molecular Brain Research* **6**, 159-166.

Stress: Neuroendocrine and Molecular Approaches
Edited by R. Kvetnansky, R. McCarty and J. Axelrod

1992 Gordon and Breach Science
Publishers S.A., New York, USA.
Photocopying permitted by license only.

STRESSORS REGULATE mRNA LEVELS OF TYROSINE HYDROXYLASE AND DOPAMINE ß-HYDROXYLASE IN ADRENALS IN VIVO AND IN PC12 CELLS

E. L. Sabban[1], R. Kvetnansky[2,3], A. McMahon[1]
K. Fukuhara[2], E. Kilbourne[1] and I. J. Kopin[2]

[1]Department of Biochemistry and Molecular Biology, New York
Medical College, Valhalla, New York USA
[2]National Institute of Neurological Disorders and Stroke,
National Institutes of Health, Bethesda, MD USA
[3]Institute of Experimental Endocrinology, Slovak Academy of Sciences,
Bratislava, Czecho-Slovakia

INTRODUCTION

The role of the sympathoadrenal medullary system, as the primary site of response to emotional as well as physiological stress, is well recognized (reviewed in Kopin *et al*, 1989). During immobilization stress there is a large rise in urinary and plasma catecholamines and corticosterone (Kvetnansky and Mikulaj, 1970; Kvetnansky *et al*, 1977, 1978). The activities of the catecholamine biosynthetic enzymes, tyrosine hydroxylase(TH) and dopamine ß-hydroxylase (DBH) and to a lesser extent phenylethanolamine N-methyltransferase (PNMT) were found to be greatly elevated in rat adrenals following repeated immobilization stress. For example, tyrosine hydroxylase and dopamine ß-hydroxylase activities were elevated about three- to four-fold following seven repeated immobilizations (Kvetnansky *et al*, 1970ab).

Tyrosine hydroxylase is subject to short-term regulation of its enzymatic activity by phosphorylation of various sites often reflected as increased affinity of TH for its pterin cofactor (reviewed in Zigmond *et al*, 1989). Long-term regulation of TH can occur by changes in enzyme levels as a result of increased mRNA levels. Dopamine ß-hydroxylase activity, in contrast to that of TH, has not been shown to be regulated by changes in kinetic properties. However DBH mRNA levels can be changed by several effectors which alter TH activity, such as treatment with reserpine (McMahon *et al*, 1990). Therefore, we investigated whether the regulation of TH and DBH in repeated immobilizations stress is reflected by changes in their mRNA levels.

In this study, the effect of various stressors on the mRNA levels of TH and DBH was examined both *in vivo*, in adrenals following a single and repeated immobilization stress, and in PC12 pheochromocytoma cell cultures. The results indicate that various stressors can regulate TH and DBH mRNA levels; however, the mechanism and/or timing of their regulation are not identical.

METHODS

Immobilization

Male, murine pathogen-free Sprague-Dawley rats (280-320 g) were obtained from Taconic Farm (Germantown, NY). At least 7 days after arrival, animals were subjected daily to a 2 hour period of immobilization stress. Immobilization stress was done by taping all four limbs of the rats to metal mounts attached to a board as previously described (Kvetnansky and Mikulaj, 1970). This protocol was approved by the NINDS Animal Care and Use Committee.

PC12 Cultures and Treatments

The PC12 cells were grown as previously described (Sabban *et al*, 1983). PC12 cells were depolarized with 50 mM KCl or 150 μM veratridine as described in Kilbourne and Sabban (1990). Treatments with dexamethasone or cyclic AMP were in cultures containing dialyzed media as described by McMahon and Sabban (1991).

Isolation of RNA and Northern Blots

RNA from adrenals and from PC12 cells was isolated and analyzed on Northern blots as previously described (McMahon *et al*, 1990). For the immobilization studies, each Northern gel was loaded with 20 μg of RNA from individual adrenals as well as from 2-4 control adrenals or from pooled controls. The same Northern blots were hybridized with probes to rat DBH and TH mRNAs and 18S rRNA as described in McMahon *et al* (1990).

RESULTS

Effect of acute and chronically repeated immobilization stress

The levels of adrenal TH and DBH mRNA were examined after single and repeated daily immobilizations (Figure 1). A single 2 hour immobilization greatly increased TH mRNA levels about eight- fold above control levels (Figure 1A). However, a single immobilization had no significant effect on DBH mRNA levels (Figure 1B). Repeated daily immobilizations markedly increased both TH and DBH mRNA levels. TH mRNA levels were about 6-fold, and DBH mRNA about 4-fold over untreated controls following seven daily 2 hour immobilizations. Adrenals of adapted controls, which underwent six daily 2 hour immobilizations and were untreated during the 7th 2 hour time period, also showed elevated TH and DBH mRNA levels.

TH mRNA was isolated from rat adrenals immediately, or 24 hours, after varying numbers of daily repeated immobilization (Figure 2). Immediately following a single immobilization, the increase in TH mRNA levels was similar to that observed for seven repeated daily immobilizations. However, when examined one day later, TH mRNA levels had returned towards control levels. A second two hour immobilization stress re-elevated TH mRNA to a similar extent. However, this time TH mRNA levels remained elevated even when examined 24 hours later and were similar to those after six daily immobilizations. Thus, two immobilizations were sufficient to lead to sustained TH mRNA levels a day later. Examination of DBH mRNA at various times indicated that while a single immobilization did not significantly alter DBH mRNA levels, after two immobilizations DBH mRNA levels were greatly elevated and were sustained at high levels when examined one day later (McMahon, 1991).

Since a large increase in TH mRNA was observed immediately

E.L. SABBAN *et al*

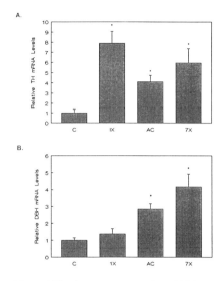

FIGURE 1.   Effect of immobilization stress on adrenal TH and DBH mRNA levels. Relative TH (A) and DBH (B) mRNA levels in adrenals of untreated controls (C); immediately following a single immobilization (1X); adapted controls, 6 daily immobilizations and untreated at time of 7th immobilization (AC). *p< 0.001 different from untreated control by Student's t-test.

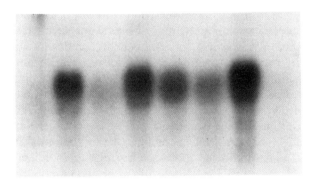

FIGURE 2.   Effect of varying number of immobilizations on TH mRNA levels. Northern blot of RNA from rat adrenals from untreated (C) or immediately following 1, 2 or 7 immobilizations for 2 hours daily (1X, 2X, 7X) or 1 day (1d) later.

after a two hour immobilization, we examined the effect of shorter immobilizations. Immobilization of rats for 30 minutes did not raise TH mRNA (Figure 3), yet immediately after a single 2 hour treatment there was an increase of 7.91 ± 1.15 fold (Mean ± SEM, n=12, p≤ 0.001 compared to controls).

Regulation of TH and DBH mRNA Levels in PC12 Cells

Immobilization stress has been shown to elevate markedly circulating glucocorticoids (Kvetnansky *et al*, 1977, 1978) as well as to require trans-synaptic mechanisms for elevation of TH and DBH activities (Kvetnansky *et al*, 1970ab). Therefore, we examined several parameters that are likely to be modified during stress on TH and DBH mRNA levels in PC12 cells, a rat pheochromocytoma cell line. Increased trans-synaptic activity would be expected to result in depolarization of the post-synaptic membrane and elevation of intracellular cAMP levels. Therefore, the effect of elevated glucocorticoids, prolonged membrane depolarization and increased cAMP levels were examined.

Depolarization of PC12 cells with either 50 mM KCl or 150 μM veratridine for 2 hours increased TH mRNA levels (Kilbourne and Sabban, 1990 and Figure 4). A 30 minute depolarization, however, was not sufficient to elevate TH mRNA. The TH mRNA remained elevated for up to 12 hours of continuous depolarization (Kilbourne and Sabban, 1990). The same depolarization conditions did not, however, induce DBH mRNA levels (Fig 4).

In contrast to the differential induction of TH mRNA, and not DBH mRNA, with membrane depolarization, treatment with dexamethasone elevated both TH and DBH mRNA levels. Following 2 days of treatment with 1 μM dexamethasone, TH mRNA levels were elevated 3.77 ± .3 fold and DBH mRNA levels 3.6 ± .34 fold (mean ± SEM) (Figure 5).

The effect of elevated cAMP on TH and DBH mRNAs in PC12 cells is shown in Figure 6. Both mRNAs were elevated following treatment of cells for 8 hours with 1 mM 8-bromo cAMP. Following 2 days of exposure to 1 mM 8-bromo cAMP, TH mRNA levels had returned to control values whereas DBH declined to below control values. Thus, elevated cAMP had a bimodal effect on DBH mRNA levels in PC12 cells.

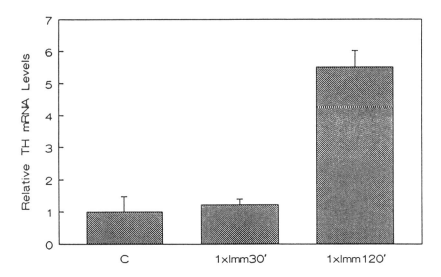

FIGURE 3. TH mRNA levels during a single immobilization. RNA was isolated from adrenals of controls (C) or immediately following a single half- hour (1/2) or 2 hour (2) immobilization, analyzed by Northern blots and hybridized to the TH cDNA.

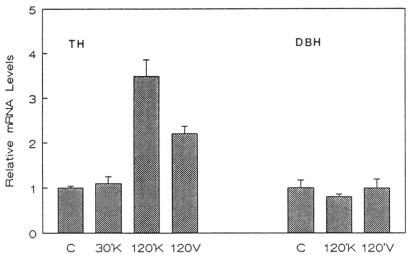

FIGURE 4. Effect of membrane depolarization of PC12 cells on TH and DBH mRNA levels. PC12 cells were treated with 50 mM KCl (K) or with 150 µM veratridine (V) for 30 or 120 minutes as described in Kilbourne and Sabban (1990) and RNA isolated and analyzed on Northern blots.

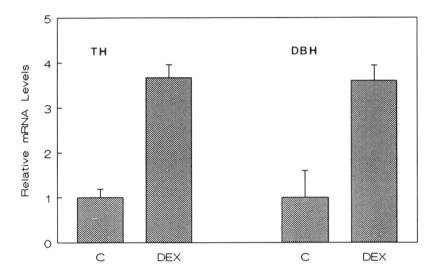

FIGURE 5. Effect of dexamethasone on TH and DBH mRNA levels in PC12 cells. PC12 cells were treated for two days with 1μM dexamethasone as described in McMahon *et al*, 1991. RNA was analyzed on Northern blots and hybridized with probes for TH and DBH mRNAs and 18S rRNA.

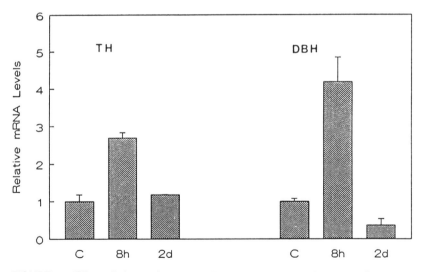

FIGURE 6. Effect of elevated cAMP on TH and DBH mRNA levels in PC12 cells. TH and DBH mRNA levels were determined in PC12 cells after 8 hours or 2 days (2d) with 1mM 8-bromo cAMP.

DISCUSSION

The results of this study indicate that immobilization stress of rats leads to alterations of not only the activities of TH and DBH but to changes in the mRNA levels for these enzymes. TH, but not DBH, mRNA levels increased following acute stress whereas both mRNAs were increased with repeated immobilization. The increases in the activities of TH and DBH in immobilization stress require trans-synaptic mechanisms (Kvetnansky *et al*, 1970). Various changes that are likely to occur with increased trans-synaptic activity were examined in PC12 cells. Membrane depolarization in PC12 cells differentially induced TH but not DBH mRNA levels. In contrast, dexamethasone and cyclic AMP induced both TH and DBH mRNA levels in PC12 cell cultures, although the latter had a bimodal effect on DBH.

Although dexamethasone increased TH and DBH mRNA levels, it is unlikely that glucocorticoids directly mediate the response to immobilization stress since trans-synaptic activity has been shown to be required for the elevation of the activities of catecholamine biosynthetic enzymes with stressful stimulation. Thus, denervation of adrenals by sectioning of the splanchnic nerve abolished most of the induction of both DBH (Kvetnansky *et al*, 1970a) and TH (Kvetnansky *et al*, 1970b) by immobilization stress, and by cold stress (Chuang *et al*, 1975).

Glucocorticoids may be more important in maintaining basal levels of adrenal TH and DBH. In hypophysectomized rats, the activities of TH, DBH and phenylethanolamine N-methyltransferase (PNMT) were decreased relative to controls and DBH and PNMT activity could be restored to near normal values by administration of ACTH or glucocorticoids (Wurtman and Axelrod, 1966; Ciaranello *et al*, 1975). TH mRNA levels increase in tissue culture in response to glucocorticoids and this increase has been shown to involve transcriptional regulation of the gene (Lewis *et al*, 1987), although a functional GRE has not been identified definitively in the TH gene.

The mechanism of regulation of TH by cyclic AMP has been shown to be mediated by the cyclic AMP regulatory element (CRE) in the 5' upstream region of the TH gene (Lewis *et al*, 1987). Characterization of the upstream region of human (Kobayashi *et al*, 1989) and rat DBH genes (McMahon *et al*, 1991) has identified one or more potential cyclic AMP regulatory elements. It remains to be determined whether any of them are functional and involved in the changes in DBH mRNA levels as a consequence of chronic intermittent stress.

An important issue is how much stress is necessary for long-term effects. The results presented here indicate that two repeated stress sessions give a similar elevation of TH and DBH mRNA as does seven repeated immobilizations. Moreover, the increase in these mRNAs were sustained when examined one day later, in contrast to the single immobilization which raised TH mRNA at least as high as the repeated stress but did not maintain this level.

The effects of membrane depolarization on TH mRNA levels have some similarities to the effects of acute immobilization stress. In fact, membrane depolarization in culture may serve as a convenient model for the cellular response to acute stress. Thus, both 2 hours of membrane depolarization in PC12 cells or immobilization stress *in vivo* had a differential effect and induced TH mRNA levels markedly without affecting DBH mRNA levels. Moreover, neither immobilization nor membrane depolarization for 30 minutes was sufficient to elevate TH mRNA, yet by 2 hours a maximal induction occurred. An additional high molecular weight TH cDNA hybridizing band, that probably represents a differentially processed or spliced form of TH, was also observed both in immobilization stress and in membrane depolarization of PC12 cells but not in any of the other treatments which induce TH (McMahon *et al*, 1991; Kilbourne *et al*, 1991). Interestingly, membrane depolarization has been proposed to serve as a model for memory, or long-term potentiation, since membrane depolarization elicits changes in gene expression of cells that subsequently modify the cell's ability to respond to other stimulations. Acute immobilization may be similar, since it "primes" the adrenal to respond differently to the second immobilization with a more sustained response.

Several studies have begun to define changes in response to membrane depolarization and specifically the induction of immediate early genes such as *c-fos* and *jun-B* (Greenberg *et al*, 1986, Bartel *et al*, 1989). The induction of *c-fos* by depolarization has been shown to involve two calcium responsive elements, one of which is indistinguishable from the cAMP response element (CRE) (Sheng *et al*, 1990).

The elevation of TH mRNA with membrane depolarization proceeds by transcriptional activation of the TH gene (Kilbourne *et al*, 1991). Work in our laboratory indicates that the cAMP responsive element is not responsible for the increased transcription of the TH gene by membrane depolarization (Kilbourne *et al*, 1991). In fact a depolarization responsive element in the TH gene has been mapped to a region containing the AP-1 site. Analysis of the DBH gene has identified many putative regulatory elements but not a classical AP-1 consensus sequence (McMahon *et al*, 1991).

It will be intriguing to determine if different transcription factors are found upstream of TH and DBH genes with repeated compared to acute immobilization, especially since the medical consequences of stress are likely to be manifestations of changes in gene expression following repeated exposure to stress.

ACKNOWLEDGEMENTS

This work was supported by grants from National Institutes of Health (NS28869), the Smokeless Tobacco Research Council and a grant in aid from the New York Heart Association. Esther L. Sabban is a recipient of an NIH Research Career Development Award NS01121.

REFERENCES

Bartel, D.P., Sheng, M., Lau, L.F., and Greenberg, M.E. (1989). Growth factors and membrane depolarization activate distinct programs of early response gene expression: dissociation of *fos* and *jun* induction. *Genes and Development* **3**, 304-313.

Ciaranello, R.D., Wooten, G.F., and Axelrod, J. (1975). Regulation of dopamine ß-hydroxylase in rat adrenal glands. *Journal of Biological Chemistry* **250**, 3204-3211.

Chuang, D., Zsila, G., and Costa, E., (1975). Turnover rate of tyrosine hydroxylase in rat adrenal medulla after exposure to cold. *Molecular Pharmacology* **11**, 784-794.

Greenberg, M.E., Ziff, E.B., and Greene, L.A. (1986). Stimulation of neuronal acetylcholine receptors induces rapid gene expression. *Science* **234**, 80-83.

Kilbourne, E.J., Osaka, H., Lewis, E., McMahon, A., and Sabban, E. (1991). The mechanism of induction of the tyrosine hydroxylase gene by membrane depolarization in PC12 cells. *Journal of Biological Chemistry*, submitted.

Kilbourne, E.J., and Sabban, E.L. (1990). Differential effect of membrane depolarization on levels of tyrosine hydroxylase and dopamine ß-hydroxylase mRNA in PC12 pheochromocytoma cells. *Molecular Brain Research* **8**, 121-127.

Kobayashi, K., Karosawa, Y., Fujita, K., and Nagatsu, T. (1989). Human dopamine ß-hydroxylase gene: two mRNA types having different 3'-terminal regions are produced through alternative polyadenylation. *Nucleic Acid Research* 17, 1089-1102.

Kopin, I.J., Eisenhofer, G., and Goldstein, D. (1989). Adrenergic response following recognition of stress. In: S. Breznitz and O. Zinder (Eds.), " Molecular Biology of Stress," pp. 123-132. Alan R. Liss: New York.

Kvetnansky, R., Gewirtz, G.P., Weise, V.K., and Kopin, I.J. (1970a). Enhanced synthesis of adrenal dopamine ß-hydroxylase induced by repeated immobilization in rats. *Molecular Pharmacology* 7, 81-86.

Kvetnansky, R., and Mikulaj, L. (1970). Adrenal and urinary catecholamines in rats during adaptation to repeated immobilization stress. *Endocrinology* 87, 738-743.

Kvetnansky, R., Sun, C.L., Lake, C.R., Thoa, N.B., Torda, T., and Kopin, I.J. (1978). Effect of handling and forced immobilization on rat plasma levels of epinephrine, norepinephrine, and dopamine ß-hydroxylase. *Endocrinology* 103, 1868-1874.

Kvetnansky, R., Sun, C.L., Torda, T., and Kopin, I.J. (1977). Plasma epinephrine and norepinephrine levels in stressed rats - effect of adrenalectomy. *Pharmacologist* 19, 241.

Kvetnansky, R., Weise, V.K., and Kopin, I.J. (1970b). Elevation of adrenal tyrosine hydroxylase and phenyethanolamine-N-methyl transferase by repeated immobilization of rats. *Endocrinology* 87, 744-749.

Lewis, E.J., Harrington, C.A., and Chikaraishi, D.M. (1987). Transcriptional regulation of the tyrosine hydroxylase gene by glucocorticoid and cyclic AMP. *Proceedings of the National Academy of Sciences* (USA) 84, 3550-3554.

McMahon, A., Geertman, R., and Sabban, E.L. (1990). Rat dopamine ß-hydroxylase: Molecular cloning and characterization of the cDNA and regulation of the mRNA by reserpine. *Journal of Neuroscience Research* 25, 395-404.

McMahon, A., and Sabban, E.L. (1991). Rat dopamine ß-hydroxylase: Regulation of mRNA levels in PC12 cells and sequence of the 5' upstream region of the gene. *Journal of Biological Chemistry*, submitted.

McMahon, A., Kvetnansky, R., Fukuhara, K., Kopin, I.J., and Sabban, E.L. (1991). Regulation of tyrosine hydroxylase and dopamine ß-hydroxylase mRNA levels in rat adrenals by a single and chronically repeated immobilization stress. *Journal of Neurochemistry*, submitted.

Sabban, E., Greene, L.A., and Goldstein, M. (1983). Mechanism of biosynthesis of soluble and membrane-bound forms of dopamine ß-hydroxylase in PC12 pheochromocytoma cells. *Journal of Biological Chemistry* 258, 7812-7818.

Sheng, M., McFadden, G., and Greenberg, M.E. (1990). Membrane depolarization and calcium induce *c-fos* transcription via phosphorylation of transcription factor CREB. *Neuron* 4, 571-582.

Wurtman, R.J., and Axelrod, J. (1966). Control of enzymatic synthesis of adrenaline in the adrenal medulla by adrenal cortical steroids. *Journal of Biological Chemistry* 241, 2301-2305.

Zigmond, R.E., Schwarzchild, M.A., and Rittenhouse, A.R. (1989). Acute regulation of tyrosine hydroxylase by nerve activity and by neurotransmitters via phosphorylation. *Annual Review of Neuroscience* 12, 415-461.

Stress: Neuroendocrine and Molecular Approaches
Edited by R. Kvetnansky, R. McCarty and J. Axelrod

1992 Gordon and Breach Science
Publishers S.A., New York, USA.
Photocopying permitted by license only.

DIFFERENTIAL REGULATION OF PHENYLETHANOLAMINE N-METHYLTRANSFERASE (PNMT) IN ADRENAL MEDULLA AND MEDULLA OBLONGATA BY GLUCOCORTICOIDS

M. C. Bohn

Department of Neurobiology and Anatomy
University of Rochester School of Medicine
Rochester, New York USA

INTRODUCTION

Epinephrine (EPI) plays a crucial role in regulating both the peripheral and central autonomic responses to stress. The release of EPI from the adrenal medulla and its effect on the cardiovascular system are well recognized. In the brain, EPI-synthesizing neurons are located in the C1-C3 cell groups of the medulla oblongata (Howe *et al*, 1980; Hokfelt, 1988; Kalia *et al*, 1985ab; Ruggiero *et al*, 1985) and project to regions in the hypothalamus and spinal cord that are involved in cardiovascular regulation and neurohormonal control of stress responses (Hokfelt *et al*, 1974, 1984; Koslow and Schlumpf, 1974; Kvetnansky *et al*, 1978; Ross *et al*, 1981, 1984; Goodchild *et al*, 1984; Granata *et al*, 1985; Cunningham *et al*, 1990). Rostrally, these adrenergic brainstem neurons project to the parvicellular regions of the hypothalamic paraventricular nucleus where they synapse on neurons that synthesize corticotropin releasing factor (CRF; Hokfelt *et al*, 1974, 1984; Swanson *et al*, 1981; Liposits *et al*, 1986; Sawchenko and Bohn, 1989). The C1 neurons also send a major projection to the locus coeruleus (Pieribone *et al*, 1988; Milner *et al*, 1989; Pieribone and Aston-Jones, 1991). Caudally, these adrenergic brainstem neurons project to the intermediolateral and central autonomic nuclei of thoracic spinal cord where they synapse on sympathetic preganglionic neurons, including those that innervate the adrenal medulla (Tucker *et al*, 1987; Bernstein-Goral and Bohn, 1988, 1989; Milner *et al*, 1988; Wesselingh *et al*, 1989; Minson *et al*, 1990).

In addition to the anatomical relationships that EPI-synthesizing neurons have with components of the hypothalamic-pituitary-adrenal axis, the enzyme that is required for EPI synthesis in the adrenal, phenylethanolamine N-methyltransferase (PNMT; E.C. 1.14.17.1), requires glucocorticoids produced by the adrenal cortex to maintain its activity (Wurtman and Axelrod, 1966). Interestingly, as will be reviewed in this chapter, PNMT in the brain is not regulated by glucocorticoids even though PNMT-immunoreactive (IR) neurons in the brain express glucocorticoid receptors. In addition to the differential regulation of PNMT in brain and adrenal by glucocorticoids, there are differences in the early development of PNMT in brain and periphery (Bohn, 1986).

PNMT EXPRESSION AND DEVELOPMENT

The concept that glucocorticoids trigger the expression of PNMT stems from two observations. In adult rats, PNMT expression in cells derived from the neural crest is confined to the adrenal medulla, a tissue exposed to high local levels of glucocorticoids. Secondly, PNMT in adult rat adrenal requires glucocorticoids to maintain its activity (Wurtman and Axelrod, 1966). However, there are a number of observations suggesting that the expression of PNMT requires factors other than simply high levels of glucocorticoids. First, not all cells in the adrenal medulla express PNMT, but remain noradrenergic, even though they are also exposed to high levels of glucocorticoids (Hillarp and Hokfelt, 1965). Secondly, in fetal and neonatal rats, PNMT is expressed outside the adrenal in cell derivatives of neural crest that are not exposed to high levels of glucocorticoids present in the adrenal, i.e. extra-adrenal chromaffin tissue and sympathetic ganglia with small, intensely fluorescent (SIF) cells (Ciaranello *et al*, 1973; Gianutsos and Moore, 1977; Bohn *et al*, 1982; Eranko *et al*, 1982; Paivarinta *et al*, 1989). PNMT expression later disappears from these regions, presumably as a consequence of cell death (Lempinen *et al*, 1964; Ciaranello *et al*, 1973; Bohn *et al*, 1982). Thirdly, PNMT is also expressed in brain and is expressed on embryonic day 14 prior to the developmental increase in glucocorticoid levels in the fetus (Figure 1; Milkovic *et al*, 1985; Bohn *et al*, 1986). Although the fetus is exposed to glucocorticoids of maternal origin, this can not account for the expression of PNMT in specific subpopulations of catecholaminergic neurons.

Experimental evidence also supports the hypothesis that factors in addition to glucocorticoids are required for PNMT expression. The initial expression of PNMT in adrenal occurs between E16-E17, later than PNMT expression in the brain, but still earlier than the surge in adrenal glucocorticoids (Teitelman *et al*, 1979; Bohn *et al*, 1981; Ehrlich *et al*, 1989). Furthermore, treatment of embryos with glucocorticoids does not cause precocious expression of PNMT (Teitelman *et al*, 1979; Bohn *et al*, 1981). Conversely, an experimental decrease in glucocorticoid levels in the fetus does not prevent adrenal PNMT expression (Bohn *et al*, 1981). PNMT is also expressed in explants of embryonic adrenal cultured in the absence of exogenous glucocorticoids (Teitelman *et al*, 1982). Also, functional glucocorticoid receptors and micromolar levels of corticosterone are present in rat adrenal on E14.5, two to three days prior to PNMT expression (Anderson and Michelsohn, 1989; Anderson, D.J., personal communication). Studies by Anderson and Michelsohn of early E14 sympatho-adrenal precursor cells grown in culture suggest that glucocorticoids are required for PNMT expression, but not for the precise timing of this event, i.e., some other process needs to occur before PNMT expression becomes glucocorticoid responsive. (Anderson, personal communication).

PNMT expression in bulbospinal neurons may play a role in early development of the autonomic nervous system. Initial expression of PNMT which occurs as early as E14 in soma located in the wall of the brainstem is followed by rapid outgrowth of PNMT-IR fibers that have reached thoracic levels of spinal cord by E17 (Bernstein-Goral and Bohn, 1988). The density of PNMT-IR fibers in the intermediolateral nucleus increases dramatically during the neonatal period so that by the second postnatal week this region is hyperinnervated. At older ages, the PNMT-IR fibers recede to the fairly sparse innervation pattern observed in adult spinal cord (Figure 2; Bernstein-Goral and Bohn, 1988). Interestingly, this period of

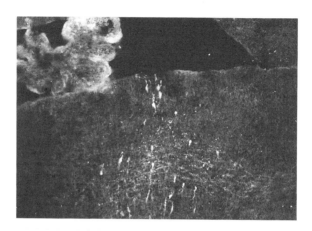

FIGURE 1. Initial expression of PNMT in a longitudinal section of E15 rat medulla oblongata. Note PNMT-IR fluorescent cells that appear to be migrating ventrally from the ventricular zone. Also note PNMT-IR beaded fibers running longitudinally.

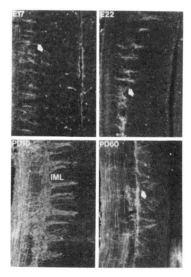

FIGURE 2. Darkfield micrograph of PNMT-IR in horizontal spinal cord sections through the n. intermediolateralis pars principalis (IML) in embryonic (E16 and E22), postnatal day (PD)10, and adult (PD60) rat. Reprinted from Bernstein-Goral and Bohn (1988) with permission.

PNMT-IR fiber exuberance, which is accompanied by a peak in spinal cord EPI levels (Bernstein-Goral and Bohn, 1988), coincides with the period when the adrenal is undergoing maturation of neurogenic mechanisms involved in stress-induced catecholamine release (Seidler and Slotkin, 1985, 1986; Kirby and McCarty, 1987; LeGamma and Adler, 1988). Although it is known that descending spinal pathways are required for normal development of sympathetic pre- and post-ganglionic neurons (Hamill *et al*, 1977), a specific role for EPI in this process has not been elucidated.

GLUCOCORTICOID REGULATION OF PNMT IN THE PERIPHERY

Early studies showed that PNMT activity in the adrenal declines following the removal of the source of adrenocorticotropin (ACTH) by hypophysectomy (HPX) in adult rats and is restored by glucocorticoid replacement (Wurtman and Axelrod, 1966). In addition, glucocorticoid treatment of newborn rats increases the number of SIF cells in sympathetic ganglia and prolongs the disappearance of extra-adrenal chromaffin tissue, as well as increases PNMT activity (Figure 3;

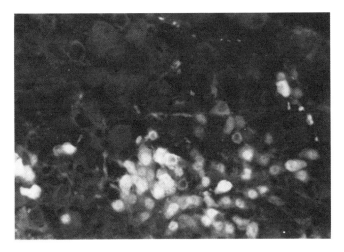

FIGURE 3.   Immunofluorescence of PNMT in small cells in SCG of 7-day old rat treated from birth with dexamethasone.   Large principal neurons are devoid of staining.   Note positively stained fibers emanating from some PNMT-positive cells. Reprinted from Bohn *et al* with permission.

Eranko *et al*, 1972; Koslow *et al*, 1975; Moore and Phillipson, 1975; Luizzi *et al*, 1977; Bohn *et al*, 1982; Paivarinta *et al*, 1985). The studies of PNMT in sympathetic ganglia are unique in that the glucocorticoid involvement in PNMT regulation in derivatives of the neural crest can be studied in the absence of high local levels of glucocorticoids as are present in the adrenal. PNMT expression in the superior cervical ganglion (SCG) coincides with PNMT expression in the adrenal, again suggesting that the milieu of the adrenal does not underlie specification of the adrenergic phenotype (Bohn *et al*, 1982). While PNMT expression apparently does not require high levels of glucocorticoids, once PNMT is expressed it can be experimentally manipulated by altering glucocorticoid levels. Quantitative studies using ^{3}H-thymidine labeling have demonstrated that the increase in SIF cells that occurs in the SCG following treatment of newborn rats with glucocorticoids is not due to an increase in the rate of cell proliferation (Bohn, 1987). Consequently, the increased number of SIF cells may result from a diminished degree of cell death, or from an increase in catecholamine phenotypic markers in precursor cells. There are experimental data supporting both possibilities and both effects may underlie the SIF cell increase. As mentioned above, glucocorticoids prolong the disappearance of extra-adrenal chromaffin tissue in neonatal rat (Lempenin, 1964; Ciaranello *et al*, 1973). In addition, glucocorticoids are required *in vitro* for differentiation of chromaffin and SIF cells (Doupe *et al*, 1985ab; Anderson and Axel, 1986). They also increase PNMT in cultured chromaffin cells (Grothe *et al*, 1985; Kelner and Pollard, 1985). Secondly, the increase in PNMT activity in explant cultures of SCG elicited by glucocorticoids requires RNA and protein synthesis (Bohn *et al*, 1984). This latter observation suggests that glucocorticoids may regulate PNMT at the mRNA level.

Regulation of PNMT gene expression by glucocorticoids is also supported by Northern blot analysis. At 2 weeks following HPX of adult rats, the levels of mRNA coding for PNMT fall to 18% of sham levels (Table 1; Jiang *et al*, 1989). Levels of PNMT mRNA are increased 2.5-fold 24 hours after implanting HPX rats with a subcutaneous pellet of corticosterone; however, they are not restored to sham levels even after a week of corticosterone treatment. Partial restoration of adrenal PNMT activity and PNMT mRNA levels following ACTH treatment of HPX rats has also been reported (Evinger and Joh, 1986). In contrast, full restoration of adrenal PNMT mRNA levels has been reported in a paradigm using dexamethasone

Table 1. Effects of hypophysectomy and corticosterone (cort.) replacement on PNMT activity and mRNA levels.

Treatment	Plasma Cort (ng/ml)	PNMT activity	PNMTmRNA
SHAM	20 (5)	11.90 (1.75)	62 (8)
HPX	not detectable	0.10 (0.06)*	11 (2)*
1 day cort.	274 (25)* +	0.53 (0.21)*	26 (5)* +
3 day cort.	211 (43)* +	1.98 (0.47)* +	15 (1)*
7 day cort.	107 (35)* +	1.73 (0.53)* +	22 (2)* +

Values are expressed as means (S.E.M.). PNMT activity is expressed as pmoles of N-methylphenylethanolamine formed/adrenal/hr. PNMT mRNA levels are expressed in arbitrary density units corrected for the amount of RNA loaded per lane. All rats were killed at 14 days after HPX or sham surgery and pellets were implanted 1, 3 or 7 days prior to killing. *P<0.05; differs from shams in Mann-Whitney test; +P<0.05; differs from HPX in Mann-Whitney test. Reprinted from Jiang et al (1989) with permission.

treatment of HPX rats at 2 weeks (Stachowiak et al, 1988). These discrepancies suggest that the effects of HPX and glucocorticoid treatment on adrenal chromaffin cells requires more study. Studies need to be done to determine the chronic effect of HPX on chromaffin cell number since prolonged HPX may result in adrenal chromaffin cell death so that full restoration in the adrenal as a whole is not possible. It would also be interesting to determine the level of mRNA in individual cells following HPX and glucocorticoid replacement. Furthermore, dexamethasone may elicit its effects through a combination of membrane and intracellular actions since it has been reported that dexamethasone has presynaptic actions and effects that are not always mimicked by corticosterone (Otten and Thoenen, 1976; Sze and Hedrick, 1983). In this regard, several studies have shown that PNMT activity and PNMT mRNA levels are also affected by splanchnic nerve activity and stress (Thoenen et al, 1970; Ciaranello and Black, 1971; Kaplan et al, 1988; Weisberg et al, 1989).

Glucocorticoid regulation of PNMT mRNA probably occurs at the gene level through a mechanism in which the glucocorticoid receptor complex interacts with the PNMT promoter to increase the rate of transcription (for a review of glucocorticoid receptor action, see Muller and Renkawitz, 1991). Cloning studies have shown that the bovine PNMT gene contains multiple consensus sequences that recognize glucocorticoid receptors (GRE; Batter *et al*, 1988). A GRE has also been reported in the 5' region of the rat PNMT gene which confers dexamethasone inducibility on a reporter gene (Ross *et al*, 1990). Taken together, the experimental evidence suggests that, in derivatives of the neural crest, a cell lineage is determined in which PNMT can be expressed independently of glucocorticoid action and independently ot cell migration into the milieu of the adrenal. Following the initial expression of the PNMT gene, the promoter region becomes vulnerable to glucocorticoid receptor action. Thereafter, both during development and in the adult, full expression of the PNMT gene requires physiological levels of glucocorticoids. Interestingly, glucocorticoid treatment can induce PNMT above basal levels in immature rat adrenal, but not in adult adrenal. The mechanism underlying this lack of further induction by glucocorticoids in adults is not clear since PNMT can be increased above basal levels in adults by other means.

GLUCOCORTICOID EFFECTS ON BRAIN PNMT

Since the majority of PNMT-IR neurons in brain stem also express glucocorticoid receptors (Harfstrand *et al*, 1986; Sawchenko and Bohn, 1989), one might assume that PNMT in the brain is also regulated by glucocorticoids. However, most experimental evidence suggests that this is not the case. PNMT activity is unaffected by HPX in adult rats (Bohn *et al*, 1986). Furthermore, direct removal of endogenous glucocorticoids following adrenalectomy leaves brain PNMT activity unaffected for up to 5 weeks (Bohn *et al*, 1986). Adrenalectomy of neonatal rats on postnatal day 7 also has no affect on brain PNMT activity on postnatal day 30 (Bohn *et al*, 1986). Conversely, treatment of neonatal rats with various synthetic and natural steroids does not increase brain PNMT activity, except at high doses of dexamethasone (Moore and Phillipson, 1975; Bohn *et al*, 1986). During fetal life, PNMT in fetal brain also is unaffected by markedly increased levels of natural glucocorticoids in the dam. Although treatment of pregnant rats with dexamethasone, specifically on gestational days 18.5-21.5, produces a small increase in PNMT activity in fetal brain, studies in explant cultures of embryonic medulla oblongata show that the effect of dexamethasone is not reproduced by other glucocorticoids and does not involve intracellular glucocorticoid receptors (Bohn *et al*, 1986, 1987).

The lack of glucocorticoid regulation of central PNMT is intriguing and the mechanisms underlying the unresponsiveness of the gene to glucocorticoids probably can not be accounted for by assuming that the gene is already maximally induced. Not only do PNMT levels increase in brain during development, but stress has been shown to increase PNMT activity and EPI levels in brain (Kvetnansky et al, 1978; Turner et al, 1978). Consequently, it appears that the promoter region containing the GRE is blocked in cells that differentiate from the neural tube, but not the neural crest.

ACKNOWLEDGEMENTS

The author is grateful for the support of this work from the National Institutes of Health (NS20832) and the Dysautonomia Foundation. This review was written while the author was on leave and working at the National Science Foundation.

REFERENCES

Anderson, D.J., and Axel, R. (1986). A bipotential neuroendocrine precursor whose choice of cell fate is determined by NGF and glucocorticoids. *Cell* **74**, 1079-1090.

Anderson, D.J., and Michelson, A. (1989). Role of glucocorticoids in the chromaffin-neuron developmental decision. *International Journal of Developmental Neuroscience* **7**, 475-487.

Batter, D.K, D'Mello, S.R., Turzai, L.M., Hughes, H.B., Gioio, A.E., and Kaplan, B.B. Complete nucleotide sequence and structure of the gene encoding bovine adrenal phenylethanolamine N-methyltransferase. *Journal of Neuroscience Research* **19**, 367-376.

Bernstein-Goral, H., and Bohn M.C. (1988). Ontogeny of adrenergic fibers in rat spinal cord in relationship to adrenal preganglionic neurons. *Journal of Neuroscience Research* **21**, 333-351.

Bernstein-Goral, H., and Bohn, M.C. (1989). Phenylethanolamine N-methyltransferase immunoreactive terminals synapse on adrenal preganglionic neurons in the rat spinal cord. *Neuroscience* **32**, 521-537.

Bohn, M.C. (1986). Expression and development of phenylethanolamine N-methyltransferase (PNMT): Role of glucocorticoids. In: P. Panula, H. Paivarinta and S. Soinila (Eds.), "Neurohistochemistry: Modern Methods and Applications", pp. 245-271. New York: A.R.Liss, Inc.

Bohn, M.C. (1987). Division of small intensely fluorescent cells in neonatal rat superior cervical ganglion is inhibited by glucocorticoids. *Neuroscience* **20**, 885-894.

Bohn, M.C., Bloom, E., Goldstein, M., and Black, I.B. (1984). Glucocorticoid regulation of phenylethanolamine N-methyltransferase (PNMT) in organ culture of superior cervical ganglia. *Developmental Biology* **105**, 130-136.

Bohn, M.C., Goldstein, M., and Black, I.B. (1981). Role of glucocorticoids in expression of the adrenergic phenotype in rat embryonic adrenal gland. *Developmental Biology* **82**, 1-10.

Bohn, M.C., Goldstein, M., and Black, I.B. (1982). Expression of phenylethanolamine N-methyltransferase in rat sympathetic ganglia and extra-adrenal chromaffin tissue. *Developmental Biology* **89**, 299-308.

Bohn, M.C., Goldstein, M., and Black, I.B. (1986). Expression and development of phenylethanolamine N-methyltransferase (PNMT) in rat brain stem: Studies with glucocorticoids. *Developmental Biology* **114**, 180-193.

Bohn, M.C., Dreyfus, C.F., Friedman, W.J., and Markey, K.A. (1987). Glucocorticoid effects on phenylethanolamine N-methyltransferase in explants of embryonic rat medulla oblongata. *Developmental Brain Research* **37**, 257-266.

Ciaranello, R.D., Jacobowitz, D., and Axelrod, J. (1973). Effect of dexamethasone on phenylethanolamine N-methyltransferase in chromaffin tissue of the neonatal rat. *Journal of Neurochemistry* **20**, 799-805.

Ciaranello, R.D., and Black, I.B. (1971). Kinetics of the glucocorticoid-mediated induction of phenylethanolamine N-methyltransferase in the hypophysectomized rat. *Biochemical Pharmacology* **20**, 3529-3532.

Cunningham, E.T., Jr., Bohn, M.C., and Sawchenko, P.E. (1990). Organization of adrenergic inputs to the paraventricular and supraoptic nuclei of the hypothalamus in the rat. *Journal of Comparative Neurology* **292**, 651-667.

Doupe, A.J., Landis, S.C., and Patterson, P.H. (1985). Environmental influences on the development of neural crest derivatives: Glucocorticoids, growth factors and chromaffin cell plasticity. *Journal of Neuroscience* **5**, 2119-2142.

Doupe, A.J., Patterson, P.H., and Landis, S.C. (1985). Small intensely fluorescent cells in culture: role of glucocorticoids and growth factors in their development and interconversions with other neural crest derivatives. *Journal of Neuroscience* **5**, 2143-2160.

Ehrlick, M.C., Evinger, M.J., Joh, T.H., and Teitelman, G. (1989). Do glucocorticoids induce adrenergic differentiation in adrenal cells of neural crest origin? *Developmental Brain Research* **50**, 129-137.

Evinger, M., and Joh, T.H. (1986). Dexamethasone stimulates synthesis of phenylethanolamine N-methyltransferase (PNMT) mRNA. *Neuroscience Abstracts* **12**, 367.

Eranko, O., Eranko, L., Hill, C.E., and Burnstock, G. (1972). Hydrocortisone-induced increase in the number of small intensely fluorescent cells and their histochemically demonstrable catecholamine content in cultures of sympathetic ganglia of the newborn rat. *Histochemistry Journal* **4**, 49-58.

Eranko, O., Pickel, V.M., Harkonen, M., Eranko, L., Joh, T.H., and Reis, D.J. (1982). Effect of hydrocortisone on catecholamines and the enzymes synthesizing them in the developing sympathetic ganglion. *Histochemistry Journal* **14**, 461-478.

Foster, G.A., Shultzberg, M., Goldstein, M., and Hokfelt, T. (1985). Ontogeny of phenylethanolamine N-methyltransferase- and tyrosine hydroxylase-like immunoreactivity in presumptive adrenaline neurons of the foetal rat central nervous system. *Journal of Comparative Neurology* **236**, 348-381.

Gianutsos, G., and Moore, K.E. (1977). Effects of pre- or postnatal dexamethasone, adrenocorticotropic hormone and environmental stress on phenylethanolamine N-methyltransferase activity and catecholamines in sympathetic ganglia of neonatal rats. *Journal of Neurochemistry* **28**, 935-940.

Goodchild, A.K., Moon, E.A., Dampney, R.A.L., and Howe, P.R.C. (1984). Evidence that adrenaline neurons in the rostral ventrolateral medulla have a vasopressor function. *Neuroscience Letters* **45**, 267-272.

Granata, A.R., Ruggiero, D.A, Park, D.H., Joh, T.H., and Reis, D.J. (1985). Brain stem area with C1 epinephrine neurons mediates baroreflex vasodepressor responses. *American Journal of Physiology* **248**, H547-H567.

Grothe, C., Hofmann, D.-D., Verhofstad, A.A.J., and Unsicker, K. (1985). Nerve growth factor and dexamethasone specify the catecholaminergic phenotype of cultured rat chromaffin cells: Dependence on developmental stage. *Developmental Brain Research* **21**, 125-132.

Hamill, R.W., Bloom, E.M., and Black, I.B. (1977). The effect of spinal transection on the development of cholinergic and adrenergic sympathetic neurons. *Brain Research* **134**, 269-278.

Harfstrand, A., Fuxe, K., Cintra, A., Agnati, L.F., Zini, I., Wikstrom, A.-C., Okret, S., Yu, Z.-Yl., Goldstein, M., Steinbusch, H., Verhofstad, A., and Gustafsson, J.-A. (1986). Glucocorticoid receptor immunoreactivity in monoaminergic neurons of rat brain. *Proceedings of the National Academy of Sciences* (USA) **83**, 9779-9783.

Hillarp, N.-A., and Hokfelt, T. (1965). Evidence of adrenaline and noradrenaline in separate adrenal medullary cells. *Acta Physiologica Scandanavica* **30**, 55-68.

Hokfelt, T., Fuxe, K., Goldstein, M., and Johansson, O. (1974). Immunohistochemical evidence for the existence of adrenaline neurons in the rat brain. *Brain Research* **66**, 235-251.

Hokfelt, T., Foster, G.A., Johansson, O., Schultzberg, M., Holets, V., Ju., G., Skagerberg., G., and Palkovits, M. (1988). Central phenylethanolamine N-methyltransferase-immunoreactive neurons: Distribution, projections, fine structure, ontogeny, and coexisting peptides. In: J. Stolk, D. U'Prichard, and K. Fuxe (Eds.), "Epinephrine in the Central Nervous System", pp. 10-45. New York: Oxford University Press.

Hokfelt, T., Johansson, O., and Goldstein, M. (1984). Central catecholamine neurons as revealed by immunohistochemistry with special reference to adrenaline neurons. In: A. Bjorklund and T. Hokfelt (Eds.), "Handbook of Chemical Neuroanatomy, Vol.2: Classical Transmitters in the CNS, Part I," pp. 157-276. Amsterdam: Elsevier.

Howe, P.R.C., Costa, M., Furness, J.B., and Chalmers, J.P. (1980). Simultaneous demonstration of phenylethanolamine N-methyltransferase immunofluorescent and catecholamine fluorescent nerve cell bodies in the rat medulla oblongata. *Neuroscience* **5**, 2229-2238.

Jiang, W., Uht, R., and Bohn, M.C. (1989). Regulation of phenylethanolamine N-methyltransferase (PNMT) mRNA in the rat adrenal medulla by corticosterone. *International Journal of Developmental Neuroscience* 7, 513-520.

Kalia, M., Fuxe, K., and Goldstein, M. (1985). Rat medulla oblongata. III. Adrenergic (C1 & C2) neurons, nerve fibers and presumptive terminal processes. *Journal of Comparative Neurology* 233, 333-349.

Kalia, M., Woodward, D.J., Smith, D.J., and Fuxe, K. (1985). Rat medulla oblongata IV. Topographical distribution of catecholamine neurons with quantitative three-dimensional computer reconstruction. *Journal of Comparative Neurology* 233, 350-364.

Kaplan, B.B., Stachowiak, M.R., Weisberg, E.P., Baruchin, A., Stricker, E.M., and Zigmond, M.J. (1989). Stress induced alterations in adrenal tyrosine hydroxylase and phenylethanolamine N-methyltransferase gene expression. In: G.R. Van Loon, R. Kvetnansky, R. McCarty and J. Axelrod (Eds.), "Stress: Neurochemical and Humoral Mechanisms." New York: Gordon and Breach.

Kelner, K., and Pollard, H.B. (1985). Glucocorticoid receptors and regulation of phenylethanolamine N-methyltransferase activity in cultured chromaffin cells. *Journal of Neuroscience* 5, 2161-2168.

Kvetnansky, R., Kopin, I.J., and Saavedra, J.M. (1978). Changes in epinephrine in individual hypothalamic nuclei after immobilization stress. *Brain Research* 155, 387-390.

Kirby, R.F., and McCarty, R. (1987). Ontogeny of functional sympathetic innervation to the heart and adrenal medulla in the preweanling rat. *Journal of the Autonomic Nervous System* 19, 67-75.

Koslow, S.H., and Schlumpf, M. (1974). Quantitation of adrenaline in rat brain nuclei and areas by mass fragmentography. *Nature* 251, 530-531.

Koslow, S.H., Bjegovic, M., and Costa, E. (1975). Catecholamines in sympathetic ganglia of rat: Effects of dexamethasone and reserpine. *Journal of Neurochemistry* 24, 277-283.

LaGamma, E.F., and Adler, J.H. (1988). Development of transynaptic regulation of adrenal enkephalin. *Developmental Brain Research* 39, 177-182.

Lempinen, M. (1964). Extra-adrenal chromaffin tissue of the rat and the effect of cortical hormones on it. *Acta Physiologica Scandanavica* 62, (Supplement 232), 1-91.

Liposits, Zs., Phelix, D., and Paull, W.K. (1986). Adrenergic innervation of corticotropin releasing factor (CRF)- synthesizing neurons in the hypothalamic paraventricular nucleus of the rat. *Histochemistry* 84, 201-205.

Luizzi, A., Foppen, F.H., Saavedra, J.M., Jacobowitz, D., and Kopin, I.J. (1977). Effect of NGF and dexamethasone on phenylethanolamine N-methyltransferase (PNMT) activity in neonatal rat superior cervical ganglia. *Journal of Neurochemistry* 28, 1515-1520.

Milkovic, S., Milkovic, K., and Paunovic, J. (1973). The initiation of fetal adrenocorticotrophic activity in the rat. *Endocrinology* 93, 380-384.

Milner, T.A., Morrison, S.F., Abate, D., and Reis, D.J. (1988). Phenylethanolamine N-methyltransferase containing terminals synapse directly on sympathetic preganglionic neurons in the rat. *Brain Research* 448, 205-222.

Milner, T.A., Abate, C., Reis, D.J., and Pickel, V.M. (1989). Ultrastructural localization of phenylethanolamine N-methyltransferase-like immunoreactivity in the rat locus coeruleus. *Brain Research* **478**, 1-15.

Minson, J., Llewellyn-Smith, I., Neville, A., Somogyi, P., and Chalmers, J. (1990). Quantitative analysis of spinally projecting adrenaline-synthesizing neurons of C1, C2 and C3 groups in rat medulla oblongata. *Journal of the Autonomic Nervous System* **30**, 209-220.

Moore, E.K., and Phillipson, O.T. (1975). Effects of dexamethasone on phenylethanolamine N-methyltransferase and adrenaline in the brains and superior cervical ganglia of adult and neonatal rats. *Journal of Neurochemistry* **25**, 289-294.

Muller, M., and Renkawitz, R. (1991). The glucocorticoid receptor. *Biochimica and Biophysica Acta* **1088**, 171-182.

Otten, U., and Thoenen, H. (1976). Selective induction of tyrosine hydroxylase and dopamine beta-hydroxylase in sympathetic ganglia in organ culture-role of glucocorticoids as modulators. *Molecular Pharmacology* **12**, 353-361.

Paivarinta, H., Soinila, S., Eranko, O., and Joh, T.H. (1985). Phenylethanolamine-N-methyltransferase-immunoreactive cells in developing rat superior cervical ganglion and the effect of hydrocortisone on their number. *International Journal of Developmental Neuroscience* **3**, 8-18.

Paivarinta, H., Pickel, V.M., Eranko, L., and Joh, T.H. (1989). Glucocorticoid-induced PNMT-immunoreactive sympathetic cells in the superior cervical ganglion of the rat. *Journal of Electron Microscopic Techniques* **12**, 389-396.

Pieribone, V.A., and Aston-Jones, G. (1991). Adrenergic innervation of the rat nucleus locus coeruleus arises predominantly from the C1 adrenergic cell group in the rostral medulla. *Neuroscience* **41**, 525-542.

Pieribone, V. A., Aston-Jones, G., and Bohn, M.C. (1988). Adrenergic and non-adrenergic neurons of the C1 and C3 areas project to locus coeruleus: a fluorescent double labeling study. *Neuroscience Letters* **85**, 297-303.

Ross, C.A., Armstrong, D.M., Ruggiero, D.A., Pickel, V.M., Joh, T.H., and Reis, D.J. (1981). Adrenaline neurons in the rostral ventrolateral medulla innervate thoracic spinal cord: A combined immunocytochemical and retrograde transport demonstration. *Neuroscience Letters* **25**, 257-262.

Ross, C. A., Ruggiero, D.A., Park, D.H., Joh, T.H., Sved, A.F., Fernandez-Pardal, J., Saavedra, J.M., and Reis, D.J. (1984). Tonic vasomotor control by the rostral ventrolateral medulla: Effect of electrical or chemical stimulation of the area containing C1 adrenaline neurons on arterial pressure, heart rate, and plasma catecholamines and vasopressin. *Journal of Neuroscience* **4**, 474-494.

Ross, M.E., Evinger, M.J., Hyman, S.E., Carroll, J.M., Mucki, L., Comb, M., Reis, D.J., Joh, T.H., and Goodman, H.M. (1990). Identification of a functional glucocorticoid response element in the phenylethanolamine N-methyltransferase promoter using fusion genes introduced into chromaffin cells in primary culture. *Journal of Neuroscience* **10**, 520-530.

Ruggiero, D.A., Ross, C.A., Anwar, M., Park, D.H., Joh, T.H., and Reis, D.J. (1985). Distribution of neurons containing phenylethanolamine N-methyltransferase in medulla and hypothalamus of rat. *Journal of Comparative Neurology* **239**, 127-154.

Sawchenko, P.E., and Bohn, M.C. (1989). Glucocorticoid receptor immunoreactivity in C1, C2 and C3 adrenergic neurons that project to the hypothalamus or to the spinal cord in the rat. *Journal of Comparative Neurology* **285**, 107-116.

Seidler, F.J., and Slotkin, T.A. (1985). Adrenomedullary function in the neonatal rat: responses to acute hypoxia. *Journal of Physiology* (London) **358**, 1-16.

Seidler, F.J., and Slotkin, T.A. (1986). Ontogeny of adrenomedullary responses to hypoxia and hypoglycemia: Role of splanchnic innervation. *Brain Research Bulletin* **16**, 11-14.

Stachowiak, M.K., Rigual, R.J., Lee, P.H.K., Viveros, O.H., and Hong, J.S. (1988). Regulation of tyrosine hydroxylase and phenylethanolamine N-methyltransferase mRNA levels in the sympathoadrenal system by the pituitary-adrenocortical axis. *Molecular Brain Research* **3**, 275-286.

Swanson, L.W., Sawchenko, P.E., Berod, A., Hartman, B.K., Helle, K.B., and Vanorden, D.W. (1981). An immunohistochemical study of the organization of catecholaminergic cells and terminal fields in the paraventricular and supraoptic nuclei of the hypothalamus. *Journal of Comparative Neurology* **196**, 271-285.

Sze, P.Y., and Hedrick, B.J. (1983). Effects of dexamethasone and other glucocorticoid steroids on tyrosine hydroxylase activity in the superior cervical ganglion. *Brain Research* **265**, 81-86.

Teitelman, G., Baker, H., Joh, T.H., and Reis, D.J. (1979). Appearance of catecholamine synthesizing enzymes during development of rat embryo sympathetic nervous system: Possible role of tissue environment. *Proceedings of the National Academy of Sciences* (USA) **76**, 509-513.

Teitelman, G., Joh, H., Park, D., Brodsky, M., New, M., and Reis, D.J. (1982). Expression of the adrenergic phenotype in cultured fetal adrenal medullary cells: role of intrinsic and extrinsic factors. *Developmental Biology* **89**, 450-459.

Thoenen, H., Mueller, R.A., and Axelrod, J. (1970). Neuronally dependent induction of adrenal phenylethanolamine N-methyltransferase by 6-hydroxydopamine. *Biochemical Pharmacology* **19**, 669-673.

Tucker, D.C., Saper, C.B., Ruggiero, D.A., and Reis, D.J. (1987). Organization of central adrenergic pathways: I. Relationships of ventrolateral medullary projections to the hypothalamus and spinal cord. *Journal of Comparative Neurology* **259**, 591-603.

Turner, B. B., Katz, R.J., Roth, K.A., and Carroll, B.J. (1978). Central elevation of phenylethanolamine N-methyltransferase activity following stress. *Brain Research* **153**, 419-422.

Wesselingh, S.L., Li, Y.W., and Blessing, W.W. (1989). PNMT-containing neurons in the rostral medulla oblongata (C1, C3 groups) are transneuronally labelled after injection of herpes simplex virus type 1 into the adrenal gland. *Neuroscience Letters* **106**, 99-104.

Weisberg, E.P., Baruchin, A., Stachowiak, M.K., Stricker, E.M., Zigmond, M.J., and Kaplan, B.B. (1989). Isolation of a rat adrenal cDNA clone encoding phenylethanolamine N-methyltransferase and cold-induced alterations in adrenal PNMT mRNA and protein. *Molecular Brain Research* **6**, 159-166

Wurtman, R.J., and Axelrod, J. (1966). Control of enzymatic synthesis of adrenaline in the adrenal medulla by adrenal cortical steroids. *Journal of Biological Chemistry* **241**, 2301-2305.

Stress: Neuroendocrine and Molecular Approaches
Edited by R. Kvetnansky, R. McCarty and J. Axelrod

A PITUITARY CYTOTROPIC FACTOR MODULATES TYROSINE HYDROXYLASE EXPRESSION IN CULTURED DOPAMINERGIC CELLS

J. C. Porter, W. Kedzierski, N. Aguila-Mansilla, S. M. Ramin and G. P. Kozlowski

Departments of Obstetrics and Gynecology and Physiology
The University of Texas Southwestern Medical Center
Dallas, Texas USA

INTRODUCTION

Since their discovery (Fuxe, 1963) and characterization (Fuxe and Hökfelt, 1966; Hökfelt, 1967), dopamine (DA) secreting neurons of the brain have been shown to be influenced by many substances, including hormones. The synthesis of L-dihydroxyphenylalanine (DOPA) and DA secretion by DAergic neurons of the hypothalamus are increased by ovarian hormones (González et al, 1988, 1989; Wang and Porter, 1986), thyroid hormones (Reymond et al, 1987; Wang et al, 1989), and prolactin (Aguila-Mansilla et al, 1991; Hökfelt and Fuxe, 1972; González and Porter, 1988; González et al, 1989; Moore, 1987; Perkins et al, 1979). On the other hand, testosterone suppresses the activity of tyrosine hydroxylase (TH) in the hypothalamus (Gunnet et al, 1986; Kizer et al, 1974), as well as its immunocytochemical expression (Brawer et al, 1986). Orchiectomy increases the mass and *in situ* activity of TH in the median eminence, corpus striatum, and substantia nigra (Aguila-Mansilla et al, 1991). González et al (1986) suggested that the pituitary secreted a substance–that was not prolactin, which stimulated dopaminergic (DAergic) neurons.

Whether these hormones have a *direct* action on DAergic neurons or involve intermediate cells, such as interneurons, is unclear. Neuritic interconnections among brain cells make it difficult to address this dilemma. However, it may be possible to resolve this issue through the use of dissociated neurons as the maintenance of dissociated neurons in culture is now possible. Growth of non-neuronal cells can be suppressed by use of serum-free medium (Faive-Bauman *et al*, 1981). The culture is aided by the coating of fixed surfaces with poly-D-lysine (Rettman and Louis, 1979) and the use of a chemically defined medium (Ahmed *et al*, 1983). In the present study, we shall consider the evidence of the existence of a pituitary factor that stimulates secretion by DAergic cells.

MATERIALS AND METHODS

Brain cells from fetal rats were cultured in serum-free medium (DAWF medium) according to the procedure of Ahmed *et al* (1983) with minor adaptation (Porter *et al*, 1990). On the 20th day of gestation, the combined hypothalamus-midbrain or only the hypothalamus of each fetus was excised. (We have found that almost all of the surviving DAergic cells are of hypothalamic origin.) The tissue was digested with trypsin (Porter *et al*, 1990). The dispersed cells were distributed among poly-D-lysine-coated wells of 24-well plates (area per well = 2 cm^2), and incubated at 37°C in a humidified atmosphere of 7% CO_2 and air.

TH mRNA was analyzed by solution hybridization using an S1 nuclease protection assay (Melton *et al*, 1984) as described earlier (Kedzierski and Porter, 1990). TH was analyzed by an immunoblot assay (Porter, 1986a; Aguila-Mansilla *et al*, 1991). DOPA and DA were assayed by HPLC with electrochemical detection (Felice *et al*, 1978) with slight modification (Porter, 1986b).

To prepare pituitary cytotropic factor (CTF), rat pituitaries (800-1000 glands) were disrupted in 0.1 M acetic acid (5 glands/ml) using a Brinkmann Polytron homogenizer. The mixture was centrifuged at 2-4°C at 10,000 g for 10 minutes. The supernate was stored at -80°C. Before use, the supernate was filtered through a micropore membrane (0.22 μm), mixed with DAWF medium, and adjusted to pH 7.4.

RESULTS

DAergic neurons in cultures of hypothalamic cells constitute a minor cellular component, but are identifiable immunocytochemically by the presence of TH. A representative TH-containing neuron is shown in Figure 1. An extensive growth cone and beaded fibers can ·be discerned.

DAergic neurons in culture secrete both DOPA and DA. At 37°C in air, the halftimes of DOPA and DA are 1.5 hours. Steady state levels of DOPA are a function of the *in situ* activity of TH. The rate of secretion of DOPA is a function of the number of cells in a well.

To ascertain the relationship of TH mRNA, TH, and DOPA secretion by unstimulated cells, cells from 140 fetuses were distributed among 140 wells. Between the first and fourth weeks of culture, TH mRNA increased from 1.6 amoles (10^{-18} moles) to 2.8 amoles per well (Figure 2). Between the fourth and ninth weeks, the amount of TH mRNA was fairly constant, after which there was a decline. The amount of TH increased from 13 fmoles/well at the end of the first week to 105 fmoles/well at the end of the seventh week. The secretion of DOPA increased from 2 pmoles/well × 18 hours after the first week to 47 pmoles/well × 18 hours by the sixth week.

When the cells were incubated with various concentrations of CTF (ranging from 0 to 1 pituitary equivalent per well), there was a dose-dependent increase in the secretion of DOPA (Figure 3). When CTF was heated in a boiling water bath for 10 minutes, its activity was destroyed (Figure 4). The action of CTF was not attributable to a non-specific effect of protein since BSA in DAWF medium is no more effective on DOPA secretion than DAWF medium alone.

Prolactin in the intact animal stimulates DAergic neurons of the tuberoinfundibular system (c.f., Moore, 1987). Moreover, prolactin is present in pituitary extracts at a concentration of 1.2×10^{-7} M. However, when cells were cultured for 9 days in the presence of 1×10^{-8} M prolactin and 1×10^{-6} M prolactin, no appreciable effect was seen on DOPA secretion, TH mRNA, or TH mass (Figure 5).

Inasmuch as short-term exposure (24 hours) of DAergic neurons to CTF stimulated DOPA secretion (Figure 3), it was of interest to ascertain whether long-term exposure to CTF would be equally or more effective. To address this issue, cells were incubated with CTF for 96 hours. Results of a representative experiment are shown in

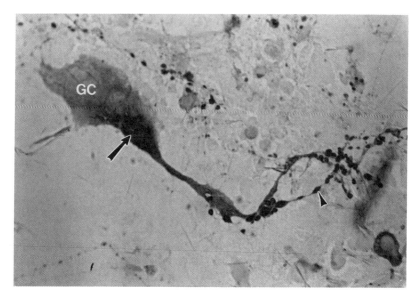

FIGURE 1.   A photomicrograph of a TH-containing neuron (×1600) in hypothalamic cell cultures. GC denotes the growth cone; PK signifies perikaryon; and BF identifies a beaded fiber.

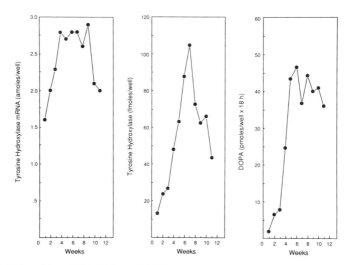

FIGURE 2.   TH mRNA, TH, and DOPA secretion by dissociated neurons. Cell density was 1.5×10^6 cells/well (0.75×10^6 cells/cm^2). The molecular weight of TH was taken as 60,000. Reproduced from Porter *et al* (1990) with permission.

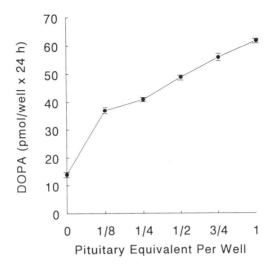

FIGURE 3.   Dose-response relationship between CTF and DOPA secretion by DAergic neurons. The values are means and SE; n=6. The cultures were 2 weeks old. Cells were exposed to CTF for 24 hours. Reproduced from Porter *et al* (1990) with permission.

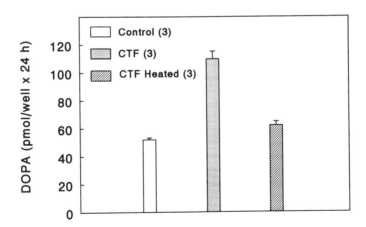

FIGURE 4.   Effect of heat on activity of CTF. The values are means and SE; n=3. The key is shown in the inset. The cultures were 3 weeks old. Cells were exposed to CTF for 24 hours. Reproduced from Porter *et al* (1990) with permission.

Figure 6. A day-to-day increase in the secretion of DOPA is evident. At the end of the incubation, cells incubated with CTF had significantly (P < 0.01) more TH mRNA than did controls. In this study, CTF did not affect the amount of TH in the cells. However, we have subsequently found that in cells incubated for two weeks with CTF, there was a marked increase in TH mass. Presumably, a 96 hour incubation is too short to reveal an action of CTF on TH mass.

We next asked whether removal of CTF from CTF-stimulated cells would lead to a reduction in DOPA secretion. Cells were incubated with CTF for 72 hours, and during the succeeding 24 hours were incubated in the absence of CTF. As shown in Figure 7, at 72 hours, cells incubated with CTF secreted 107 ± 4.9 (mean and SE) pmoles DOPA/well × 24 hours. Then, half of the CTF-stimulated cells were incubated for another 24 hours without CTF, and half were incubated another 24 hours with CTF. In the absence of CTF, DOPA secretion decreased to 32 ± 0.9 pmoles DOPA/well × 24 hours. Cells incubated with CTF for 96 hours secreted 118 ± 4.1 pmoles DOPA/well × 24 hours. In the absence of CTF for 96 hours, DOPA was secreted at a low level throughout the 96 hours, and these cells contained 3.5 ± 0.4 amoles TH mRNA/well, cells incubated with CTF for 96 hours contained 4.6 ± 0.2 amoles TH mRNA, and those incubated for 72 hours with CTF followed by 24 hours without CTF had 3.6 ± 0.1 amoles TH mRNA/well.

DISCUSSION

Control of the secretion of tuberoinfundibular DAergic neurons has long been of interest, not only for the inherent importance of these neurons but also as a model system for all brain catecholaminergic neurons. There are hormones that stimulate DA secretion as well as those that depress its secretion. Most of these agents have been shown to affect DA secretion in living animals. Nonetheless, caution is required in any attempt to generalize from the intact model to the dispersed cell model. In the case of intact models, it can seldom be said that the active agent acts on the DAergic neuron *per se*. The involvement of intermediate cells such as interneurons may have essential roles in the response. Dispersed cells would seem to reduce the likelihood of an interaction with an intermediate cell. At the same time, dispersed neurons may be quite dissimilar to those of an intact

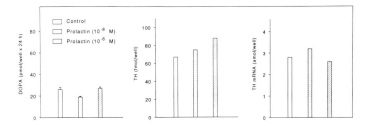

FIGURE 5. Effect of prolactin on DOPA secretion, TH mass, and TH mRNA of DAergic cells. The key is shown in the inset. The age of the cultures was 2 weeks when the study began. The cells were cultured for 9 days with prolactin. The molecular weight of TH was taken as 60,000. Reproduced from Porter *et al* (1990) with permission.

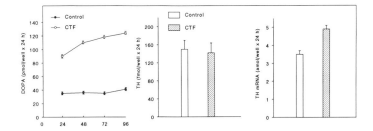

FIGURE 6. Trophic action of CTF on DAergic neurons. The values are means and SE. The keys are shown in the insets. For DOPA secretion, n = 10; for TH mRNA, n = 5; and for TH, n = 15. The molecular weight of TH was taken as 60,000. Reproduced from Porter *et al* (1990) with permission.

model in some important but unrecognized manner.

These caveats seem relevant when the actions of CTF and prolactin are considered. Prolactin stimulates DA secretion into portal blood of the rat (Gudelsky and Porter, 1980), DOPA synthesis in the median eminence (González and Porter, 1988), as well as DA turnover in the median eminence (Annunziato and Moore, 1978; Fuxe et al, 1975; Gudelsky and Moore, 1977; Gudelsky et al, 1976; Hökfelt and Fuxe, 1972; Kizer et al, 1978). Yet, prolactin does not stimulate the secretion of DOPA by dispersed DAergic neurons in cell culture, whereas CTF does do so. If prolactin were to affect DA secretion by way of an intermediate cell, a result such as this would be expected.

The stimulatory action of combined estrogen and progesterone on these DAergic neurons requires the presence of the pituitary (González et al, 1989). These observations speak to the existence of a factor from the pituitary that simulates tuberoinfundibular DAergic neurons. The data reviewed here provide evidence for such a factor, i.e., cytotropic factor (CTF).

CTF not only acts acutely to stimulate DOPA secretion by DAergic neurons in cell culture but also has a trophic action. CTF stimulates TH gene expression, as reflected by an increase in the mass of TH mRNA in cells. The longer the exposure of cells to CTF, the greater the secretion of DOPA. This is probably due to an increase in the mass of TH in the cells. Withdrawal of CTF from cells previously exposed to CTF results in a prompt reduction in the secretion of DOPA.

We speculate that one action of CTF is to activate TH, perhaps by phosphorylation, which has been shown to increase the enzymatic activity of TH (Ames et al, 1978; Edelman et al, 1978; Joh et al, 1978). Several protein kinases can phosphorylate TH. These include cAMP-dependent protein kinase (Joh et al, 1978; Vulliet, 1980; Yamauchi and Fujisawa, 1979), Ca^{2+}/calmodulin-dependent protein kinase (Vulliet et al, 1984; Yamauchi and Fujisawa, 1981), Ca^{2+}/phospholipid-dependent protein kinase (Albert et al, 1984), cGMP-dependent protein kinase (Roskoski et al, 1987), protein kinase in NGF-treated PC12 cells (kinase N) (Rowland et al, 1987), and TH kinase (Campbell et al, 1986).

If CTF has an important role in the intact animal in regulating DAergic cells, it is of interest to know how CTF reaches the brain. CTF could be secreted into the general circulation and thereby reach appropriate brain cells. It is also conceivable that CTF reaches the

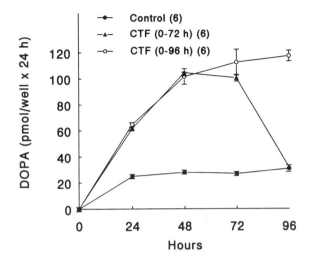

FIGURE 7.   Effect of withdrawal of CTF from DAergic cells on the secretion of DOPA. The values are means and SE; n=6. The key is shown in the inset.

brain by retrograde flow in the vasculature of the hypothalamic-hypophysial vasculature (Oliver *et al*, 1977). Regardless of the answer to these questions, it seems clear that the pituitary contains a CTF that stimulates DAergic cells in culture. The importance of CTF in intact animals remains to be established.

ACKNOWLEDGEMENTS

We thank Kay Stanley, Nhu-Y Dong, Sharyn Monroe, Jodie Roberts, Robert Lipsey and Nasir Sadatyar for assistance. This work was supported by U.S. Public Health Service Grants DK-01237, AG-08173, and AG-04344.

REFERENCES

Aguila-Mansilla, Jorquera, B.A., N., Kedzierski, W., and Porter, J.C. (1991). Effect of cerebroventricular anterior pituitary grafts on *in situ* expression of tyrosine hydroxylase in dopaminergic neurons of the aged animal. *Endocrinology* **128**, 1303-1309.

Ahmed, Z., Walker, P.S., and Fellows, R.E. (1983). Properties of neurons from dissociated fetal rat brain in serum-free culture. *Journal of Neuroscience* **3**, 2448-2462.

Albert, K.A., Helmer-Matyjek, E., Nairn, A.C., Muller, T.H., Haycock, J.W., Greene, L.A., Goldstein, M., and Greengard, P. (1984). Calcium/phospholipid-dependent protein kinase (protein kinase C) phosphorylates and activates tyrosine hydroxylase. *Proceedings of the National Academy of Sciences* (USA) **81**, 7713-7717.

Ames, M.M., Lerner, P., and Lovenberg, W. (1978). Tyrosine hydroxylase. Activation by protein phosphorylation and end product inhibition. *Journal of Biological Chemistry* **253**, 27-31.

Annunziato, L., and Moore, K.E. (1978). Prolactin in CSF selectively increases dopamine turnover in the median eminence. *Life Sciences* **22**, 2037-2042.

Brawer, J., Bertley, J., and Beaudet, A. (1986). Testosterone inhibition of tyrosine hydroxylase expression in the hypothalamic arcuate nucleus. *Neuroscience Letters* **67**, 313-318.

Campbell, D.G., Hardie, D.G., and Vulliet, P.R. (1986). Identification of four phosphorylation sites in the N-terminal region of tyrosine hydroxylase. *Journal of Biological Chemistry* **261**, 10489-10492.

Edelman, A.M., Raese, J.D., Lazar, M.A., and Barchas, J.D. (1978). *In vitro* phosphorylation of a purified preparation of bovine corpus striatal tyrosine hydroxylase. *Communications in Psychopharmacology* **2**, 461-465.

Faivre-Bauman, A., Rosenbaum, E., Puymirat, J., Grouselle, D., and Tixier-Vidal, A. (1981). Differentiation of fetal mouse hypothalamic cells in serum-free medium. *Developmental Neuroscience* **4**, 118-129.

Felice, L.J., Felice, J.D., and Kissinger, P.T. (1978). Determination of catecholamines in rat brain parts by reverse-phase ion-pair liquid chromatography. *Journal of Neurochemistry* **31**, 1461-1465.

Fuxe, K. (1963). Cellular localization of monoamines in the median eminence and in the infundibular stem of some mammals. *Acta Physiologica Scandanavica* **58**, 383-384.

Fuxe, K., Agnati, L.F., Hökfelt, T., Jonsson, G., Lidbrink, P., Ljungdahl, A., Löfström, A., and Ungerstedt, U. (1975). The effect of dopamine receptor stimulating and blocking agents on the activity of supersensitive dopamine receptors and on the amine turnover in various dopamine nerve terminal systems in the rat brain. *Journal of Pharmacology* (Paris) **6**, 117-129.

Fuxe, K., and Hökfelt, T. (1966). Further evidence for the existence of tubero-infundibular dopamine neurons. *Acta Physiologica Scandanavica* **66**, 245-246.

González, H.A., Kedzierski, W., Aguila-Mansilla, N., and Porter, J.C. (1989). Hormonal control of tyrosine hydroxylase in the median eminence: demonstration of a central role for the pituitary gland. *Endocrinology* **124**, 2122-2127.

González, H.A., Kedzierski, W., and Porter, J.C. (1988). Mass and activity of tyrosine hydroxylase in the tuberoinfundibular dopaminergic neurons of the aged brain:

control by prolactin and ovarian hormones. *Neuroendocrinology* **48**, 663-667.

González, H.A., and Porter, J.C. (1988). Mass and *in situ* activity of tyrosine hydroxylase in the median eminence: effect of hyperprolactinemia. *Endocrinology* **122**, 2272-2277.

Gudelsky, G.A., and Moore, K.E. (1977). A comparison of the effects of haloperidol on dopamine turnover in the striatum, olfactory tubercle and median eminence. *Journal of Pharmacology and Experimental Therapeutics* **202**, 149-156.

Gudelsky, G.A., and Porter, J.C. (1980). Release of dopamine from tuberoinfundibular neurons into pituitary stalk blood after prolactin or haloperidol administration. *Endocrinology* **106**, 526-529.

Gudelsky, G.A., Simpkins, J., Mueller, G.P., Meites, J., and Moore, K.E. (1976). Selective actions of prolactin on catecholamine turnover in the hypothalamus and on serum LH and FSH. *Neuroendocrinology* **22**, 206-215.

Gunnet, J.W., Lookingland, K.J., and Moore, K.E. (1986). Effects of gonadal steroids on tuberoinfundibular and tuberohypophysial dopaminergic neuronal activity in male and female rats. *Proceedings of the Society for Experimental Biology and Medicine* **183**, 48-53.

Hökfelt, T. (1967). The possible ultrastructural identification of tubero-infundibular dopamine-containing nerve endings in the median eminence of the rat. *Brain Research* **5**, 121-123.

Hökfelt, T., and Fuxe, K. (1972). Effects of prolactin and ergot alkaloids on the tubero-infundibular dopamine (DA) neurons. *Neuroendocrinology* **9**, 100-122.

Joh, T.H., Park, D.H., and Reis, D.J. (1978). Direct phosphorylation of brain tyrosine hydroxylase by cyclic AMP-dependent protein kinase: mechanism of enzyme activation. *Proceedings of the National Academy of Sciences* (USA) **75**, 4744-4748.

Kedzierski, W., and Porter, J.C. (1990). Quantitative study of tyrosine hydroxylase mRNA in catecholaminergic neurons and adrenals during development and aging. *Molecular Brain Research* **7**, 45-51.

Kizer, J.S., Humm, J., Nicholson, G., Greeley, G., and Youngblood, W. (1978). The effect of castration, thyroidectomy and haloperidol upon the turnover rates of dopamine and norepinephrine and the kinetic properties of tyrosine hydroxylase in discrete hypothalamic nuclei of the male rat. *Brain Research* **146**, 95-107.

Kizer, J.S., Palkovits, M., Zivin, J., Brownstein, M., Saavedra, J.M., and Kopin, I.J. (1974). The effect of endocrinological manipulations on tyrosine hydroxylase and dopamine-ß-hydroxylase activities in individual hypothalamic nuclei of the adult male rat. *Endocrinology* **95**, 799-812.

Melton, D.A., Krieg, P.A., Rebagliati, M.R., Maniatis, T., Zinn, K., and Green, M.R. (1984). Efficient *in vitro* synthesis of biologically active RNA and RNA hybridization probes from plasmids containing a bacteriophage SP6 promoter. *Nucleic Acid Research* **12**, 7035-7056.

Moore, K.E. (1987). Interactions between prolactin and dopaminergic neurons. *Biology of Reproduction* **36**, 47-58.

Oliver, C., Mical, R.S., and Porter, J.C. (1977). Hypothalamic-pituitary vasculature: evidence for retrograde blood flow in the pituitary stalk. *Endocrinology* **101**,

598-604.

Perkins, N.A., Westfall, T.C., Paul, C.V., MacLeod, R.M., and Rogol, A.D. (1979). Effect of prolactin on dopamine synthesis in medial basal hypothalamus: evidence for a short loop feedback. *Brain Research* **160**, 431-444.

Porter, J.C. (1986a). Relationship of age, sex, and reproductive status to the quantity of tyrosine hydroxylase in the median eminence and superior cervical ganglion of the rat. *Endocrinology* **118**, 1426-1432.

Porter, J.C. (1986b). *In situ* activity and phosphorylation of tyrosine hydroxylase in the median eminence. *Molecular and Cellular Endocrinology* **46**, 21-27.

Porter, J.C., Kedzierski, W., Aguila-Mansilla, N., and Jorquera, B.A. (1990). Expression of tyrosine hydroxylase in cultured brain cells: stimulation with an extractable pituitary cytotropic factor. *Endocrinology* **126**, 2474-2481.

Rettman, B., and Louis, J.C. (1979). Morphological and biochemical maturation of neurons cultured in the absence of glial cells. *Nature* **281**, 378-380.

Reymond, M.J., Benotto, W., and Lemarchand-Berand, T. (1987). The secretory activity of the tuberoinfundibular dopaminergic (TIDA) neurons is modulated by the thyroid status in the adult rat: consequence of prolactin secretion. *Neuroendocrinology* **46**, 62-68.

Roskoski, Jr., R., Vulliet, P.R., and Glass, D.B. (1987). Phosphorylation of tyrosine hydroxylase by cyclic GMP-dependent protein kinase. *Journal of Neurochemistry* **48**, 840-845.

Rowland, E.A., Muller, T.H., Goldstein, M., and Greene, L.A. (1987). Cell-free detection and characterization of a novel nerve growth factor-activated protein kinase in PC12 cells. *Journal of Biological Chemistry* **262**, 7504-7513.

Vulliet, P.R., Langan, T.A., and Weiner, N. (1980). Tyrosine hydroxylase: a substrate of cyclic AMP-dependent protein kinase. *Proceedings of the National Academy of Sciences* (USA) **77**, 92-96.

Vulliet, P.R., Woodgett, J.R., and Cohen, P. (1984). Phosphorylation of tyrosine hydroxylase by calmodulin-dependent multiprotein kinase. *Journal of Biological Chemistry* **259**, 13680-13683.

Wang, P.S., González, H.A., Reymond, M.J., and Porter, J.C. (1989). Mass and *in situ* molar activity of tyrosine hydroxylase in the median eminence: effect of thyroidectomy and thyroid replacement. *Neuroendocrinology* **49**, 659-663.

Wang, P.S., and Porter, J.C. (1986). Hormonal modulation of the quantity and *in situ* activity of tyrosine hydroxylase in neurites of the median eminence. *Proceedings of the National Academy of Sciences* (USA) **83**, 9804-9806.

Yamauchi, T., and Fujisawa, H. (1981). Tyrosine 3-monooxygenase is phosphorylated by Ca^{2+}-calmodulin-dependent protein kinase, followed by activation by activator protein. *Biochemical and Biophysical Research Communications* **100**, 807-813.

PART FIVE

NEUROENDOCRINE RESPONSES TO STRESS

Stress: Neuroendocrine and Molecular Approaches
Edited by R. Kvetnansky, R. McCarty and J. Axelrod

1992 Gordon and Breach Science
Publishers S.A., New York, USA.
Photocopying permitted by license only.

REGULATION OF THE HYPOTHALAMIC-PITUITARY-ADRENAL AXIS DURING STRESS: ROLE OF NEUROPEPTIDES AND NEUROTRANSMITTERS

G. Aguilera, A. Kiss, R. Hauger
and Y. Tizabi

Section on Endocrine Physiology, Developmental Endocrinology
Branch, National Institute of Child Health and Human Development,
NIH, Bethesda, Maryland USA

INTRODUCTION

Adaptation to stress involves behavioral, visceral and neuroendocrine responses directed to overcome disturbances of homeostasis caused by physical or psychological stressors. The main endocrine response is the activation of the hypothalamic-pituitary-adrenal (HPA) axis with rapid stimulation of ACTH and glucocorticoid secretion. However, during continuous or chronic intermittent exposure to stressful stimuli, activation of the pituitary-adrenal system is transient, and plasma ACTH levels return to near basal levels by 24 hours despite the persistence of the stressor (Selye, 1976; Keller-Wood and Dallman, 1984). This adaptative response can be attributed to a number of mechanisms, including negative feedback of glucocorticoids, decreased hypothalamic secretion of corticotropin releasing factor(s), and pituitary desensitization to ACTH regulators (Keller-Wood and Dallman, 1984; Rivier and Plotsky, 1986; Aguilera *et al*, 1987). However, regulatory changes in the hypothalamic-pituitary-adrenal axis essential for adaptation to chronic stress are not as yet fully understood. For example, in a number of chronic stress situations,

365

while plasma ACTH levels decrease with time, pituitary ACTH content and responsiveness to a novel stress are increased (Gann *et al*, 1985; Vernikos *et al*, 1982; Hauger *et al*, 1988). This paradoxical response suggests that chronic stress may reduce the sensitivity of the brain and pituitary to glucocorticoid feedback, or that activation of distinct neural pathways by different stressors determines changes in the rate of secretion and type of regulator released from the hypothalamus. In addition, in several chronic stress situations, the secretion of glucocorticoids remains elevated despite falling plasma ACTH levels, suggesting that the sensitivity of the adrenal is modified during prolonged stress (Dallman and Jones, 1973; Engeland *et al*, 1981; Hauger *et al*, 1988).

This review will discuss mechanisms involved in the adaptation to chronic stress, including regulatory actions at different levels of the axis, the hypothalamus, the pituitary and the adrenal.

Role of Corticotropin Releasing Hormone

CRH is the major physiological regulator of ACTH release in the anterior pituitary. The peptide is a potent stimulator of ACTH release *in vitro* and *in vivo* in both experimental animals and in humans (Rivier and Plotsky, 1986; Chrousos *et al*, 1989). *In vivo* studies have shown that inhibition of endogenous CRH by either hypothalamic lesions, immunoneutralization, or CRH antagonist administration reduces ACTH secretion in response to adrenalectomy and acute stress (Rivier and Plotsky, 1986; Vermes *et al*, 1981).

CRH is synthesized in the cell bodies of the parvicellular zone of the paraventricular nucleus (PVN) and released into the portal circulation from nerve endings in the external zone of the median eminence (Swanson, 1986; Antoni, 1986). Studies based on direct measurement of CRH in portal blood or the content of the peptide in the median eminence by radioimmunoassay and immunocytochemistry have shown that CRH secretion is increased during stress (Rivier and Plotsky, 1986). An example is shown in Figure 1, which illustrates the marked decrease in immunoreactive CRH in the median eminence following acute immobilization stress. This effect is more evident in rats pretreated with icv colchicine (Figure 1), which prevents the replenishment of the nerve terminals following release by blocking the axonal transport of the peptide from the cell body.

In vitro and *in vivo* studies have shown that the synthesis and

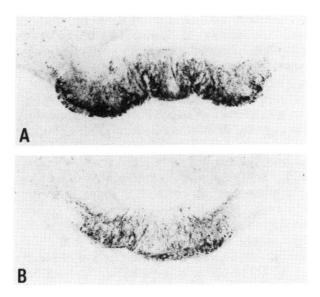

FIGURE 1. Immunoperoxidase staining of CRH in the median eminence of control rats (A) or rats subjected to 120 minutes immobilization stress (B). Rats received colchicine (75 ug, i.c.v.) 6 hours before immobilization.

secretion of CRH are under the influence of a number of stimulatory and inhibitory factors. These factors include catecholamines, serotonin, acetylcholine, angiotensin II (AII), interleukins, GABA and glucocorticoids (Assenmacher *et al*, 1987; Tsagarakis *et al*, 1989). Receptors for several of these factors, including AII, alpha and beta adrenergic ligands, and glucocorticoids have been demonstrated in the PVN (Mendelsohn *et al*, 1984; Leibowitz *et al*, 1982; Cummings *et al*, 1988; Reul and de Kloet, 1986). Although it is clear that catecholamines have a primary role in the regulation of hypothalamic corticotropic releasing factors, the type of adrenergic receptor involved and the precise regulatory effect are still controversial (Al-Damluji, 1988). The PVN receives noradrenergic innervation from the ventral noradrenergic bundle which contains nerve fibers from the areas A_1, A_2 and A_6 in the brain stem (Swanson, 1986). In addition, adrenergic and noradrenergic pathways innervate the amygdala, one of the sites of origin of the stria terminalis, which influences the PVN through peptidergic fibers (Gray *et al*, 1989). A number of studies indicate that catecholamines have an inhibitory effect on the activity of the hypothalamic-pituitary-adrenal axis during stress. Thus, depletion of endogenous catecholamines by central or peripheral injection of synthesis inhibitors is followed by increases in the activity of the hypothalamic-pituitary-adrenal axis, which has been interpreted to reflect a central catecholaminergic tonic inhibition of the system (Scampagni *et al*, 1975; Mezey *et al*, 1984; Murakami *et al*, 1989). In anesthetized dogs, icv injections of clonidine and norepinephrine inhibited cortisol secretion, an effect which was abolished by the alpha$_2$ adrenergic antagonist, yohimbine (Ganong *et al*, 1976). In the rat, yohimbine elevates basal ACTH secretion, suggesting an inhibitory role of alpha$_2$ activity, but the failure of ether stress to further increase ACTH levels is inconsistent with this argument (Murakami *et al*, 1989).

On the other hand, several studies support the proposal that adrenoreceptor activation stimulates hypothalamic secretion of corticotropin releasing factors. Selective lesions of the ventral noradrenergic bundle result in a reduction of the activity of the hypothalamic-pituitary-adrenal axis (Feldman *et al*, 1984; Alonso *et al*, 1986). Also, ACTH responses to ether stress, and CRH responses to induced hypotension can be partially inhibited in a non-additive manner by icv injection of alpha$_1$- or beta- adrenergic antagonists (Szafarczyk *et al*, 1987). Recent studies in our laboratory using immunocytochemical techniques demonstrate that alpha adrenergic activation is critical for the synthesis and release of CRH during acute and prolonged stress. In these experiments, icv injection of the alpha adrenergic agonist, methoxamine, caused marked depletion of CRH in the median eminence, and the characteristic CRH depletion following acute stress was partially prevented by icv injection of the alpha$_1$ adrenergic antagonist, prazosin (Kiss, 1991). Continuous icv minipump infusion of the alpha$_2$-adrenergic antagonist yohimbine also prevented the release of CRH from the ME 48 hours following colchicine administration. This finding is in disagreement with the belief that alpha$_2$-adrenergic receptor activation is inhibitory (Murakami *et al*, 1989). Since axonal transport in the present experiments was blocked by colchicine, the protective effect of the alpha antagonist on the release of CRH can be the result of blockade of stimulatory activity in the median eminence, which contains alpha$_2$-adrenoreceptors. In contrast, the stimulatory effect in the former reports could be due to blockade of an inhibitory effect in the cell bodies of the PVN.

In addition to CRH, other hypothalamic peptides such as vasopressin, oxytocin, and AII, are secreted into the portal circulation during stress and influence pituitary ACTH secretion by modulating the effect of CRH (Rivier and Plotsky, 1986). The relative secretion of CRH and other modulators into the portal circulation may be an important determinant for the changes in responsiveness of the corticotrophs during prolonged stress. For example, the activity of vasopressin neurons in the parvicellular system is increased during chronic stress (Whitnall, 1989). This is illustrated in Figure 2, showing a marked increase in the content of vasopressin in the external zone of the median eminence following 14 days of daily immobilization for 2.5 hours. As previously shown in adrenalectomized rats (Holmes *et al*, 1985), the ratio of CRH/vasopressin secretion from the median eminence following *in vitro* stimulation with 40 mM KCl is markedly reduced in rats subjected to 2.5 hours immobilization for 14 days compared with controls (Table 1). Recent immunocytochemical studies have shown changes in the rate of CRH and vasopressin secretion during stress, with initial release of CRH followed by a decline and increased vasopressin secretion (Goeij *et al*, 1991). The above observations indicate that during the course of chronic stress the proportional release of the various hypothalamic regulators changes, and this may be a determinant of pituitary responsiveness.

Catecholamines are also important mediators of the release of vasopressin. Pharmacological studies have shown that increases in ACTH secretion following catecholamine administration are due to alpha$_1$- adrenergic stimulation of vasopressin release (Al-Damluji, 1988). Consistent with the studies noted above is our recent observation that icv injection of prazosin, 2 hours prior to decapitation, in colchicine treated rats results in markedly higher vasopressin content in the median eminence. Future studies on the role of other regulatory neurotransmitters and peptides in the control of hypothalamic function will be critical for understanding the relative contribution of CRH and other regulators during adaptation to chronic stress.

Pituitary Effects of CRH

CRH activates pituitary corticotrophs through interaction with plasma membrane receptors coupled to a cyclic AMP-dependent signalling transduction system (Abou-Samra *et al*, 1987). The affinity of the pituitary CRH receptor is in the nanomolar range, which is consistent with the concentrations of the peptide required to elicit cAMP and ACTH release from pituitary cells *in vitro* (Aguilera *et al*, 1986). The

TABLE 1. Effect of chronic intermittent restraint stress on the potassium-stimulated release of CRH and vasopressin (VP) from the median eminence (ME) *in vitro.*

| | PEPTIDE RELEASE (fmol/ME) | | RATIO VP/CRH |
	VP	CRH	
CONTROL	11.9 ± 0.8	5.4 ± 0.2	2.1 ± 0.1
STRESS	23.8 ± 1.6**	6.7 ± 0.3	3.6 ± 0.4*

Median eminences from control rats and rats subjected to immobilization for 2 hours/day for 14 days were preincubated for 60 minutes and then incubated for two consecutive periods of 10 minutes in medium 199 containing 5 or 50 mM KCl, respectively. Values are means and SEM of 6 determinations in pools of 3 median eminences. **p<0.01, *p<0.05 with respect to control.

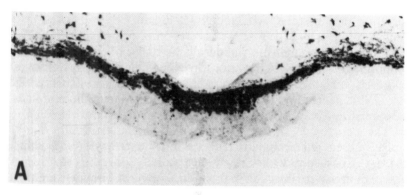

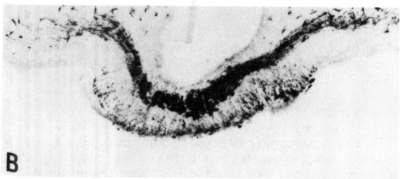

FIGURE 2. Immunoreactive vasopressin content in the median eminence of control (A) and chronically stressed rats (B). Rats were subjected daily to 120 minutes immobilization for 14 days.

number of pituitary CRH receptors undergoes marked regulatory changes during manipulations of the hypothalamic-pituitary- adrenal axis (Aguilera *et al*, 1987; DeSouza and Kuhar, 1986). However, a number of studies have shown that CRH receptor changes cannot account for the changes in responsiveness of the corticotrophs. For example, following adrenalectomy, ACTH responses to CRH are increased despite losses in CRH receptors of about 80% (Aguilera *et al*, 1987). In other situations such as prolonged continuous immobilization stress, the return of plasma ACTH to near basal levels after the initial hours of stress is accompanied by decreases in pituitary CRH receptors (Hauger *et al*, 1989) (Figure 3). Although in this stress model, pituitary CRH receptor downregulation could explain the transient increase in ACTH, plasma ACTH responses to exogenous CRH administration or to a novel stressor are augmented, indicating that full ACTH responses can be obtained with partial receptor occupancy (Figure 4). In other stress paradigms, such as prolonged cold exposure, the elevation in plasma ACTH levels is transient as in immobilization stress, but pituitary CRH receptor content is unchanged (Hauger *et al*, 1988). Thus, it is likely that interaction of CRH with other factors at a postreceptor level may be more important than receptor regulation in determining pituitary responsiveness.

Interaction Between CRH and Other Regulators

The pituitary effects of CRH are modulated by the inhibitory effect of glucocorticoids and a number of stimulatory factors such as vasopressin, oxytocin, angiotensin II, catecholamines and others (Rivier and Plotsky, 1986; Aguilera *et al*, 1986). The inhibitory effect of glucocorticoids, the major negative modulator of corticotroph activity, is biphasic with a rapid and a delayed component (Keller-Wood and Dallman, 1984). The rapid feedback inhibition occurs within minutes of exposure to glucocorticoids while the delayed effect which involves nuclear interaction of the steroid receptor complex occurs within hours and is sensitive to the higher concentrations of the steroid seen during stress (Abou-Samra *et al*, 1987). The stimulatory factors, of which the most important is vasopressin, have a minor effect on their own, but they potentiate the effect of CRH. In contrast to CRH, which stimulates ACTH secretion through stimulation of cAMP production, vasopressin, angiotensin II and alpha-adrenergic agonists exert their effects through calcium/phospholipid-dependent mechanisms (Aguilera

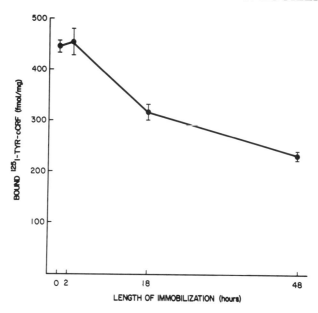

FIGURE 3.   Time course of the effect of continuous immobilization stress on anterior pituitary CRH receptors in the rat.

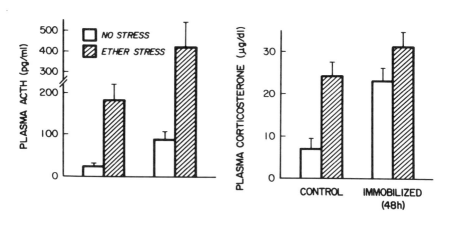

FIGURE 4.   Plasma ACTH and corticosterone levels after 5 minutes ether exposure in controls and rats subjected to continuous immobilization stress for 48 hours.

et al, 1990). Vasopressin alone does not stimulate cAMP production, but the synergism with CRH involves protein kinase C-dependent potentiation of CRH-stimulated adenylate cyclase activity as well as effects at more distal loci (Abou-Samra, 1987). It is noteworthy that in some experimental conditions, vasopressin can overcome the inhibitory effect of glucocorticoids on CRH-stimulated ACTH production (Abou-Samra *et al*, 1987). As shown in Figure 5, preincubation of cultured rat anterior pituitary cells with 1 uM corticosterone markedly inhibits maximum release of ACTH by CRH. Addition of 1 nM vasopressin during the incubation with CRH partially reverses the inhibition, restoring ACTH production to levels observed in control cells.

Role of CRH and Vasopressin in Pituitary Sensitivity During Stress

A number of studies have shown that CRH causes desensitization of the pituitary to a subsequent exposure to the peptide (Aguilera *et al*, 1990). Prolonged CRH administration in rats results in down-regulation of pituitary CRH receptors and decreases in cAMP and ACTH responses to the peptide in pituitary cells isolated from these animals. CRH infusion into rats or humans causes an early peak in plasma ACTH followed by a decrease to lower levels, similar to the pattern seen in chronic stress (Chrousos *et al*, 1989). Despite the partial desensitization, prolonged CRH exposure has trophic effects in the pituitary corticotroph, including increases in cell size and number (Childs *et al*, 1987) and stimulation of POMC transcription (Levin *et al*, 1989). Since the release of CRH into the portal circulation is increased during stress, it is possible that the priming effects of the peptide may account for the increased secretory capacity of the pituitary to a novel stressor. To study this possibility, the pituitary responsiveness to acute stimulation was studied in rats receiving chronic administration of CRH (Tizabi and Aguilera, 1990). Infusion of 50 ng/min of CRH for 48 hours with osmotic minipumps increased plasma CRH to levels similar to those reported in the portal circulation. As seen after chronic stress, adrenal weight and basal plasma ACTH and corticosterone were also increased. However, the plasma ACTH and corticosterone responses to immobilization stress were markedly lower than those in control rats (Figure 6). As observed following continuous CRH infusion, transient elevation in plasma CRH by single daily injections of CRH for 10 days also

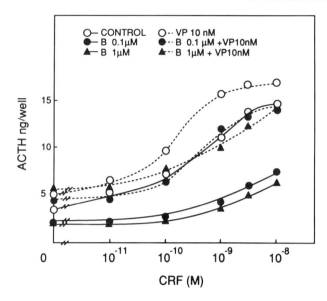

FIGURE 5.   Effect of vasopressin (VP) on the inhibitory action of corticosterone on CRH-stimulated ACTH production in cultured anterior pituitary cells.

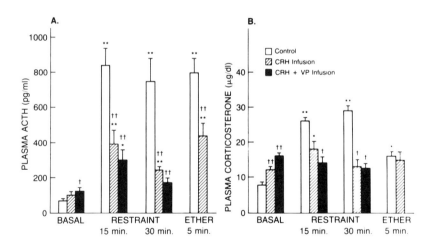

FIGURE 6.   Effect of minipump infusion of CRH or CRH plus vasopressin on basal and stress-stimulated plasma ACTH (A) and corticosterone (B). *$p < 0.05$ compared to basal, **$p < 0.01$ compared to basal, +$p < 0.05$ compared with non-infused controls, + +$p < 0.01$ compared with, non-infused controls.

resulted in diminished ACTH responses to stress.

Since vasopressin is an important modulator of the pituitary response to CRH and the secretion of vasopressin into the portal circulation appears to increase during prolonged stress, the possibility that combined exposure to CRH and vasopressin mimics the effect of chronic stress was tested in rats receiving simultaneous infusions of both peptides. However, the combination of CRH and vasopressin caused further pituitary desensitization rather than increasing the responses to acute stimulation. Another experimental model used to study the interaction of CRH with vasopressin, water restriction, has provided conflicting results. Following 60 hours of water deprivation, basal ACTH levels in plasma were slightly lower than in controls (41.4 $\pm$ 4.5 and 31.9 $\pm$ 2.5 pg/ml in controls and water restriction, respectively, n = 7-8), despite marked increases in plasma vasopressin (1.8 $\pm$ 0.7 and 45.1 $\pm$ 3.8 pg/ml for controls and water restriction, respectively). Since immunocytochemical and *in situ* hybridization studies have shown decreases in immunoreactive CRH in the parvicellular region of the paraventricular nucleus following increased plasma osmolarity, it is possible that the lack of ACTH response is the result of low CRH secretion (Jessop *et al*, 1990; Dohanics *et al*, 1990). However, CRH administration failed to increase plasma ACTH levels, suggesting that the levels of vasopressin at which the pituitary was exposed during dehydration are different from those during stress, or that there is an additional factor which is released as a result of inhibition of the parvicellular system. In addition, it has been shown that prolonged CRH administration in control rats causes dowregulation of anterior pituitary CRH receptors, an effect which is markedly accentuated by simultaneous vasopressin administration (Aguilera *et al*, 1987). On the other hand, the decrease in CRH receptors following 48 hours of minipump administration of CRH was identical in control and water deprived rats, suggesting that stimulation of the corticotrophs requires higher vasopressin levels than the renal effects of the peptide.

Although it is not possible to rule out that different patterns of administration of CRH and vasopressin may reproduce the pituitary effects of stress, it is more likely that corticotroph sensitivity during chronic stress may require more complex interactions with additional regulators such as cytokinins, catecholamines and others which are known to modulate ACTH secretion.

Regulation of Adrenal Glucocorticoid Secretion During Stress

There is evidence indicating that adrenocortical secretion of glucocorticoids depends not only on the prevailing levels of circulating ACTH, but also on the adrenal sensitivity to ACTH. Repeated stress has been shown to have facilitory or inhibitory influences on the adrenal responses to subsequent stimulation depending on the type and time of application of the stimulus (De Souza and Van Loon, 1982; Gann *et al*, 1985; Keller-Wood and Dallman, 1984). Inhibition of the corticosterone response independent of plasma ACTH levels has been observed when a second stressor is applied 60, but not 90 minutes after a 2 minute restraint stress (De Souza and Van Loon, 1982). On the other hand, prior surgical stress in dogs causes sensitization of the adrenal cortisol response to hemorrhage (Gann *et al*, 1985) or hypoxia (Raff *et al*, 1983). In rats, electrical shock, laparotomy or immobilization resulted in equal or larger corticosterone responses to a second stressor, an effect not seen when the effect of the primary stressor was mimicked by ACTH injections (Dallman and Jones, 1973). Similarly, during prolonged immobilization (Hauger *et al*, 1989) there is a dissociation between ACTH and plasma corticosterone levels, with a marked decline in ACTH after the first 6 hours of stress, but a slight decline in corticosterone levels after 48 hours. It is unlikely that the increased corticosterone response is due to adrenal hypertrophy, because CRH or ACTH administration also causes increases in adrenal weight, but corticosterone responses to acute stress are markedly reduced (Figure 6-B).

Recent studies in our laboratory using isolated adrenal fasciculata cell, provide evidence for increased adrenocortical responsiveness to ACTH following stress. Corticosterone responses to increasing ACTH concentrations were significantly increased in cells from rats subjected to 2.5 hours immobilization for 14 days compared with those from control rats (Figure 7). ACTH-stimulated cyclic AMP accumulation was also higher in cells from stressed rats (Figure 7), suggesting that ACTH receptors or their coupling to adenylate cyclase was increased. Corticosterone responses to cholera toxin and the conversion of progesterone to corticosterone were also increased in cells from chronically stressed rats, indicating that the increased responses are partially due to postreceptor events, such as an increase in the activity of steroidogenic enzymes.

The mechanism by which the adrenal sensitivity to ACTH is

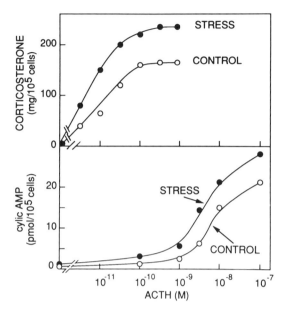

FIGURE 7. Corticosterone and cAMP responses to ACTH in isolated adrenal fasciculata cells from control and chronically stressed rats. Rats were subjected daily to 120 minutes immobilization for 14 days.

modified during stress could involve neural or humoral mechanisms. Several peptides such as vasopressin and POMC products have been shown to modify adrenal responses to ACTH *in vitro* (Pederson *et al*, 1980; Lowry, 1985; Payet and Lehoux, 1982), and changes in their local or systemic release during stress may modify the adrenal sensitivity to ACTH. Alternatively, neural mechanisms such as those involved in compensatory adrenal growth following unilateral adrenalectomy (Dallman *et al*, 1976; Engeland and Gann, 1989) may affect the sensitivity to ACTH during stress.

CONCLUSIONS

The neuroendocrine adaptation to stress involves complex regulatory actions at three different levels, the hypothalamus, the pituitary and the adrenal. During chronic stress, plasma ACTH responses are only transiently increased, but pituitary responsiveness to a novel stimulation is enhanced. Accumulated evidence indicates that the major regulators of pituitary ACTH secretion are CRH and vasopressin. At the hypothalamic level, the secretion of CRH and vasopressin is under the influence of stimulatory and inhibitory factors, of which alpha$_1$-adrenergic stimulation is of primary importance during stress. CRH and vasopressin exert their actions by interacting with plasma membrane receptors located in the pituitary corticotrophs. Although CRH receptors undergo marked regulatory changes during alterations of the hypothalamic-pituitary-adrenal axis, CRH receptor regulation cannot account for the changes in pituitary sensitivity to CRH or to a novel stressor. Vasopressin markedly potentiates the stimulation of ACTH by CRH and partially overcomes the feedback inhibition by corticosterone. However, continuous or episodic administration of CRH and vasopressin in rats results in pituitary desensitization rather than increased sensitivity to a novel stressor, suggesting that during stress, other factors may modulate the sensitivity of the pituitary to the effects of the main regulators, CRH and vasopressin.

Plasma corticosterone levels during chronic stress are often disproportionately elevated with respect to the increases in plasma ACTH, indicating that the sensitivity of the adrenal cortex to ACTH is also increased during stress. Future studies will elucidate the precise role of the different regulators and the mechanisms which determine the responsiveness of the hypothalamic-pituitary-adrenal axis during stress.

REFERENCES

Abou-Samra, A.-B., Harwood, J.P., Catt, K.J., and Aguilera, G. (1987). Mechanisms of action of CRF and other regulators of ACTH release in pituitary corticotrophs. *Annals of the New York Academy of Sciences* **512**, 67-84.

Aguilera, G., Flores, M., Carvallo, P., Harwood, J.P., Millan, M., and Catt, K.J. (1990). Receptors for corticotropin-releasing factor. In: E.B. De Souza and C.B. Nemeroff (Eds.), "Corticotropin-Releasing Factor: Basic and Clinical Studies of a Neuropeptide," pp. 154-171. Boca Raton, Florida: CRC Press, Inc.

Aguilera, G., Millan, M.A., Hauger, R.L., and Catt, K.J. (1987). Corticotropin-releasing factor receptors: Distribution and regulation in brain, pituitary, and peripheral tissues. *Annals of the New York Academy of Sciences* **512**, 48-66.

Aguilera, G., Wynn, P.C., Harwood, J.P., Hauger, R.L., Millan, M.A., Grewe, C., and Catt, K.J. (1986). Receptor-mediated actions of corticotropin-releasing factor in pituitary gland and nervous system. *Neuroendocrinology* **43**, 79-88.

Al-Damluji, S. (1988). Adrenergic mechanisms in the control of corticotrophin secretion. *Journal of Endocrinology* **119**, 5-14.

Alonso, G., Szafarczyk, A., Balmefrezol, M., and Assenmacher, I. (1986). Immunocytochemical evidence for stimulatory control by the ventral noradrenergic bundle of parvicellular neurons of the paraventricular nucleus secreting corticotropin releasing hormone and vasopressin in rats. *Brain Research* **397**, 297-307.

Antoni, F.A. (1986). Hypothalamic control of adrenocorticotropin secretion: Advances since the discovery of 41-residue CRF. *Endocrine Reviews* **7**, 351-378.

Assenmacher, I., Szafarczyk, A., Alonso, G., Ixart, G., and Barbanel, G. (1987). Physiology of neural pathways affecting CRH secretion. *Annals of the New York Academy of Sciences* **512**, 149-161.

Childs, G.V. (1987). Cytochemical studies of the regulation of ACTH secretion. *Annals of the New York Academy of Sciences* **512**, 248-274.

Chrousos, G.P., Udelsman, R., Gold, P.W., Margioris, A.N., Oldfield, E.H., Schurmeyer, T.H., Schulte, H.M., Doppman, J., and Loriaux, D.L. (1989). Corticotropin releasing factor (hormone): Physiological and clinical implications. In: E.E. Muller and R.M. Macleod (Eds.), "Neuroendocrine Perspectives," Volume 7, pp.49-83. New York: Springer-Verlag.

Cummings, S., and Seybold, V. (1988). Relationship of alpha-1 and alpha-2 adrenergic binding sites to regions of the paraventricular nucleus of the hypothalamus containing corticotropin releasing factor and vasopressin neurons. *Neuroendocrinology* **47**, 523-530.

Dallman, M.F., and Jones, M.T. (1973). Corticosteroid feedback control of ACTH secretion: Effect of stress-induced corticosterone secretion on subsequent stress responses in the rat. *Endocrinology* **103**, 1367-1375.

De Souza, E.B., and Kuhar, M.J. (1986). Corticotropin releasing factor receptors in the pituitary gland and central nervous system: Methods and overview. *Methods in Enzymology* **124**, 560-590.

De Souza, E.B., and Van Loon, G.R. (1982). Stress-induced inhibition of the plasma corticosterone response to a subsequent stress in rats: A nonadrenocorticotropin-mediated mechanism. *Endocrinology* **110**, 23-33.

Dohanics, J., Kovacs, K.J., and Makara, G.B. (1990). Oxytocinergic neurons in rat hypothalamus. *Neuroendocrinology* **51**, 515-522.

Engeland, W.C., Byrnes, G.J., Presnell, K., and Gann, D.S. (1981). Adrenocortical sensitivity to adrenocorticotropin (ACTH) in awake dogs changes as a function of the time of observation and after hemorrhage independently of changes in ACTH. *Endocrinology* **108**, 2149-2153.

Feldman, S., Siegel, R.A., Weidenfeld, J., Conforti, N., and Melamed, E. (1984). Adrenocortical responses to ether stress and neural stimuli in rats following the injection of 6-hydroxydopamine into the medial forebrain bundle. *Experimental Neurology* **83**, 215-220.

Gann, D.S., Bereiter, D.A., Carlson, D.E., and Thrivikraman, K.V. (1985). Neural interaction in control of adrenocorticotropin. *Federation Proceedings* **44**, 161-167.

Ganong, W.F., Kramer, N., Salmon, J., Reid, I.A., Lovinger, R., Scapaginini, U., Boryczka, A.T., and Shackelford, R. (1976). Pharmacological evidence for inhibition of ACTH secretion by central adrenergic system in the dog.

Neurosciences 1, 167-174.

Goeij, D.C.E., Kvetnansky, R., Whitnall, M.H., Jezova, D., Berkenbosch, F., and Tilders, F.J.H. (1991). Repeated stress induced activation of corticotropin releasing factor (CRF) neurones enhances vasopressin stores and colocalization with CRF in the median eminence of rats. *Neuroendocrinology* 53, 150-159.

Gray, T.S., Carney, M.E., and Magnuson, D.J. (1989). Direct projections from the central amygdaloid nucleus to the nucleus paraventricularis: Possible role in stress adrenocorticotropin release. *Neuroendocrinology* 50, 433-446.

Hauger, R.L., Millan, M.A., Lorang, M., Harwood, J.P., and Aguilera, G. (1988). Corticotropin-releasing factor receptors and pituitary adrenal responses during immobilization stress. *Endocrinology* 123, 396-405.

Holmes, M.C., Antoni, F.A., Catt, K.J., and Aguilera, G. (1985). Predominant release of vasopressin vs corticotropin-releasing factor from the isolated median eminence after adrenalectomy. *Neuroendocrinology* 43, 245-251.

Jessop, D.S., Chowdrey, H.S., and Lightman, S.L. (1990). Inhibition of rat corticotropin-releasing factor and adrenocorticotropin secretion by an osmotic stimulus. *Brain Research* 523, 1-4.

Keller-Wood, M.E., and Dallman, M.F. (1984). Corticosteroid inhibition of ACTH secretion. *Endocrine Reviews* 5, 1-24.

Kiss, A. (1991). Evidence for a role of alpha-1-adrenergic stimulation in the release of corticotropin releasing hormone during acute immobilization stress. *The Endocrine Society: 73rd annual meeting,* Washington, DC.

Leibowitz, S.F., Jhanwar-Uniyal, M., Dvorkin, B., and Makman, M.H. (1982). Distribution of alpha-adrenergic, beta-adrenergic and dopaminergic receptors in discrete hypothalamic areas of rat. *Brain Research* 233, 97-114.

Levin, N., Blum, M., and Roberts, J.L. (1989). Modulation of basal and corticotropin-releasing factor-stimulated proopiomelanocortin gene expression by vasopressin in rat anterior pituitary. *Endocrinology* 125, 2957-2966.

Lowry, P.J. (1985). Pro-opiocortin peptides and adrenal mitogenesis. In: F. Mantero, E.G. Biglieri, J. Funder, and B.A. Scoggins (Eds.) "The Adrenal Gland and Hypertension," pp 1-6. New York: Raven Press.

Mendelsohn, F.A.O., Quirion, R., Saavedra, J., Aguilera, G., and Catt, K.J. (1984). Autoradiographic localization of angiotensin II receptors in rat brain. *Proceedings of the National Academy of Sciences* (USA) 81, 1575-1579.

Mezey, E., Kiss, J.Z., Skirboll, L.R., Goldstein, M., and Axelrod, J. (1984). Increase of corticotropin-releasing factor staining in rat paraventricular nucleus neurones by depletion of hypothalamic adrenaline. *Nature* 310, 140-141.

Murakami, K., Akana, S., Dallman, M.F., and Ganong, W.F. (1989). Correlation between the stress-induced transient increase in corticotropin-releasing hormone content of the median eminence of the hypothalamus and adrenocorticotropic hormone secretion. *Neuroendocrinology* 49, 233-241.

Payet, N., and Lehoux, J-G. (1982). Aldosterone and corticosterone stimulation by ACTH in isolated rat adrenal glomerulosa cells: Interactions with vasopressin. *Journal of Physiology* 78, 317-321.

Reul, J.M.H.M., and de Kloet, E.R. (1986). Anatomical resolution of two types of corticosterone receptor sites in rat brain with two types of corticosterone receptor sites in rat brain with *in vitro* autoradiography and computerized image analysis. *Journal of Steroid Biochemistry* **24**, 269-272.

Rivier, C.L., and Plotsky, P.M. (1986). Mediation by corticotropin releasing factor (CRF) of adenohypophysial hormone secretion. *Annual Review of Physiology* **48**, 475-494.

Scampagni, U., Annunziato, L., Lombardi, G., Oliver, C.H., and Preziosi, P. (1975). Time-course of the effect of alpha-methyl-p-tyrosine on ACTH secretion. *Neuroendocrinology* **18**, 272-276.

Selye, H. (1976). "Stress in Health and Disease." Boston, Massachusetts: Little Brown.

Swanson, L.W. (1986). Organization of mammalian neuroendocrine system. In: F. Bloom (Ed.), "Handbook of Physiology, Section 1: The Nervous System," pp.317-363. Bethesda, MD: American Physiological Society.

Szafarczyk, A., Malaval, F., Laurent, A., Gibaud, R., and Assenmacher, I. (1987). Further evidence for a central stimulatory action of catecholamines on adrenocorticotropin release in the rat. *Endocrinology* **121**, 883-892.

Tizabi, Y., and Aguilera, G. (1990). Desensitization of the hypothalamic-pituitary-adrenal axis following prolonged administration of corticotropin releasing hormone and vasopressin. *Neuroscience Abstracts* **16**, 853.

Tsagarakis, S, Gillies, G., Rees, L.H., Besser, M., and Grossman, A. (1989). Interleukin-1 directly stimulates the release of corticotropin releasing factor from rat hypothalamus. *Neuroendocrinology* **49**, 98-101.

Vermes, I., Berkenbosch, F., Tilders, F.J.H., and Smelik, P.G. (1981). Hypothalamic deafferentation in the rat appears to discriminate between the anterior lobe and intermediate lobe response to stress. *Neuroscience Letters* **27**, 89-93.

Vernikos, J., Dallman, M.F., Bonner, C., Katzen, A., and Shinsako, J. (1982). Pituitary-adrenal function in rats chronically exposed to cold. *Endocrinology* **110**, 413-424.

Whitnall, M.H. (1989). Stress selectivity activates the vasopressin-containing subset of corticotropin-releasing hormone neurons. *Neuroendocrinology* **50**, 702-707.

Stress: Neuroendocrine and Molecular Approaches
Edited by R. Kvetnansky, R. McCarty and J. Axelrod

ROLE OF CATECHOLAMINERGIC AND SELECTED PEPTIDERGIC SYSTEMS IN THE CONTROL OF HYPOTHALAMIC-PITUITARY-ADRENOCORTICAL RESPONSES TO STRESS

I. Assenmacher, A. Szafarczyk, G. Barbanel, G. Ixart,
P. Siaud, S. Gaillet and F. Malaval

Endocrinological Neurobiology Laboratory, URA 1197-CNRS
Department of Physiology, University of Montpellier-2
Montpellier, France

INTRODUCTION

During the past several decades, the hypothalamic-pituitary-adrenocortical (HPA) axis has maintained a dominant position in neuroendocrinological studies, and frequent revisions of concepts have occurred. This may be explained not only by the rapidly developing and increasingly powerful panoply of technical approaches that now enable deeper insight into the various components of this regulatory system but also by a few updated physiological concepts concerning the HPA axis. Among other innovations, the theory of pulsatile patterns of basal hormonal release, originally demonstrated within the gonadotropic axis (Knobil, 1978; Levine *et al*, 1980), was recently extended to all levels of the HPA axis, including CRH-41 (Ixart *et al*, 1991), ACTH (Carnes *et al*, 1986) and corticosterone (Reynolds *et al*, 1980). Moreover, the physiological role of the HPA axis response to stressful stimulation, as proposed in the 1930s by Selye, has been challenged recently by a broader concept involving a moderating role of post-stress corticosteroid output on a number of more specific physiological responses to various challenges, including antidiuresis, hypoglycemia and above all, activation of inflammatory and immune systems (Munk *et al*, 1984). This revised theory of the HPA axis stress response, as part of a complex adaptive response of the organism to a variety of stress situations, has boosted recent interest in the nature and role of the CNS pathways controlling activation of the corticotropic axis.

Among the neuronal systems involved in central control of the HPA axis, catecholaminergic systems originating from the brain stem have been the subject of considerable interest over the past few years (for a review refer to Rose, 1989). The present contribution was designed to discuss some new aspects in this rapidly expanding field.

BASAL PULSATILITY IN THE HPA AXIS

For technical reasons, the pulsatile patterns of hormone release were first detected peripherally in the HPA system, based upon measures of plasma corticosteroids and ACTH (Krieger, 1979; Reynolds *et al*, 1980). Regarding CRF release in the median eminence (ME), the push-pull cannulation technique in the ME, which allows slow local perfusion (13 ml/minute) with an artificial fluid close to the junction of the hypophysial portal system in unanesthetized rats, has provided clear evidence of a pulsatile pattern of CRH-41 release (Ixart *et al*, 1991), as previously shown for gonadotropin releasing hormone (Levine *et al*, 1980) and somatostatin (Arancibia *et al*, 1984). We used statistical methods specially designed for the analysis of rhythmic processes, such as "Ultra" or "Pulsar" algorithms (review in Merriam, 1986), to determine the rhythmic parameters of CRH-41 release under basal levels of HPA axis activity, as demonstrated by low morning levels of plasma ACTH (41 ± 7 pg/ml). Under these conditions the frequency was 2.7 ± 0.1 cycles/hour and the amplitude was 4.4 ± 0.3 pg (Ixart *et al*, 1990). Interestingly, in a similar quantitative study of the pulsatile pattern of plasma ACTH based on the same mathematical model, Carnes *et al* (1989) obtained a frequency value very close to that of CRH-41 in our study (3.2 ± 0.4 cycles/hour), thus providing evidence that some temporal information may be encoded in the pulsatile frequency within the corticotropic system. We recently obtained additional evidence of a close correlation between CRH-41 pulsatility and ACTH release. By investigating CRH-41 pulsatility throughout the circadian rhythm of HPA axis activity, we discovered a 4-fold rise in plasma ACTH occurring in late afternoon *versus* morning basal levels to be associated with a 55 % increase in the amplitude of CRH-41 pulsatility, with the frequency remaining unchanged. In addition, we noted that experimental removal of corticosteroid negative feedback modulation of the HPA system by adrenalectomy was associated at the ME level with a 40 % increase in both the amplitude and frequency of CRH-41 pulsatility, leading to a 10-fold rise in plasma ACTH (Ixart *et al*, 1990). Even though in target cells hormonal pulsatility might be an effective biological process to sustain adequate receptor sensitization and an unstable cybernetic state well suited for swift adjustments of the system to randomly occurring input, the data clearly indicate that even moderate alterations in pulsatility, amplitude and/or frequency lead to changes in ACTH secretion.

Stimulation of the HPA axis brought about by stress or experimental administration of specific CRF secretagogues results in transient perturbation of the pulsatile pattern of CRH-41 release which is immediately transmitted to the corticotrophs (Figure 1). Here again,

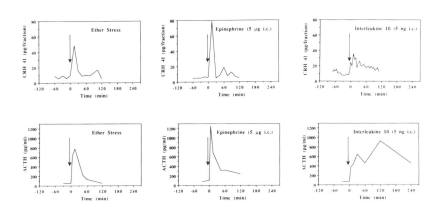

FIGURE 1. Representative individual examples of quantitative correlations between shape and time course of HPA axis responses to ether stress or to secretagogues injected (arrow) unilaterally into the PVN (i.c.): EPI (5 μg) or IL-1ß (5 ng). CRH-41 samples were collected from push-pull cannulae in the ME, and ACTH samples from an intra-carotid cannula.

it is interesting to note that there was a clear correlation between shape and duration of the surge in both the hypothalamic and pituitary signalling systems. Two minutes of ether stress was shown to induce a shortlived CRH-41 surge using the push-pull cannula system, with a corresponding increase in plasma ACTH. An intra-paraventricular nucleus (intra-PVN) infusion of 5 μg of epinephrine led to an intense CRH-41 surge that was followed by a series of after-discharges lasting for about 2 hours, and was associated with a prolonged state of ACTH hypersecretion for more than 2 hours. An intra-PVN injection of 5 ng interleukin-1β induced an even longer series of perturbed CRH-41 pulses (> 150 minutes) along with increased, biphasic and long-lasting (> 3 hours) ACTH release (Barbanel *et al*, 1990). A quantitative comparative analysis of the disturbed pulsatile patterns of CRH-41 and ACTH release in longitudinal time series studies of single animals might therefore provide a new methodological approach to the different mechanisms involved in responses of the HPA axis to various stressors.

MAJOR NEURONAL PATHWAYS TO CRH-41 NEURONS

The aim of this section is to delineate a few major connections between the PVN, arcuate nucleus and catecholaminergic nuclei located in the lower medulla which are organized in control loops providing the functional substrates for CNS control of the HPA axis.

The noradrenergic ascending bundle (NAB) bilaterally provides major direct innervation to the PVN, and particularly to CRH-41 secretory neurons. This pathway arises from two main neuronal groups that produce norepinephrine (NE) and epinephrine (EPI), both co-secreted with neuropeptide Y (NPY): (1) the ventral A1/C1 group supplies the most ventral section of the NAB innervating the hypothalamus, and (2) more dorsally, the A2/C2 group along with reduced participation of the A1/C1 group, forms the dorsal portion of the NAB leading to the hypothalamus. Moreover, there is a more dorsal branch of the NAB, unrelated to CRH neurons, which includes axons of the A2/C2 group and a substantial contribution from the locus coeruleus. This tract is directed to superior brain areas including the cerebral cortex (Sawchenko and Swanson, 1982). In addition to the main direct NAB targets in the PVN, at least two other catecholaminergic NAB pathways seem to be indirectly involved in the

HPA axis: one is through synaptic connections in the arcuate nucleus (Chromwall, 1985), and the second through possible contacts in the median eminence (Palkovits *et al*, 1980).

The arcuate nucleus itself, provides further major innervation of CRH-41 neurons, with the main neurotransmitters being POMC-derived neuropeptides, in particular ß-endorphin, and NPY (Liposits *et al*, 1988). In addition, two neuronal pathways are directed back to the medullary A1/C1 and A2/C2 cells groups arising from the arcuate nucleus (POMC and NPY neurons) and the parvocellular PVN (oxytocinergic neurons), respectively (Siaud *et al*, 1989). Interestingly, both medullary groups of catecholaminergic neurons also provide descending innervation to the preganglionic neurons of the autonomic nervous system. This indicates that the PVN-medullary circuitry is a major coupling mechanism controlling the overall corticotropic and neurovegetative responses to stress (Saper *et al*, 1976).

CATECHOLAMINERGIC ACTIVATING SYSTEMS

Since Ganong's pioneering studies in the 1970s on dogs, the prevailing theory in the literature was that catecholaminergic innervation of the hypothalamus had a predominantly inhibitory role on corticotropic response to stressors (review in Ganong, 1984). However, in the past several years, a number of important studies have favored the alternative concept that major NAB catecholaminergic pathways innervating the PVN actually convey the most potent stimulatory signals to CRH-41 neurons. Data from our group strongly supported this revised theory. Bilateral destruction of the major catecholaminergic pathways to the PVN (NAB-X) by direct microinjection of the neurotoxin 6-hydroxydopamine (6-OHDA; 4 μg in 0.2 μl vehicle) into this tract, which caused a 70% and 85% drop in the hypothalamic content of NE and EPI, respectively, induced: (i) an immunocytochemical accumulation of rCRH-41 in ME nerve terminals (Alonso *et al*, 1986); (ii) a 90% decrease in plasma concentrations of rCRH-41 in cannulated portal vessels of anesthetized rats, with a 65% fall in plasma ACTH (Guillaume *et al*, 1987); and (iii) a dramatic blockade of the ACTH surge following ether stress, which was reversed by 0.5 mg icv infusions of NE or EPI (Szafarczyk *et al*, 1985). In addition, icv or intra-PVN infusions of NE or EPI induced, as was mentioned above, a rapid CRH-41 surge in the ME of push-pull cannulated rats (Barbanel et al, 1991) that was associated with a dose-dependent rise in plasma ACTH (Szafarczyk et al, 1987). This latter effect was blunted by immunoneutralization with anti-rCRH-41 (Szafarczyk et al, 1987). A series of pharmacological studies showed that (i) the ACTH response to ether was blocked by icv pretreatment with either an α_1 adrenergic antagonist (prazosin) or a ß-adrenergic antagonist (propranolol), and that the former antagonized the stimulatory effect of both NE and EPI, whereas the latter was only effective in the blockade of EPI (Szafarczyk et al, 1987) and; (ii) after destruction of direct catecholaminergic input to CRH-41 neurons by NAB-X, an ACTH surge could also be produced by icv infusion of the α_2-adrenergic agonist, clonidine and this effect was blocked by the α_2-adrenergic antagonist, idazoxan (Szafarczyk et al, 1990).

These data led us to propose that the main catecholaminergic pathways to CRH-41 neurons act as stimulatory components in CNS control of the CRF-ACTH axis across postsynaptic α_1, α_2 and ß-adrenergic receptors. This conclusion has been further confirmed by a large body of recent experimental work in this field (Assenmacher et al, 1987; Plotsky et al, 1989).

More recent investigations concerning the specific medullary origins of this facilitory catecholaminergic regulation of the HPA axis have shown that near total destruction of the entire NAB blocked these regulatory processes. However, Feldman et al (1983) first demonstrated a similar effect after discrete lesions of axons stemming from the A2/C2 cell group (i.e., of the dorsal portion of the medullary-hypothalamic NAB tract). This result was recently confirmed in our new series of discrete 6-OHDA lesions within the dorsal or ventral sections of this tract or within both (Gaillet et al, in preparation).

PEPTIDERGIC ACTIVATING SYSTEMS

In accordance with the morphological data detailed above, we have also concentrated our investigations on two peptidergic neurotransmitters acting on CRH-41 secretory neurons, including Neuropeptide Y (NPY) arising from the arcuate nucleus and, possibly as a co-localized signal via the NAB and ß-endorphin, one of the major POMC cleavage products derived from the arcuate nucleus. To avoid possible interference of both neuropeptides with the NAB neurotransmitter systems, we studied their effects on ACTH secretion in NAB-X rats. Under these conditions, an icv infusion of NPY or ß-endorphin induced a rapid 10-fold increase in plasma ACTH concentrations above sham-injected controls. For both peptides, the effect on ACTH was long lasting, since one hour later ACTH levels were still 4- to 5-fold higher than levels in their respective controls (Gaillet *et al*, in press).

ß-endorphin, which is involved in direct innervation of CRH-41 neurons by the arcuate nucleus, appears to be a quite potent secretagogue in the CRF-ACTH axis, as previously reported when given by other routes (Suemaru *et al*, 1989). NPY displays a very similar stimulatory effect on ACTH release which seems to be independent of noradrenergic hypothalamic innervation since it persists in NAB-X rats, contrary to previous reports (Haas and George, 1989). It can thus be tentatively proposed that, in addition to the NAB, peptidergic innervation of CRH-41 neurons from the arcuate nucleus may also activate the HPA axis.

According to an emerging concept that bidirectional functional interactions link cellular and biochemical agents of both the HPA axis and the immune system (Weigent and Blalock, 1987), special interest has been devoted to interleukin-1ß (IL-1ß), a monokine known to be produced by macrophages in response to bacterial infections. With respect to the HPA axis, IL-1ß was shown to stimulate CRH-41 release when given iv or ip (Berkenbosch *et al*, 1987). As mentioned above, IL-1ß infused unilaterally into the PVN (5 ng in 0.25 ml vehicle) quickly induced (within 5 minutes) a 3-fold rise in CRH-41 release in the push-pull cannulated ME of unanesthetized rats, followed by a prolonged decrease (about 2 hours). In the periphery, plasma ACTH responded with a biphasic rise, with an initial 5-fold increase after 5 minutes, and a secondary increase reaching 15- to 20-fold stimulation 2 hours later (Barbanel *et al*, 1990). When given iv, IL-1ß (100 ng) also stimulated plasma ACTH. However, the time course of this surge differed from that of the intra-PVN route, by its monophasic profile that culminated (within 5 minutes) in a 10-fold increase above baseline levels before declining to normal levels 2 hours later (Barbanel *et al*, 1990).

Whatever the route of administration, interleukin-1ß is clearly a potent secretagogue of the HPA axis. However, marked differences in the shape and time course of ACTH responses, depending on the mode of peptide infusion, and its weak diffusion across the blood-brain barrier, suggest that the two experiments may have produced different physiological conditions, in which CNS macrophages (microglia) or their peripheral counterparts monocytes or histiocytes could have released IL-1ß. While IL-1ß release centrally could exert a direct influence on CRH-41 neurons, IL-1ß given peripherally would trigger the HPA axis via an unknown "reflexive arc" (Tazi et al, 1988), as is the case for other local inflammatory agents.

CATECHOLAMINERGIC INHIBITORY SYSTEMS

The controversy over the stimulatory versus inhibitory role of the major catecholaminergic pathways to the anterior hypothalamus has been resolved recently although the results of in vitro catecholamine perifusions of hypothalami are still open to question (Suda et al, 1987). Indeed, we have recently obtained evidence in at least three experimental conditions, that catecholaminergic pathways appear to be involved as inhibitory regulators of the HPA axis. For example, although a dramatic inhibition of the ACTH response to ether stress has been reported in NAB-X rats, we recently demonstrated that this blockade could be reversed entirely with an icv infusion of the α_2 antagonist, idazoxan (10 nmol), which per se had no effect in intact controls (Szafarczyk et al, 1990). Interestingly, this idazoxan-induced restoration of the normal post-stress ACTH surge was itself blunted by icv pretreatment with naloxone (10 nmol). This strongly suggests that under stressful conditions, α_2 adrenergic receptors participate in an inhibitory control of the HPA axis (see also Al-Damluji et al, 1990) that is mediated indirectly via a stimulatory endorphinergic component (Gaillet et al, in press). This role may tentatively be attributed to catecholaminergic (NAB) innervation to POMC-producing neurons of the arcuate nucleus.

An icv or intra-PVN injection of EPI or NE consistently resulted in a significant stimulation of the HPA axis, while direct infusion of EPI into the ME by means of the "push" cannula of a push-pull system never induced CRH-41 release. Instead, higher doses (10^{-4}M) of EPI led to decreased CRH-41 output during EPI perfusion (Barbanel et al, 1991). Considering that push-pull cannulation of the ME perfuses about 10% of the ME, which might explain why in such experiments only higher doses of agonists (or antagonists) are effective when injected directly into the system, this study suggests that catecholaminergic innervation of the ME may be another inhibitory component for CRH-41 release into the hypophysial portal system.

Finally, the strong monophasic stimulation of ACTH by peripheral (iv), but not central (intra-PVN) injections of IL-1ß was recently shown to be doubled in NAB-X rats (Barbanel et al, 1990). This finding clearly indicates that this particular route of stimulation of the HPA axis by IL-1ß may also involve an inhibitory catecholaminergic component, which may well be one of the two pathways discussed earlier. On the other hand, NAB-X dramatically decreased the ACTH response to IL-1ß infused into the PVN, which again suggests the possible involvement of different pathways in the response of the HPA axis to this cytokine.

CONCLUSIONS

Based on our present knowledge of several biochemically characterized neuronal pathways involved in the innervation of CRH-41 secretory neurons, a wealth of experimental data has demonstrated the presence of a few major neuronal routes, which, under stressful conditions, are responsible for transient alterations of the basal pulsatile release of CRH-41. The shape and time course of this release seem to be dependent on the nature of the CRF secretagogue involved in the stress response. These pathways include: (1) direct innervation of CRH-41 neurons in the PVN by the NAB pathway partially originating from the A1/C1, and even more from the A2/C2 catecholaminergic and NPY neurons of the medulla. This is a major stimulatory component in the HPA axis stress response which functions across postsynaptic α_1, α_2 and ß-adrenergic receptors. (2) The POMC (ß-endorphin) and NPY neurons of the arcuate nucleus provide two other stimulatory pathways to CRH-41 neurons. (3) Interleukin-1ß, a cytokine known to be produced by CNS and peripheral macrophages, is a potent stimulator of the CRH-ACTH axis when given either centrally (icv or intra-PVN) or peripherally (iv). Even though the pattern of the CRH-ACTH response to IL-1ß and its modulation after NAB-X differed depending on the route of cytokine administration, these particular interactions may be part of the complex intercellular signalling between the HPA axis and the immune system triggered along with inflammatory processes. (4) In addition to the major facilitory role played by NAB innervation of the PVN in the HPA axis response to stress, a few secondary catecholaminergic pathways appear to be involved in a series of inhibitory modulation processes. At least two occur at the hypothalamic level, one via the ß-endorphinergic system of the arcuate nucleus involving α_2 receptors, and another at the terminals of CRH-41 neurons in the ME.

This is clearly only a partial view of the complex neuronal mechanisms which may be set into action along the HPA axis during stressful conditions, presumably in a variety of configurations depending on the nature of the stress situation.

REFERENCES

Al-Damluji, S., Bouloux, P., White, A., and Besser, M. (1990). The role of alpha-2-adrenoreceptors in the control of ACTH secretion: Interaction with the opioid system. *Neuroendocrinology* **51**, 76-81.

Alonso. G., Szafarczyk, A., Balmefrezol, M., and Assenmacher, I. (1986). Immunocytochemical evidence for stimulatory control by the ventral noradrenergic bundle of parvocellular neurons of the paraventricular nucleus secreting corticotropin releasing hormone and vasopressin in rats. *Brain Research* **397**, 297-307.

Arancibia, S., Epelbaum, J., Boyer, R., and Assenmacher, I. (1984). *In vivo* release of somatostatin from rat median eminence after K+ infusion or delivery of nociceptive stress. *Neuroscience Letters* **50**, 97-102.

Assenmacher, I., Szafarczyk, A., Alonso, G., Ixart, G., and Barbanel, G. (1987). Physiology of neural pathways affecting CRH secretion. *Annals of the New York Academy of Sciences* **512**, 149-161.

Barbanel, G., Ixart, G., Szafarczyk, A., Malaval, F., and Assenmacher.I. (1990). Intra-hypothalamic infusion of interleukin-1ß increases the release of corticotropin releasing hormone (CRH-41) and adrenocorticotropin hormone (ACTH) in free moving rats bearing a push-pull cannula in the median eminence. *Brain Research* **516**, 31-36.

Barbanel, G., Ixart, G., and Assenmacher, I. (1991). *In vivo* infusion of adrenaline stimulates corticotropin-releasing hormone-41 producing neurons when given centrally but not distally. *Journal of Neuroendocrinology* **3**, 145-148.

Berkenbosch, F., Van Oers, J., Del Rey, A., Tilders, F., and Besedovsky, H. (1987). Corticotropin releasing factor-producing neurons in the rat activated by interleukin 1. *Science* **238**, 524-526.

Chromwall., B.H. (1985). Anatomy and physiology of the endocrine arcuate nucleus. *Peptides* **6**, 1-11.

Carnes, M., Brownfield, M.S., Kalin, N.M., Lent, S.T., and Barksdale, C.M. (1986). Episodic secretion of ACTH in rats. *Peptides* **7**, 219-223.

Feldman, S., Siegel, R.A. Weidenfeld, J., Conforti, N., and Melamed, E. (1983). Adrenocortical responses to ether stress and neural stimuli in rats following the injection of 6-hydroxydopamine into the medial forebrain bundle.

Experimental Neurology **83**, 215-220.

Gaillet, S., Malaval, F., Barbanel, G., Pelletier, G., Assenmacher, I., and Szafarczyk, A. (1991). Inhibitory interactions between α_2-adrenergic, but not NPY-ergic mechanisms controlling the CRH-ACTH axis in the rat. *Regulatory Peptides, in press.*

Ganong, W.F. (1984). Neurotransmitter mechanism underlying stress responses. In G.A. Brown, S.H. Koplow and S. Reichlin (Eds.). "Neuroendocrinology and Psychiatric Disorders." pp. 133-143. New York: Raven Press.

Guillaume, V., Conte-Devolx, B., Szafarczyk, A., Malaval, F., Pares-Herbuté, N., Grino, M., Alonso, G., Assenmacher, I., and Oliver, C. (1987). The corticotropin-releasing factor release in rat hypophysial portal blood is mediated by brain catecholamines. *Neuroendocrinology* **46**, 143-146.

Haas, D. A., and George, S.R. (1989). Neuropeptide Y-induced effects on hypothalamic corticotropin-releasing factor content and release are dependent on noradrenergic/ adrenergic neurotransmission. *Brain Research* **498**, 333-338.

Ixart, G., Barbanel, G., Nouguier, J., and Assenmacher, I. (1990). Diurnal and adrenalectomy-induced variations in the pulsatile CRH-41 release measured in the push-pull cannulated rat median eminence. *Proceedings of the Second International Congress of Neuroendocrinology.*

Ixart, G., Barbanel, G., Nouguier-Soulé, J., and Assenmacher, I. (1991). A quantitative study of the pulsatile parameters of CRH-41 secretion in unanesthetized free-moving rats. *Experimental Brain Research, in press.*

Jones, M.T., Gillham, B., Campbell, E.A., Al-Taher, A.R.H., Chuang, T.T., and Di Sciullo, A. (1987). Pharmacology of neural pathways affecting CRH secretion. *Annals of New York Academy of Sciences* **512**, 162-175.

Knobil, E. (1980). The neuroendocrine control of the menstrual cycle. *Recent Progress in Hormone Research* **36**, 53-88.

Krieger, D.T. (1979). Rhythms in CRF, ACTH and corticosteroids. In: D.T. Krieger (Ed.), "Endocrine Rhythms," pp. 123-142. New York: Raven Press.

Levine, J.E., and Ramirez, V.D. (1980). *In vivo* release of LHRH estimated with push-pull cannula from the mediobasal hypothalami of ovariectomized, steroid-primed rats. *Endocrinology* **107**, 1782-1790.

Liposits, Zs., Phelix, C., and Paull W.K. (1986). Adrenergic innervation of corticotropin releasing factor (CRF) synthesizing neurons in the hypothalamic paraventricular nucleus of the rat. A combined light and electron microscopic study. *Histochemistry* **84**, 201-205.

Liposits, Zs., Sievers, L., and Paull, W.K. (1988). Neuropeptide - Y and ACTH - immunoreactive innervation of corticotropin releasing factor (CRF) synthesizing neurons in the hypothalamus of the rat. *Histochemistry* **88**, 227-234.

Merriam, G. (1986). Methods for the characterization of episodic hormone secretion. In: W.F. Crowley, Jr. "Episodic Hormone Secretion," pp. 47-65. New York: Wiley.

Munk, A., Guyre, P.M., and Holbrook, N.J. (1984). Physiological functions of glucocorticoids in stress and their relation to pharmacological actions. *Endocrine Reviews* **5**, 25-44.

Palkovits, M., Zaborsky, A., Feminger, E., Mezey, E., Fekete, M.I., Herman, J.P., and

Szab, D.O. (1980). Noradrenergic innervation of the rat hypothalamus: Experimental biochemical and electron microscopic studies. *Brain Research* 151, 161-171.

Plotsky, P.M., Cunningham Jr, E.T., and Widmaier, E.P. (1989). Catecholaminergic modulation of corticotropin-releasing factor and adrenocorticotropin secretion. *Endocrine Reviews* 10, 437-458.

Reynolds, R.W., Keith, L.D., Harris, D.R., and Calvano, S. (1980). Rapid pulsatile corticosterone response in unanesthetized individual rats. *Steroids* 35, 305-314.

Rose, F.C. (1989). "The Control of the Hypothalamo-Pituitary Adrenal Axis." Madison: International University Press.

Saper, C.B., Loewy, A.D., Swanson, L.W., and Cowan, W.M. (1976). Direct hypothalamo-autonomic nucleus. *Brain Research* 117, 305-312.

Sawchenko, P. E., and Swanson, L. W. (1982). The organization of noradrenergic pathways from the brainstem to the paraventricular and supra optic nuclei in the rat. *Brain Research Reviews* 4, 275-326.

Siaud, P., Denoroy, L., Assenmacher, I., and Alonso, G. (1989). Comparative immunocytochemical study of the catecholaminergic and peptidergic afferent innervation of the dorsal vagal complex in the rat and the guinea pig. *Journal of Comparative Neurology* 290, 323-335.

Suda, T., Yajima, F., Tomori, N., Sumitomo, T., Nakagami, Y., Ushiyama, T., Demura, H., and Shizume, K. (1987). Inhibitory effect of norepinephrine on immunoreactive corticotropin-releasing factor release from the rat hypothalamus *in vitro*. *Life Sciences* 40, 1645-1649.

Suemaru, S., Dallman, M.F., Darlington, D., Cascio, C.S., and Shinsako, J. (1989). Role of alpha-adrenergic mechanisms in effects of morphine on the hypothalamo-pituitary- adrenocortical and cardiovascular systems in the rat. *Neuroendocrinology* 49, 181-190.

Szafarczyk, A., Alonso, G., Ixart, G., Malaval, F., and Assenmacher, I. (1985). Diurnal stimulated and stress-induced ACTH release in rats is mediated by ventral noradrenergic bundle. *American Journal of Physiology* 249, E219-E226.

Szafarczyk, A., Gaillet, S., Barbanel, G., Malaval, F., and Assenmacher, I. (1990). Alpha-2 adrenergic post synaptic receptors participate in the mechanisms of the central catecholaminergic activation of the corticotropic axis in the rat. *Comptes Rendus de l'Academie des Sciences* (Paris) 311, 81-88.

Szafarczyk, A., Malaval, F., Laurent, A., Gibaud, R., and Assenmacher I. (1987). Further evidence for a central stimulatory action of catecholamines on adrenocorticotropin release in the rat. *Endocrinology* 121, 883-892.

Tazi, A., Dantzer, R., Crestani, F., and Le Moal, M. (1988). Interleukin-1 induces conditioned taste aversion in rats: A possible explanation for its pituitary-adrenal stimulating activity. *Brain Research* 473, 369-371.

Weigent, D.A., and Blalock, J.E. (1987). Interactions between the neuroendocrine and immune systems: Common hormones and receptors. *Immunology Reviews* 100, 79-108.

Stress: Neuroendocrine and Molecular Approaches
Edited by R. Kvetnansky, R. McCarty and J. Axelrod

1992 Gordon and Breach Science
Publishers S.A., New York, USA.
Photocopying permitted by license only.

CORTICOTROPIN RELEASING FACTOR CONTROLS CORTICOSTERONE SECRETION FROM THE RAT ADRENAL GLAND

F. J. H. Tilders[1], J. W. A. M. van Oers[1] and J. P. Hinson[2]

[1]Department of Pharmacology, Medical Faculty, Free University
1081 BT Amsterdam, The Netherlands

[2]Department of Biochemistry, Faculty of Basic Medical Sciences
Queen Mary and Westfield College, University of London, UK

INTRODUCTION

It is generally accepted that glucocorticoid production and secretion by the adrenal cortex is under control of ACTH originating from the corticotroph cells in the pituitary gland. It should be noted that this view is based on an overwhelming amount of indirect indications although there is very little direct evidence from studies with ACTH receptor antagonists or neutralizing antibodies. According to the above mentioned concept, changes in plasma levels of glucocorticoids (cort) should be accompanied by changes in circulating ACTH concentrations. Although this seems to be the case in most studies, there are many reports demonstrating a dissociation between these two parameters under physiological, pathological and experimental conditions. One possible explanation for such discordant findings is that the time characteristics of the cort responses are delayed and prolonged as compared to those of ACTH. Thus, cort levels may remain elevated for some time after ACTH has fallen to resting values, giving rise to an elevated cort/ACTH ratio.

SUBSTANCES AFFECTING ADRENAL GLUCOCORTICOID SECRETION

It is conceivable that a dissociation between plasma levels of ACTH and cort might also be caused by factors that exhibit intrinsic corticotropic activity or act in synergism with ACTH. In fact, results from *in vitro* studies have demonstrated that a wide variety of substances other than ACTH can influence glucocorticoid secretion from adrenal fragments and dispersed or cultured adrenocortical cells. Several neurotransmitters, neuropeptides and other signaling substances present in the adrenal gland have been found to enhance cort secretion including acetylcholine (Hadjian *et al*, 1982), catecholamines (Holzwarth *et al*, 1987), histamine, serotonin (Hinson *et al*, 1989), prostaglandins (Winter *et al*, 1990), vasoactive intestinal polypeptide (VIP, Holzwarth *et al*, 1987), endothelin (Hinson *et al*, 1991), calcitonin gene-related peptide (Hinson and Vinson, 1990), interleukin-1 (Andreis *et al*, 1991a), vasopressin (Hinson *et al*, 1987, Schneider, 1988) and CRF (Winter *et al*, 1990, Andreis *et al*, 1991b).

It should be stressed that these observations illustrate effects but do not demonstrate physiological roles of these substances in the control of adrenal glucocorticoid secretion. Nevertheless these findings support the view that mechanisms are present within the adrenal gland whereby factors other than ACTH may affect cort secretion.

NON-ACTH MECHANISMS STIMULATING GLUCOCORTICOID SECRETION

Here, we will briefly discuss three examples of conditions in which changes in glucocorticoid secretion are not due to, or cannot be fully explained by, changes in circulating ACTH concentrations.

Electrical Stimulation of the Splanchnic Nerve. In calves and dogs, electrical stimulation of the splanchnic nerve has been reported to cause marked increases in the secretion rate of cortisol from the adrenal gland under conditions that do not lead to elevated ACTH levels. Although not mediated by increased ACTH concentrations, it was noted that cort secretion in response to splanchnic stimulation only occurred when ACTH was present in the circulation (Engeland and Gann, 1989; Edwards and Jones, 1987,1988). Conversely, splanchnic denervation reduces the cortisol response to ACTH (Edwards and Jones, 1986).

The mechanism by which splanchnic stimulation enhances cortisol secretion remains to be elucidated. It is likely that one or more neuronal, endocrine or other cellular elements present within the adrenal gland are involved (c.f. Charlton, 1990; Hinson, 1990). This hypothesis is supported by observations demonstrating that splanchnic nerve stimulation can release substances from the adrenal gland that are known to affect cort secretion, including acetylcholine, catecholamines, VIP, enkephalin and CRF (Edwards and Jones, 1987, 1988; Bloom et al, 1987,1988; Engeland and Gann, 1989; Holzwarth et al, 1987).

Alternatively, vascular responses may mediate the effects of splanchnic stimulation since adrenal blood flow markedly increases in response to electric stimulation (Breslow et al, 1987; Edwards and Jones, 1988) and decreases after sectioning of the splanchnic nerves (Engeland et al, 1985). Under various experimental conditions, changes in adrenal perfusion rate by itself have been demonstrated to induce parallel changes in cortisol secretion rate (Urquhart, 1965; Porter and Klaiber, 1965; Hinson et al, 1986; Engeland and Gann, 1989). In addition, it should be noted that many though not all substances that activate adrenal glucocorticoid secretion, including ACTH, reduce vascular resistance in the adrenal gland (c.f. Hinson et al, 1986, 1989). This has led some authors to introduce the concept of 'ACTH presentation rate', implying that with increased adrenal blood flow, the adrenal gland may 'see' more ACTH even when plasma levels are steady (e.g., Jones et al, 1990). Obviously, this will only be the case when blood flow is a limiting determinant for the ACTH concentration in the direct vicinity of ACTH receptors. Irrespective of the underlying mechanisms, these observations lead us to conclude that stimulation of the splanchnic nerve enhances the responsiveness of the adrenal gland to a given concentration of ACTH.

Small Volume Hemorrhage

An acute reduction in blood volume in dogs (10 ml/kg BW) can stimulate the adrenals to secrete cortisol despite very little or no increase in plasma ACTH levels (Engeland et al, 1981; Wood et al, 1982; Dempsher and Gann, 1983). Although it is not fully clear whether this response can occur in the absence of circulating ACTH (cf. Wood *et al*, 1982), all available data support the view that hemorrhage leads to a rapid increase in adrenal sensitivity to ACTH. Recent studies with microspheres have demonstrated that hemorrhage can enhance adrenal flow rate in rats and dogs (Sparrow and Coupland, 1987; Breslow *et al*, 1987). However, at the degree of hemorrhage mentioned above, no increase in adrenal blood flow was found (Dempsher and Gann, 1983; Engeland *et al*, 1985). By monitoring the responses of both adrenal glands in dogs subjected to unilateral adrenal denervation, Engeland *et al* (1985) found that sympathetic (T9-12) or splanchnic denervation did prevent epinephrine (EPI) and norepinephrine (NE) secretion in response to hemorrhage without affecting the increase in cortisol secretion. Taken together, these data suggest that neither the sympathetic innervation of the adrenal gland nor changes in adrenal blood flow play a mediating role in this response.

Diurnal Variation in Adrenal Cort Secretion

This phenomenon represents the most extensively documented example of a dissociation between plasma ACTH and glucocorticoid levels. As demonstrated in animals and humans, circulating ACTH levels are inadequate for the induction of diurnal peaks in plasma cort (Engeland *et al*, 1981; Fehm *et al*, 1984; Kaneko *et al*, 1980,1981; Van Oers and Tilders, 1991b). In rats, the diurnal swings in resting cort levels are primarily controlled by changes in adrenal responsiveness to ACTH (Ottenweller *et al*, 1978; Dallman *et al*, 1978; Kaneko *et al*, 1980,1981; Engeland *et al*, 1981). Even in hypophysectomized rats supplemented with ACTH in a manner that is supposed to result in constant plasma levels throughout the day, corticosterone levels showed diurnal variations with peak values around the onset of the activity period (Ottenweller and Meier, 1982). The afternoon peak in plasma cort disappeared after treatment of these animals with sodium pentobarbital or atropine or after transection of the spinal cord at T7, suggesting that neuronal mechanisms play a crucial role in these daily sensitivity changes (Ottenweller and Meier, 1982). Accordingly, autotransplantation of the adrenal glands under the kidney capsule resulted in a disappearance (Ottenweller and Meier, 1982) or in a strong reduction (Wilkinson *et al*, 1981) of the diurnal variation in resting cort levels. The persistence of a corticosterone rhythm and coupling to the light-dark cycle in rats with transplanted adrenal glands

(Gibson and Krieger, 1981) may indicate functional reinnervation of the transplants.

In addition to the examples mentioned above, a number of other conditions, including chronic stress, major depression and aging are characterized by altered cort/ACTH ratios either at rest or after challenges. Taken together, these observations clearly demonstrate that both under physiological and pathological conditions, the production of glucocorticoids by the adrenal gland is under control of circulating ACTH and of other mechanisms that can modulate adrenal responsiveness to adrenocorticotropic hormone.

EVIDENCE FOR AN INTRA-ADRENAL CRF MECHANISM

CRF immunoreactivity is present in adrenal extracts of mammals. Although the peptide has not been sequenced, the majority of the CRF immunoreactive material in extracts from dog, human and rat adrenal glands exhibits biological and chromatographic characteristics identical to that of authentic CRF (Hashimoto et al, 1984; Minamino et al, 1988; Suda et al, 1984; Bruhn et al, 1987ab). The exact cellular location of adrenal CRF stores is not fully clear and may show species specificity. CRF-immunoreactive varicose nerve fibers have been detected in various layers of the sheep adrenal cortex (Rundle et al, 1988). Although an extra-adrenal origin of these fibers cannot be excluded, they may well be intrinsic since CRF-positive neuron-like cells are present in the adrenal gland and occur in particularly high densities in the contact area between the cortex and the medulla. In dog adrenal gland, CRF appeared to be confined to cells located in the medulla and CRF positive varicose nerve fibers and cells were not detected in the adrenal cortex. These cells represent a specific population of adreno-medullary cells that are located in the contact zone between the medulla and the cortex (Bruhn et al, 1987ab). They have processes that, according to EM studies, contain storage particles and sites for release of CRF into blood (Bruhn et al, 1987ab).

Indeed, detailed studies on CRF concentrations in peripheral circulation of dogs and calves have revealed an arterio-venous gradient over the adrenal gland. In fact, Bruhn and colleagues (1987b) suggested that the rate of CRF secretion from the adrenal gland in dogs is of the same order of magnitude as that from the hypothalamus. The high CRF concentrations present in adrenal venous blood fall to undetectable levels in peripheral plasma and therefore are unlikely to contribute to the control of ACTH secretion from corticotroph cells. Release of CRF can be influenced by experimental manipulations. For instance, electrical stimulation of the splanchnic nerve was found to enhance the rate of CRF secretion from the adrenal gland (Edwards and Jones, 1988). In addition, hemorrhage in dogs has been reported to evoke massive secretion of CRF from the adrenal gland (Bruhn *et al*, 1987ab).

Relatively high densities of CRF receptors have been detected in membrane preparations of adrenal glands. Although significant numbers of CRF binding sites are present in rat and bovine adrenal cortex (Dave *et al*, 1985), binding studies and autoradiography revealed that the highest density of binding sites is present in the adrenal medulla (Dave *et al*, 1985; Udelsman *et al*, 1986; Aguilera *et al*, 1987). Accordingly, attention has been focused primarily on CRF receptors on chromaffin cells. Both in membrane preparations and in intact cultured chromaffin cells, CRF induced a dose-dependent increase in cyclase activity or cAMP accumulation (Dave *et al*, 1985; Aguilera *et al*, 1987). In view of these observations, possible effects of CRF on chromaffin cell activity were studied. As first demonstrated by Udelsman *et al* (1986), exposure of bovine chromaffin cells to CRF resulted in a dose-dependent increase in release of EPI, NE and enkephalin into the culture medium. Although it is tempting to speculate that the chromaffin cells secreting these substances exhibit CRF receptors, these responses are slow in onset and indirect effects cannot be excluded (Udelsman *et al*, 1986; Aguilera *et al*, 1987).

Taken together these observations lead us to believe that

1. CRF is produced by, stored in and secreted by certain cells in the adrenal gland in a neuronally controlled manner and

2. Although CRF can be detected in adrenal vein blood, it is likely that CRF exerts a physiological role in the local control of adrenal function.

ROLE OF CRF IN DIURNAL CHANGES IN CORT SECRETION

As discussed above, diurnal variations in plasma corticosterone concentrations in rats are due to small amplitude changes in plasma ACTH that run in phase with large amplitude changes in adrenal responsiveness to ACTH. In order to study the possible involvement of CRF in the diurnal swings in adrenal sensitivity to ACTH, we studied the effects of blockade of CRF signal transfer by the administration of a CRF antiserum (Van Oers and Tilders, 1991a). In these studies, endogenous CRF mechanisms were blocked by the administration of PFU83, a rat monoclonal antibody to r/h CRF in a dose that fully blocks ether-induced ACTH secretion in rats (Van Oers *et al*, 1989). As illustrated in Table 1, antibody administration resulted in complete blockade of the evening increase in plasma ACTH which is in accord with recent observations of others (Bagdy *et al*, 1991). Thus, it appears that in rats, resting ACTH secretion is driven by CRF only during the peak but not during the nadir of the diurnal cycle. In addition, blockade of CRF signal transfer induced a 15-fold reduction of evening cort levels which is in accord with recent findings (Bagdy *et al*, 1991) but not with earlier observations (Ixart *et al*, 1985; Carnes *et al*, 1989).

Thus, evening cort levels after pretreatment with the CRF monoclonal antibody were not different from morning cort levels in control rats. The fact that the same holds for plasma ACTH, is most surprising since the adrenal gland is expected to be more responsive or sensitive to circulating ACTH in the evening than during the nadir of the diurnal cycle. These observations lead us to postulate that blockade of endogenous CRF signal transfer not only prevents the evening rise in circulating ACTH, but more importantly inhibits the typical evening increase in adrenal sensitivity to ACTH. This observation implies that CRF directly or indirectly enhances adrenal responsiveness to ACTH.

Since the CRF antibody does not affect morning ACTH levels, but nevertheless suppresses morning corticosterone levels by 40-60%, we postulated that this CRF mechanism may operate throughout the day, although to a different extent in the morning than in the evening.

ENDOGENOUS CRF AND CONTROL OF ADRENAL RESPONSIVENESS TO ACTH

In order to test the hypothesis that endogenous CRF directly or indirectly enhances adrenal responsiveness to ACTH, we studied the adrenal response to exogenous ACTH in rats with or without blockade of endogenous CRF (Van Oers *et al*, 1991). In these studies, the secretion of endogenous ACTH was inhibited by dexamethasone

TABLE I.   Effects of immunoneutralization of endogenous CRF on morning and evening plasma levels of ACTH and corticosterone in handled male Wistar rats.

	1000 hours	1700 hours
	Plasma ACTH (pg/ml)	
Control IgG	54.3 ± 2.4	88.4 ± 8.0[a]
CRF MoAb	50.1 ± 2.0	53.0 ± 1.5[b]
	Plasma corticosterone (ng/ml)	
Control IgG	6.0 ± 0.9	87.4 ± 19.5[a]
CRF MoAb	2.6 ± 0.2[b]	6.7 ± 0.9[a,b]

[a] $p < 0.01$ morning versus evening group, [b] $p < 0.001$ CRF MoAb versus Control IgG group. Data represent means and SEM (n = 8). For details see Van Oers and Tilders (1991a).

(0.5 mg/kg sc) 30 minutes prior to the administration of the CRF monoclonal antibody or control IgG.   Rats were anesthetized by sodium pentobarbital 1.5 hours later and ACTH was given as an i.v. infusion for 60 minutes.   Preinfusion levels of ACTH were not affected by PFU83, whereas cort levels were reduced by the CRF antibody (Van Oers *et al*, 1991b) which is in harmony with the results mentioned earlier in non-dexamethasone treated rats (Table 1).   This in itself is an interesting finding indicating that the CRF mechanism involved is unlikely to be of hypothalamic origin since it appears relatively insensitive to dexamethasone.

Although the CRF monoclonal antibody did not affect steady state plasma ACTH concentrations reached during the infusion (Table 2), the cort levels reached by infusion with medial effective doses of ACTH (1 and 3 ng/kg.min) were reduced by approximately 35%.   In contrast, the response to ACTH in a dose inducing a (near) maximal physiological cort response was not affected by the CRF antibody.

These observations lead us to conclude that an endogenous CRF mechanism tonically potentiates the cort response of the adrenal gland to low circulating ACTH but not to high ACTH levels, indicating that this mechanism somehow controls the sensitivity of the adrenal gland to circulating ACTH.

SYNERGISTIC ACTION OF CRF AND ACTH ON THE ADRENAL GLAND

In view of the insensitivity to dexamethasone, it seems most unlikely that hypothalamic CRF plays a crucial role in the control of adrenal

TABLE 2. Effects of immunoneutralization of endogenous CRF on ACTH-induced corticosterone secretion in dexamethasone-treated anesthetized rats.

ACTH dose	plasma ACTH (pg/ml)		plasma corticosterone(ng/ml)	
	CNTRL	CRF MOaB	CNTRL	CRF MoAb
0	46 ± 7	38 ± 4	43 ± 24	13 ± 1
1	116 ± 21	109 ± 25	111 ± 11	72 ± 9
3	173 ± 15	188 ± 6	197 ± 161	35 ± 14
10	695 ± 111	186 ± 44	341 ± 293	43 ± 17

Results represent means and SEM (n = 4) of plasma levels of ACTH and corticosterone as measured after 60 minutes infusion with ACTH (dose 0, 1, 3 or 10 ng/kg.min). CNTRL: control IgG, CRF MoAb; PFU83. For details see Van Oers et al, 1991.

sensitivity to ACTH. For the same reason, the involvement of pituitary- derived POMC peptides is unlikely.

To test the possibility that CRF acts directly at the level of the adrenal gland, we used the perfused isolated adrenal gland *in situ* preparation as described previously (Sibley *et al*, 1981; Hinson *et al*, 1985). In this preparation, challenges with CRF alone in doses of 1-100 pmol did not affect adrenal cort output. However, in combination with a subeffective dose of ACTH, CRF induced a dose-dependent increase in corticosterone secretion to approximately 3-fold pre-challenge levels at 10 pmol. Apparently CRF only stimulates cort secretion when acting in synergism with ACTH (Van Oers *et al*, 1991). These observations are in agreement with recent findings demonstrating that CRF can stimulate adrenal cort secretion in pituitary stalk lesioned calves and hypophysectomized rats (Jones and Edwards, 1990; Andreis *et al*, 1991b).

This synergistic action of ACTH and CRF does not seem to take place at the level of the cort-producing cell itself. Studies with dispersed adrenal cells *in vitro* did not show effects of CRF alone or in combination with ACTH (Andreis *et al*, 1991b; Van Oers *et al*, 1991). Therefore, it is most likely that the effects of CRF are mediated by other cellular elements present in the adrenal gland. The exact nature of these cellular elements and of the signals involved remains to be uncovered.

　　　　Two hypotheses that are not mutually exclusive may play a role. A). The 'vascular response hypothesis' is based on our observation that CRF in combination with a subeffective dose of ACTH resulted in a marked increase in the perfusion rate of the isolated adrenal gland *in situ*. In fact, this response closely parallels the changes in cort output of the same preparations (Van Oers *et al*, 1991). This hypothesis is supported by the fact that increases in adrenal perfusion rate enhance the cort secretion rate per se (Hinson *et al*, 1986). Furthermore, the two stimuli that are known to enhance CRF secretion from the adrenal gland (hemorrhage and splanchnic stimulation) also increased adrenal perfusion rate and enhanced adrenal responsiveness to ACTH as discussed above.

B). The 'intra-adrenal CRF-ACTH-cort axis hypothesis' (Charlton, 1990) is based on the observation that small amounts of POMC derived products are present in the adrenal gland, are present in adrenal venous blood of functionally hypophysectomized animals, and are released in response to CRF challenge (Jones and Edwards, 1990). Recent findings demonstrating that CRF-induced cort secretion from adrenal fragments *in vitro* can be prevented by blockade of ACTH receptors (Andreis *et al*, 1991b) further support this hypothesis.

　　　　The results discussed in this chapter lead to the following conclusions.

(A). The mechanisms for CRF production, storage, secretion and action are all present in the adrenal gland and are possibly under control of the nervous system.

(B). Intra-adrenal CRF plays an important physiological role as a modulator of adrenal responsiveness to ACTH.

(C). Intra-adrenal CRF plays a crucial role in the control of diurnal variations in adrenal glucocorticoid secretion.

ACKNOWLEDGEMENTS

We thank Mr. R. Binnekade for his indispensable experimental contributions.

REFERENCES

Andreis, P.G., Neri, G., and Nussdorfer, G.G. (1991a). Corticotropin releasing hormone (CRH) directly stimulates corticosterone secretion by the rat adrenal gland. *Endocrinology* **128**, 1198-1200.

Andreis, P.G., Neri, G., Belloni, A.S., Mazzochi, G., Kasprzak, A., and Nussdorfer, G.G. (1991b). Interleukin-1β enhances corticosterone secretion by acting directly on the rat adrenal gland. *Endocrinology* **129**, 53-57.

Aguilera, G., Millan, M.A., Hauger, R.L., and Catt, K.J. (1987). Corticotropin releasing factor receptors: distribution and regulation in the brain, pituitary and peripheral tissues. *Annals of the New York Academy of Sciences* **512**, 48-66.

Bagdy, G., Chrousos, G.P., and Calogero. A.E. (1991). Circadian patterns of plasma immunoreactive corticotropin, beta-endorphin, corticosterone and prolactin after immunoneutralization of corticotropin-releasing hormone. *Neuroendocrinology* **53**, 573-578.

Bloom, S.R., Edwards, A.V., and Jones, C.T. (1987). Adrenal cortical responses to vasoactive intestinal polypeptide in conscious hypophysectomized calves. *Journal of Physiology* **391**, 441-450.

Bloom, S.R., Edwards, A.V., and Jones, C.T. (1988). The adrenal contribution to the neuroendocrine responses to splanchnic nerve stimulation in conscious calves. *Journal of Physiology* **397**, 513-526.

Breslow, M.J., Jordan, D.A., Thellman, S.T., and Traystman, R.J. (1987). Neural control of adrenal medullary and cortical blood flow during hemorrhage. *American Journal of Physiology* **252**, H521-H528.

Bruhn, T.O., Engeland, W.C., Anthony, E.L.P., Gann, D.S., and Jackson, I.M.D. (1987a). Corticotropin-releasing factor in the dog adrenal medulla is secreted in response to hemorrhage. *Endocrinology* **120**, 25-33.

Bruhn, T.O., Engeland, W.C., Antony, E.L.P., and Jackson, I.M.D. (1987b). Corticotropin-releasing factor in the adrenal medulla. *Annals of the New York Academy Sciences* **512**, 115-128.

Carnes, M., Lent, S.J., Erisman, S., Barksdale, C., and Feyzi, J. (1989). Immunoneutralization of corticotropin releasing hormone prevents the diurnal surge of ACTH. *Life Sciences* **45**, 1049-1056.

Charlton, B.G. (1990). Adrenal cortical innervation and glucocorticoid secretion. Journal of *Endocrinology* **126**, 5-8.

Dallman, M.F., Engeland, W.C., Rose, J.C., Wilkinson, C.W., Shinsako, J., and Siedenburg, F. (1978). Nycthemeral rhythm in adrenal responsiveness to ACTH. *American Journal of Physiology* **235**, R210-R218.

Dave, J.R., Eiden, L.E., and Eskay, R.L. (1985). Corticotropin-releasing factor binding to peripheral tissue and activation of the adenylate cyclase-adenosine 3'5'-monophosphate system. *Endocrinology* **116**, 2152-2159.

Dempsher, D.P., and Gann, D.S. (1983). Increased cortisol secretion after small hemorrhage is not attributable to changes in adrenocorticotropin. *Endocrinology* **113**, 86-93.

Edwards, A.V., and Jones, C.T. (1986). Reduced adrenal cortical sensitivity to ACTH in lambs with cut splanchnic nerves. *Journal of Endocrinology* **110**, 81-85.

Edwards, A.V., and Jones, C.T. (1987). The effect of splanchnic nerve stimulation on adrenocortical activity in conscious calves. *Journal of Physiology* **382**, 385-396.

Edwards, A.V., and Jones, C.T. (1988). Secretion of corticotropin releasing factor from the adrenal during splanchnic nerve stimulation in conscious calves. *Journal of Physiology* **400**, 89-100.

Engeland, W.C., Byrnes, G.J., Presnell, K., and Gann, D.S. (1981). Adrenocortical sensitivity to adrenocorticotropin (ACTH) in awake dogs changes as a function of the time of observation and after hemorrhage independently of changes in ACTH. *Endocrinology* **108**, 2149-2152.

Engeland, W.C., and Gann, D.S. (1989). Splanchnic nerve stimulation modulates steroid secretion in hypophysectomized dogs. *Neuroendocrinology* **50**, 124-131.

Engeland, W.C., Lilly, M.P., and Gann, D.S. (1985). Sympathetic adrenal denervation decreases adrenal blood flow without altering the cortisol response to hemorrhage. *Endocrinology* **117**, 1000-1010.

Fehm, H.L., Klein, E., Holl, R., and Voigt, H. (1984). Evidence for extrapituitary mechanisms mediating the morning peak of plasma cortisol in man. *Journal of Clinical Endocrinology and Metabolism* **58**, 410-414.

Gibson, M.J., and Krieger, D.T. (1981). Circadian rhythm and stress response in rats with adrenal autotransplants. *American Journal of Physiology* **240**, E363-366

Hadjian, A.J., Cultey, M., and Chambaz, E.M. (1982). Cholinergic muscarinic stimulation of steroidogenesis in bovine adrenal cortex fasciculata cell suspensions. *Biochimica and Biophysica Acta* **714**, 157-163.

Hashimoto, K., Murakami, K., Hattori, T., Niimi, M., Ujino, K., and Zensuke, O. (1984). Corticotropin-releasing factor (CRF)-like immunoreactivity in the adrenal medulla. *Peptides* **5**, 707-711.

Hinson, J.P. (1990). Paracrine control of adrenocortical function: a new role for the medulla? *Journal of Endocrinology* **124**, 7-9.

Hinson, J.P., and Vinson, G.P. (1990). Calcitonin gene-related peptide stimulates adrenocortical function in the isolated perfused rat adrenal gland *in situ*. *Neuropeptides* **16**, 129-133.

Hinson, J.P., Vinson, G.P., Kapas, S., and Teja, R. (1991). The role of endothelin in the control of adrenocortical function: Stimulation of endothelin release by ACTH and the effects of endothelin-1 and endothelin-3 on steroidogenesis in rat and human adrenocortical cells. *Journal of Endocrinology* **128**, in press.

Hinson, J.P., Vinson, G.P., Porter, I.D., and Whitehouse, B.J. (1987). Oxytocin and vasopressin stimulate steroid secretion by the isolated perfused adrenal gland. *Neuropeptides* **10**, 1-7.

Hinson, J.P., Vinson, G.P., Pudney, J., and Whitehouse, B.J. (1989). Adrenal mast cells modulate vascular and secretory responses in the intact adrenal gland of the rat. *Journal of Endocrinology* **121**, 253-260.

Hinson, J.P., Vinson, G.P., and Whitehouse, B.J. (1986). The relationship between perfusion medium flow rate and steroid secretion on the isolated perfused rat adrenal gland *in situ*. *Journal of Endocrinology* **111**, 391-396.

Hinson, J.P., Vinson, G.P., Whitehouse, B.J., and Price, G. (1985). Control of zona glomerulosa function in the isolated perfused rat adrenal gland *in situ*. *Journal of Endocrinology* **104**, 387-395.

Holzwarth, M.A., Cunningham, L.A., and Kleitman, N. (1987). The role of adrenal nerves in the regulation of adrenocortical functions. *Annals of the New York Academy of Sciences* **512**, 449-464.

Ixart, G., Conte-Devolx, B., Szafarczyk, A., Malaval, F., Oliver, C., and Assenmacher, I. (1985). L'immunisation passive avec une immune serum anti-oCRF 41 inhibe l'augmentation cicadienne de l'ACTH plasmatique chez le rat. *Comptes Rendu Academie Sciences* Paris III **14**, 659-664.

Jones, C.T., and Edwards, A.V. (1990). Adrenal responses to corticotropin releasing factor in conscious hypophysectomized calves. *Journal of Physiology* **430**, 25-36.

Jones, C.T., Edwards, A.V., and Bloom, S.R. (1990). The effect of changes in adrenal blood flow on adrenal cortical responses to adrenocorticotropin in conscious calves. *Journal of Physiology* **429**, 377-386.

Kaneko, M., Hiroshige, T., Shinsako, J., and Dallman, M.F. (1980). Diurnal changes in amplification of hormone rhythms in the adrenocortical system. *American Journal of Physiology* **239**, R309-R316.

Kaneko, M., Kaneko, K., Shinsako, J., and Dallman, M.F. (1981). Adrenal sensitivity to adrenocorticotropin varies diurnally. *Endocrinology* **109**, 70-75.

Minamino, N., Uehara, A., and Arimura, A. (1988). Biological and immunological characterization of corticotropin-releasing activity in the bovine adrenal medulla. *Peptides* **9**, 37-45.

Ottenweller, J.E., and Meier, A.H. (1982). Adrenal innervation may be an extrapituitary mechanism able to regulate adrenocortical rhythmicity in rats. *Endocrinology* **111**, 1334-1338.

Ottenweller, J.E., Meier, A.H., Ferrel, B.R., Horseman, N.D., and Proctor, A. (1978). Extrapituitary regulation of the circadian rhythm of plasma corticosteroid concentration in rats. *Endocrinology* **103**, 1875-1881.

Porter, J.C., and Klaiber, M.S. (1965). Corticosterone secretion in rats as a function of ACTH input and adrenal blood flow. *American Journal of Physiology* **209**, 811-814.

Rundle, S.E., Benedict, J.C., Robinson, P.M., and Funder, J.W. (1988). Innervation of the sheep adrenal cortex: an immunohistochemical study with rat corticotropin-releasing factor antiserum. *Neuroendocrinology* **48**, 8-15.

Schneider, E.G. (1988). Effect of vasopressin on adrenal steroidogenesis. *American Journal of Physiology* **255**, R806-R811.

Sibley, C.P., Whitehouse, B.J., Vinson, G.P., Goddard, C., and McCredie, E. (1981). Studies on the mechanism of secretion of corticosteroids by the isolated perfused adrenal of the rat. *Journal of Endocrinology* **91**, 313-323.

Sparrow, R.A., and Coupland, R.E. (1987). Bloodflow to the adrenal gland of the rat: Its distribution between the cortex and the medulla before and after hemorrhage. *Journal of Anatomy* **155**, 51-61.

Suda, T., Tomori, N., Tozawa, F., Demura, H., Shizume, K., Mouri, T., Miura, Y., and Sasano, N. (1984). Immunoreactive corticotropin and corticotropin releasing factor in human hypothalamus, adrenal, lung cancer and pheochromocytoma. *Journal of Clinical Endocrinology and Metabolism* **58**, 919-924.

Udelsman, R., Harwood, J.P., Millan, M.A., Chrousos, G.P., Goldstein, D.S., Zimlichman. R., Catt, K.J., and Aguilera, G. (1986). Functional corticotropin releasing factor receptors in the primate peripheral sympathetic nervous system. *Nature* **319**, 147-150.

Urquhart, J. (1965). Adrenal blood flow and the adrenocortical response to corticotropin. *American Journal of Physiology* **209**, 1162-1168.

Van Oers, J.W.A.M., and Tilders, F.J.H. (1991a). Antibodies in passive immunization studies: characteristics and consequences. *Endocrinology* **128**, 496-503.

Van Oers, J.W.A.M., and Tilders, F.J.H. (1991b). Non-adreno-corticotropin mediated effects of endogenous corticotropin releasing factor on the adrenocortical activity in the rat. *Journal of Neuroendocrinology* **3**, 119-121.

Van Oers, J.W.A.M., Tilders, F.J.H., and Berkenbosch, F. (1989). Characterization and biological activity of a rat monoclonal antibody to rat/human corticotropin-releasing factor. *Endocrinology* **124**, 1239-1246.

Van Oers, J.W.A.M., Hinson, J.P., Binnekade, R., and Tilders, F.J.H. (1991). Physiological role of corticotropin releasing factor in the control of ACTH mediated corticosterone release from the rat adrenal gland. *Endocrinology,* in press.

Wilkinson, C.W., Shinsako, J., and Dallman, M.F. (1981). Return of pituitary-adrenal function after adrenal enucleation or transplantation: Diurnal rhythms and responses to ether. *Endocrinology* **109**, 162-169.

Winter, J.S.D., Gow, K.W., Perry, Y.S., and Greenberg, A.H. (1990). A stimulatory effect of interleukin-1 on adrenocortical cortisol secretion mediated by prostaglandins. *Endocrinology* **127**, 1904-1909.

Wood, C.E., Shinsako, J., Keil, L.C., Ramsay, D.J., and Dallman, M.F. (1982). Apparent dissociation of adrenocorticotropin and corticosteroid responses to 15 ml/kg hemorrhage in conscious dogs. *Endocrinology* **110**, 1416-1421.

Stress: Neuroendocrine and Molecular Approaches
Edited by R. Kvetnansky, R. McCarty and J. Axelrod

1992 Gordon and Breach Science
Publishers S.A., New York, USA.
Photocopying permitted by license only.

RENIN AS A STRESS HORMONE

W. F. Ganong, N. C. Tkacs and E. C. Gryler

Department of Physiology, University of California,
San Francisco, California USA

INTRODUCTION

Renin secretion is increased by a number of psychological and other stimuli that most would call stressors. These include the psychological stress of immobilization (Jindra and Kvetnansky, 1982; Ganong and Barbieri, 1982) and more complex psychological stimuli such as handling and anticipation of electric shock in rats (Van de Kar *et al*, 1984), conditioned avoidance in monkeys (Blair *et al, 1976*), and mental arithmetic and IQ tests in humans (Ganong and Barbieri, 1982; Hollenberg *et al*, 1981). An increase in plasma renin activity (PRA) is produced in humans who under hypnosis imagine they are running without actually performing any physical activity (Ganong and Barbieri, 1982). In addition, renin secretion is increased during slow wave sleep and decreased during REM sleep (Brandenberger *et al*, 1990). Presumably, these stimuli act via the nervous system. Some of them also increase ACTH secretion, but others do not. Most but not necessarily all are known to produce increased sympathetic discharge to blood vessels and other viscera. Their significance in terms of vascular homeostasis and the pathogenesis of hypertension is uncertain. However, angiotensin converting enzyme inhibitors are known to lower blood pressure in hypertensive patients with normal or low circulating renin (Dzau, 1987), and it is possible that repeated increases in renin due to psychological and other stimuli could sensitize pressor mechanisms, leading to hypertension with normal levels of PRA. Therefore, it is of considerable interest to explore the areas and the transmitters in brain that are involved in the regulation of renin secretion and more broadly in the control of circulating levels of angiotensin II.

NEURAL CONTROL OF RENIN SECRETION

Role of the Raphe Nuclei

Some years ago, we commenced the task of exploring the neural pathways and transmitters by which divergent stimuli operate to increase renin secretion. One facet of this research was an investigation of the mechanism by which 5-hydroxytryptophan (5-HTP) increased renin secretion in dogs. With Zimmermann (Zimmermann and Ganong, 1980), we found that this increase was blocked by metergoline, which blocks 5-hydroxytryptamine (5-HT) receptors, particularly the $5-HT_1$ and $5-HT_2$ subtypes. It was potentiated by carbidopa, which inhibits dopa decarboxylase in the periphery but not in the brain. Therefore, the effect of 5-HTP was due to the release of 5-HT, and 5-HT release was in the central nervous system. The pathway from the central nervous system was sympathetic, since the response was blocked by renal denervation (Zimmermann and Ganong, 1980). In subsequent studies, we switched to rats and found that neurochemical or electrolytic lesions of the dorsal raphe nucleus blocked the renin response to p-chloroamphetamine (PCA), a drug which in the short term experiments we conducted acts as a 5-HT releasing drug. Lesions in other raphe nuclei have no effect on the response (Van de Kar *et al*, 1981, 1982). We then traced the efferent path from the dorsal raphe to the hypothalamus. Large lesions of the ventral hypothalamus reduced or abolished the renin response to PCA (Karteszi *et al*, 1982), as did knife cuts rostral and inferior to the dorsal raphe nucleus, which interrupted the neural connections of this nucleus to the hypothalamus and depleted hypothalamic serotonin. Control knife cuts of comparable magnitude in the anterior hypothalamus had no effect on the renin response. The final common pathway for the renin response to PCA is sympathetic, since the response was blocked by propranolol and other drugs which block ß-adrenergic receptors (Alper and Ganong, 1984). Norepinephrine (NE) released from postganglionic sympathetic fibers in the renal nerves and circulating catecholamines stimulate renin secretion primarily by acting directly on ß₁ adrenergic receptors on the renin secreting juxtaglomerular cells in the kidneys (Assaykeen and Ganong, 1971).

We next asked whether other stimuli for renin secretion act via the dorsal raphe nucleus and its serotonergic connections to the hypothalamus. Unfortunately, none of the stimuli we tested was affected; in rats with lesions of the dorsal raphe nuclei the renin responses to the psychological stimulus of immobilization, the postural stimulus of head-up tilt, and the volume stimulus of a low sodium diet were all normal. However, Van de Kar and his associates (Van de Kar *et al*, 1984) have reported that the renin response to another, more complicated psychological stimuli was blocked by dorsal raphe lesions. In their experiments, they exposed rats to unavoidable foot shock in a special cage for 3 consecutive days, then put the rats in the cage on the fourth day but did not turn on the current. This produced an increase in renin secretion that was blocked by dorsal raphe lesions. It may be that renin responses to complex psychological stimuli are mediated via the dorsal raphe nucleus. However, confirmation and extension of Van de Kar's results are needed to settle the point.

Other Stimuli

In the meantime, we conducted further research on the role of the hypothalamus in the regulation of renin secretion. The original hypothalamic lesions that blocked the PCA response were large, so we studied smaller bilateral lesions in the paraventricular nuclei, ventromedial nuclei, and dorsomedial nuclei. To determine which, if any, of these nuclei was involved in the day-to-day regulation of renin secretion, we tested PCA and 3 other stimuli in rats that were specifically selected because they would be expected to affect renin secretion by diverse pathways. These stimuli were the psychological stress of immobilization; the gravitational stimulus of head-up tilt; and the volume stimulus of a low sodium diet for 7 days.

It was first necessary to determine that these stimuli increased renin secretion via the sympathetic nervous system, since decreases in renal perfusion pressure can act directly on the kidneys to increase renin secretion via renal baroreceptor mechanism, and a decrease in NaCl in the fluid leaving the loops of Henle can increase renin secretion via a macula densa mechanism (Davis and Freeman, 1976). However, we found that the responses to immobilization and head-up tilt were readily blocked by propranolol (Golin *et al*, 1988). With prolonged or marked salt depletion, non-neural effects probably come into play, but in rats placed on a low sodium diet for 7 days, lesions in the brain totally abolished the renin response (see below). There was some residual elevation of renin secretion after propranolol in rats fed a low sodium diet for 9 days, but a marked reduction was still produced by ß-adrenergic blockade (Tkacs *et al*, 1990).

Effects of Paraventricular Lesions

We initially found that bilateral lesions of the paraventricular nuclei reduced the PRA response to PCA, immobilization, head-up tilt, and a low sodium diet. This was in contrast to dorsomedial lesions, which had no effect. However, the paraventricular lesions also lowered circulating angiotensinogen (Gotoh *et al*, 1987), and since the concentration of angiotensinogen as well as the concentration of renin are important in determining the amount of angiotensin I generated when rat plasma is incubated (Menard *et al*, 1983), the reduced PRA response could be due to angiotensinogen instead of a renin deficiency. Consequently, we re-analyzed the plasma specimens after adding excess exogenous angiotensinogen to them, i.e. we measured plasma renin concentration (PRC). The PRC response to PCA was still significantly reduced, but the PRC response to immobilization, head-up tilt, and a low sodium diet were normal (Gotoh *et al*, 1987). Thus, the effect on renin secretion of these 3 stimuli was not mediated by the paraventricular nuclei.

The observation that paraventricular lesions lowered circulating angiotensinogen led to a series of experiments designed to delineate the mechanism by which the reduction was produced. One possibility was that the effect was neural, and that nerves to the liver mediated increases in angiotensinogen secretion. However, in collaborative studies with Dr. J.P. Porter at the University of Louisville (Porter and Ganong, unpublished data), we found that electrical stimulation of the paraventricular nuclei failed to increase circulating angiotensinogen even though, as reported before (Porter, 1988), it produced a prompt increase in PRA and PRC.

Another possibility was that the effect was neuroendocrine. It has been known for many years that glucocorticoids, estrogens, and thyroid hormones increase circulating angiotensinogen by stimulating its secretion from the liver (Menard *et al*, 1983). Glucocorticoid secretion is regulated by ACTH from the anterior pituitary, and cells in the paraventricular nuclei synthesize CRH, which in turn controls the secretion of ACTH. Similarly, thyroid hormone secretion is controlled by TSH from the anterior pituitary, and neurons in the paraventricular nuclei produce the TRH which regulates TSH secretion. The secretion of estrogens is regulated by LH from the anterior pituitary and the LH-regulating hypophysiotropic hormone GnRH is produced in the hypothalamus, though not directly in paraventricular nuclei.

If the effect of paraventricular lesions on angiotensinogen secretion was neuroendocrine, it should be duplicated by removal of the pituitary. Consequently, we compared the effects of hypophysectomy to those of paraventricular lesions. Both produced comparable slow declines in plasma angiotensinogen (Kjos et al, 1991).

We also measured circulating ACTH, thyroid hormones, and LH in rats with paraventricular lesions. The lesions which lowered plasma angiotensinogen did not lower plasma LH, and lesions posterior to the paraventricular nuclei lowered plasma LH without affecting plasma angiotensinogen. Plasma ACTH was generally lower in the rats with paraventricular lesions and low plasma angiotensinogen, but the decline was not statistically significant. Previous studies had demonstrated that although paraventricular lesions prevent stress-induced increases in plasma ACTH, they do not reduce resting plasma ACTH (Ganong, 1963; Makara et al, 1986). We did not specifically avoid stress in handling and sacrificing our animals, and the somewhat lower ACTH values in the lesioned rats was probably explained on this basis. Since it remains possible that episodic increases in plasma ACTH and the stress of everyday living were necessary for normal plasma angiotensinogen levels, we did an ACTH replacement experiment, giving single doses of a short-acting ACTH preparation each day for 3 days. However, this treatment had no effect on the depressed circulating angiotensinogen in rats with paraventricular lesions.

On the other hand, replacement treatment with thyroxine (T_4) for 5 days restored plasma angiotensinogen to normal in rats with paraventricular lesions (Kjos et al, 1991), and there was a clear-cut positive correlation between plasma T_4 and plasma angiotensinogen. This indicated that the depression of plasma angiotensinogen in animals with paraventricular lesions was due to a decrease in thyroid hormone secretion produced by the lesions.

Effects of Ventromedial Lesions

Bilateral lesions of the ventromedial nuclei did not lower plasma angiotensinogen but did prevent increases in plasma PRA produced not only by PCA but by immobilization, head-up tilt, and a low sodium diet for 7 days (Gotoh *et al*, 1988). As noted above, the PRA response to all four of these stimuli was markedly inhibited by blocking ß-adrenergic receptors with propranolol, indicating they are neurally mediated via the sympathetic nervous system. Thus, it appears that the ventromedial nuclei are part of a neural system by which diverse sympathetic stimuli act to increase secretion of renin.

Recently, the role of the ventromedial nuclei has been explored further by Gryler in our laboratory. He found (Gryler, unpublished data) that electrical stimulation of many points in the ventromedial nuclei increased PRA. He subsequently sought to determine if this increase was due to stimulation of neuronal cell bodies in the ventromedial nuclei or due to stimulation of fibers of passage, presumably from more anterior areas. His results are still preliminary, but he failed to observe an increase in PRA when he injected the excitatory amino acid DL-homocysteic acid into many different locations in the ventromedial nuclei. Therefore, our preliminary impression is that the ventromedial nuclei are not an integrating center per se but that nerve fibers passing through them are part of a descending system regulating sympathetic output to the kidney.

DISCUSSION

The effects of the lesions discussed in this paper are summarized in Table 1. How many different inputs affect neural output to the juxtaglomerular cells? Serotonergic inputs mediate the response to PCA via the dorsal raphe nucleus. The inputs mediating the response to immobilization are presumably from the limbic system, whereas the inputs responsible for the increase produced by head-up tilt are presumably from the baroreceptors and the nucleus tractus solitarius. The input mediating the response to a low sodium diet is uncertain, but a good possibility would be that it is also mediated via arterial or central venous baroreceptors. Whether all the inputs converge on a single nucleus anterior to the ventromedial nuclei or whether they independently send fibers that happen to pass through the ventromedial nuclei remains to be determined by further research.

TABLE 1. Effect of brain lesions on the increase in plasma renin activity produced by various stimuli.

	Lesion of			
Stimulus:	Dorsal Raphe Nucleus	Paraventricular Nuclei	Ventromedial Nuclei	Dorsomedial Nuclei
PCA	-	-*	-	+
Immobilization	+	-**	-	+
Head-up Tilt	+	-**	-	+
Low Sodium Diet	+	-**	-	+

+ = present, - = absent. * plasma angiotensinogen depressed; PRC response reduced. ** plasma angiotensinogen depressed; PRC response normal.

REFERENCES

Alper, R.H., and Ganong, W.F. (1984). Pharmacological evidence that the sympathetic nervous system mediates the increase in renin secretion produced by para-chloroamphetamine. *Neuropharmacology* 23, 1237-1240.

Assaykeen, T.A., and Ganong, W.F. (1971). The sympathetic nervous system and renin secretion. In: L. Martini and W.F. Ganong (Eds.), "Frontiers in Neuroendocrinology, 1971," pp. 67-102. New York: Oxford University Press.

Blair, M.L., Feigl, E.O., and Smith, O.A. (1976). Elevation of plasma renin activity during avoidance performance in baboons. *American Journal of Physiology* 231, 772-776.

Brandenberger, G., Kraut, H.M., Ehrhart, J., Liebert, J.P., Simon, C., and Follenius, M. (1990). Modulation of episodic renin release during sleep in humans. *Hypertension* 15, 370-375.

Davis, J.O., and Freeman, R.H. (1976). Mechanisms regulating renin release. *Physiological Reviews* 45, 1-56.

Dzau, V.J. (1987). Implications of local angiotensin production in cardiovascular physiology and pharmacology. *American Journal of Cardiology* 59, 59A-65A.

Ganong, W.F. (1963). The central nervous system and the synthesis and release of ACTH. In: A.V. Nalbandov (Ed.), "Advances in Neuroendocrinology," pp. 92-149. Urbana, IL: University of Illinois Press.

Ganong, W.F., and Barbieri, C. (1982). Neuroendocrine components in the regulation of renin secretion. In: W.F. Ganong and L. Martini (Eds.), "Frontiers in Neuroendocrinology," Volume 7, pp. 231-262. New York: Raven Press.

Golin, R.M.A., Gotoh, E., Said, S.I., and Ganong, W.F. (1988). Pharmacological evidence that the sympathetic nervous system mediates the increase in renin secretion produced by immobilization and head up tilt in rats. *Neuropharmacology* 27, 1209-1213.

Gotoh, E., Murakami, K., Bahnson, T.D., and Ganong, W.F. (1987). Role of brain serotonergic pathways and hypothalamus in regulation of renin secretion. *American Journal of Physiology* 253, R179-R185.

Gotoh, E., Golin, R.M.A., and Ganong, W.F. (1988). Relation of the ventromedial nuclei of the hypothalamus to the regulation of renin secretion. *Neuroendocrinology* 47, 518-522.

Hollenberg, N.K., Williams, G.H., and Adams, D.S. (1981). Essential hypertension: Abnormal renal vascular and endocrine responses to a mild psychological stimulus. *Hypertension* 3, 11-17.

Jindra, A. Jr., and Kvetnansky, R. (1982). Stress-induced activation of inactive renin. *Journal of Biological Chemistry* 257, 5997-5999.

Karteszi, M., Van de Kar, L.D., Makara, G., Stark, E., and Ganong, W.G. (1982). Evidence that the mediobasal hypothalamus is involved in serotonergic stimulation of renin secretion. *Neuroendocrinology* 34, 323-326.

Kjos, T., Gotoh, E., Tkacs, N., Shackelford, R., and Ganong, W.F. (1991). Neuroendocrine regulation of plasma angiotensinogen. *Endocrinology* 129, 901-906.

Makara, G.B., Stark, E., Kapocs, G., and Antoni, F.A. (1986). Long term effects of hypothalamic paraventricular lesion on CRF content and stimulated ACTH secretion. *American Journal of Physiology* 250, E319-E324.

Menard, J., Bouhnik, J., Clauser, E., Richoux, J.P., and Corvol, P. (1983). Biochemistry and regulation of angiotensinogen. *Clinical and Experimental Hypertension* (A) 5, 1005-1019.

Porter, J.P. (1988). Electrical stimulation of paraventricular nucleus increases plasma renin activity. *American Journal of Physiology* 254, R325-R330.

Tkacs, N.T., Kim, M., Denzon, M., Hargrave, B., and Ganong, W.F. (1990). Pharmacological evidence for involvement of the sympathetic nervous system in the increase in renin secretion produced by a low sodium diet. *Life Sciences* 47, 2317-2322.

Van de Kar, L.D., Wilkinson, C.W., and Ganong, W.F. (1981). Pharmacological evidence for a role of brain serotonin in the maintenance of plasma renin activity in unanesthetized rats. *Journal of Pharmacology and Experimental Therapeutics* 219, 85-90.

Van de Kar, L.D., Wilkinson, C.W., Skrobik, Y., Brownfield, M.S., and Ganong, W.F. (1982). Evidence that serotonergic neurons in the dorsal raphe nucleus exert a stimulatory effect on the secretion of renin but not of corticosterone. *Brain Research* 235, 233-243.

Van de Kar, L.D., Lawrence, S.A., McWilliams, C.R., Kunimoto, K., and Bethea, C.L. (1984). Role of midbrain raphe in stress-induced renin and prolactin secretion. *Brain Research* 311, 333-341.

Zimmermann, H., and Ganong, W.F. (1980). Pharmacological evidence that stimulation of central serotonergic pathways increases renin secretion. *Neuroendocrinology* 30, 101-107.

Stress: Neuroendocrine and Molecular Approaches
Edited by R. Kvetnansky, R. McCarty and J. Axelrod

1992 Gordon and Breach Science
Publishers S.A., New York, USA.
Photocopying permitted by license only.

DEFINED SITES OF IMPACT OF MAGNESIUM MODULATING STRESS RESPONSES

S. Porta, W. Emsenhuber, H. G. Classen, J. Helbig,
K. Schauenstein, A. Epple and A. Ehrenberg

Institute of Functional Pathology, Endocrinological Research Unit,
University of Graz, Austria

INTRODUCTION

During the past year, we have concluded extensive experiments relating to the action of Magneium (Mg) in stress. Our work has included experiments with laboratory rats as well as many field experiments with commercially reared pigs and with Austrlian soldiers during training exercises.

Our objective was twofold. First we wished to show from an endocrinological perspective some defined and specific sites where Mg modulates parameters measured during stress. In this regard, we hope to counteract the widely spread notion of a "beneficial" effect of Mg in stress. Second, we examined the effects Mg supplement on organisms and hypothesized that such effects can only be seen when the organism is faced with an additional demand, which usually happens in an otherwise healthy organism during stressful stimulation.

MATERIALS AND METHODS

In the first experiment, 5 groups (38 animals per group) of Sprague-Dawley rats were pretreated for 10 days with 1 of 5 specially prepared

417

standardized laboratory chows (Institute for Nutritional Research, University of Hohenheim, Germany) of different Mg concentrations (Mg aspartate hydrochloride, VERLA PHARM, Tutzing, Germany). Concentrations of Mg ranged from 250 ppm (Mg deficient) to 1500 ppm (control) up to 4000, 6500 and 9000 ppm. After 5 days, 7 animals in each each diet group were sacrificed and the right femur used for Mg determination with Atomic Absorption Spectrometry (AAS). After 10 days of prefeeding, 7 more animals in each diet group were sacrificed and their right femur taken for Mg determination by AAS. Half of the remaining animals (12) were treated with a subcutaneously implantable epinephrine (EPI) tablet with a steady output rate of 1.1 microgram/ minute for 24 hours (Porta *et al*, 1978), the other half (12) received placebo tablets. After 24 hours and on the 11th day of the experiment, all test and placebo-treated animals were sacrificed and blood and tissue samples were collected.

Blood samples were processed for determination of insulin by a radioimmunoassay (RIA). Liver samples were frozen in liquid nitrogen for later glycogen determination and the adrenal glands were frozen and later assayed for content of corticosterone by RIA. Insulin content of the pancreas was measured according to the method of Ziegler *et al* (1985).

In the second experiment, 18 healthy members of the Austrian Theresianische Militärakademie (trainees for commissioned officers) were pretreated for 10 days with 5 mM Mg-aspartate hydrochloride (VERLA PHARM, Tutzing, Germany) or placebo in a double blind study. They were previously told about the purpose and hazards of the experiment according to the Helsinki Charter. In the morning of the 11th day they underwent the following regimen: cannulation of the antecubital vein, 30 minutes later the first blood sampling and after 5 minutes of exhaustive training using the Harvard-Step-Test, a second blood sample was collected. Three consecutive, evenly spaced blood samples were collected in the next 90 minutes followed by 5 minutes of the Harvard-Step-Test and a 6th and final blood sample. Plasma catecholamines (HPLC, Beckman Gold) and plasma cortisol (RIA) were determined. Additionally, mitogen stimulation (pokeweed antigen) of lymphocytes was carried out using the whole blood stimulation assay according to Mangge and co-workers (1990). The last test was performed with samples from only 10 persons, so we consider the results preliminary.

In the third experiment, two groups of 20 pigs were treated with either 5 mg Mg per day (Mg aspartate hydrochloride, VERLA PHARM, Tutzing, Germany), or placebo for 105 days of fattening in the stable. Afterwards, they were slaughtered in the usual way in the slaughterhouse. Non-cannulated blood (determination of catecholamines, HPLC, Beckman System Gold), and samples of the m. longissimus dorsi, the cutlet, (for electron microscopic investigations of mitochondria and for MAO determination) were taken and prepared accordingly (fixation or liquid nitrogen) as quickly as possible.

RESULTS

Five and 10 days of Mg treatment of rats ranging from deficient to high Mg levels resulted in a highly significant correlation between the amount of Mg in the diet and femoral Mg content. Note that there was Mg loss from bone in rats fed a Mg deficient diet (Figure 1).

Mg deficiency should increase synthesis and secretion of some endocrine glands, which is the case for pancreatic insulin production, especially during EPI application (Porta et al, 1984) (Figure 2). That the increase in pancreatic insulin content is not brought about by inhibition of insulin secretion is shown by the fact that there was no significant difference in serum insulin levels between the Mg deficient group and two groups fed diets with normal or high levels of Mg (Figure 3).

EPI treatment for 20 hours decreased liver glycogen content. Animals prefed with the highest Mg doses had triple the remaining glycogen concentration compared to normally fed rats (Figure 4).

A striking effect of high Mg pretreatment upon corticosterone levels during the 24 hours of EPI treatment was observed. Low Mg and normally fed rats showed a typical 2-fold increase of corticosterone during EPI compared to placebo-treated animals, but in animals fed the higher Mg diets (4000, 6500ppm), the effect was virtually abolished (Figure 5).

A Mg-induced curbing effect of cortisol secretion of a similar order of magnitude was seen during stress in young officer trainees, where stress-induced cortisol increases of about 100% (placebos) were nearly completely absent in Mg pretreated participants (Figure 6).

Also the highly significant increase in total (free and sulfate conjugated norepinephrine (NE) after exhaustive exercise in the placebo group did not occur in Mg pretreated subjects (Figure 7).

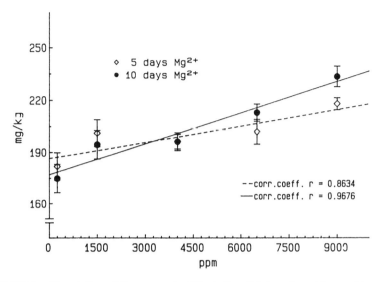

FIGURE 1.   Correlation of the amount of Mg in the diet (ppm) and femoral bone content of Mg (mg/kg) in laboratory rats. Values are means ± SEM for groups fed the diets for 5 or 10 days.

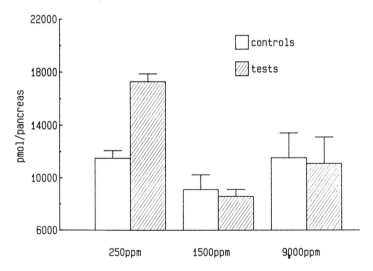

FIGURE 2.   Pancreatic insulin content (pmol/ pancreas) in rats fed deficient (250 ppm), control (1500 ppm) and supplemented (9000 ppm) Mg diets. Rats in each group served as placebo controls (open bars) or received EPI subcutaneously (tests, cross-hatched bars). Values are means and vertical lines denote 1 SEM.

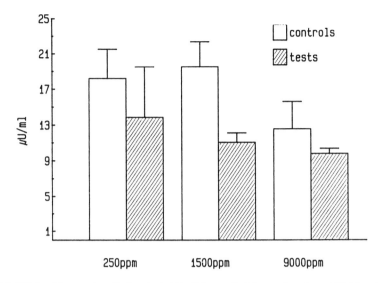

FIGURE 3.   Plasma insulin levels (μU/ml) in control (open bars) and EPI treated rats (cross-hatched bars) fed diets of varying Mg levels. Values are means and vertical lines denote 1 SEM.

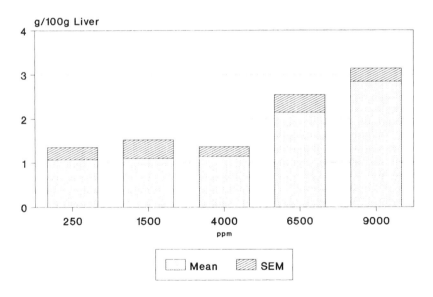

FIGURE 4.   Effects of Mg content of diet and EPI treatment on liver glycogen content (g/100g liver).

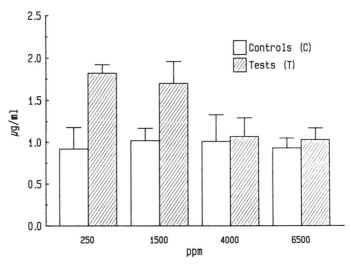

FIGURE 5.   Plasma corticosterone (μg/ml) in rats fed diets containing varying amounts of Mg (ppm). Rats in each diet group served as controls or received EPI. Values are means and vertical lines denote 1 SEM.

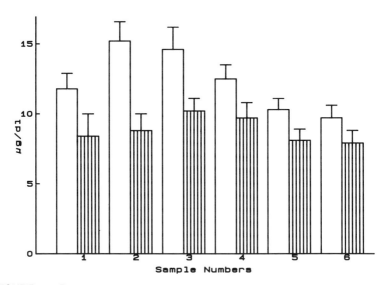

FIGURE 6.   Post-stress serum cortisol concentrations (μg/dl) in Mg pretreated officer trainees (hatched bars) compared to placebo controls (open bars). Values are means and vertical bars denote 1 SEM. Refer to text for details of the sampling protocol.

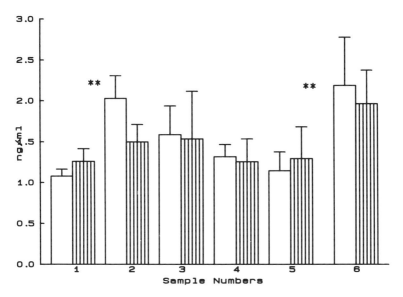

FIGURE 7.   Total plasma norepinephrine (ng/ml) in Mg pretreated officer trainees (hatched bars) compared to placebo controls (open bars). Values are means and vertical lines denote 1 SEM. Refer to text for details of the sampling protocol.

The stimulation of peripheral blood lymphocytes with pokeweed mitogen, which activates both B and T cells, did show a significant inhibition after the two stressors in the placebo group, but not in the Mg pretreated group (Figure 8, not determined in all individuals).

In pigs treated with 5mg of Mg for 105 days, a slight increase in MAO activity in skeletal muscle and a concomitant significant (-30%, $p < 0.05$) decrease in bound EPI compared to controls (from 100 ng/ml to 69 ng/ml) could be seen. There was less interstitial edema and more mitochondria of appropriate shape present in the muscle of Mg treated animals which seems to be important because of the significant negative correlation between MAO activity in skeletal muscle (but not in the liver) and plasma EPI values.

DISCUSSION

As mentioned above we set out to discover defined sites of action of Mg during stress, especially in relation to endocrine function. First of all, we had to establish that Mg depletion or supplementation actually

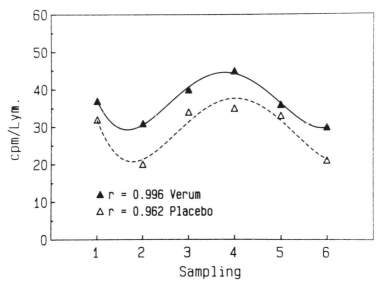

FIGURE 8. Effects of stress on lymphocyte stimulation in Mg treated (shaded triangle) and placebo control (open triangles) officer trainees. Refer to text for details of the sampling protocol.

alters Mg concentrations in organisms.   We did this by Mg determination in the femur, because serum Mg levels change only in response to exceptional high doses of Mg (Rayssiguier *et al*, 1990), and may even be high in a moment of stress-induced Mg depletion, when Mg is lost from tissues and may accumulate in blood (see below).

We established that different Mg pretreatment results in different Mg concentrations in bone, the main Mg reservoir. A 100% increase in pancreatic insulin in the Mg deficient animals together with no difference in secretion of insulin compared to the normal and high Mg groups indicated increased insulin synthesis but decreased peripheral insulin effectiveness (Rapado *et al*, 1990). This occurred even during Mg deficiency while animals underwent chronic EPI treatment, which usually inhibits synthesis and secretion of insulin (Porta *et al*, 1984).   Thus, Mg deficiency seemed to override the inhibiting effect of EPI.   The reason may lie in a down-regulation of adrenergic receptors during the 10 days of Mg deficiency and a concomittant increase in adrenal EPI secretion (Emsenhuber *et al*, 1990), so that adrenergic inhibitory effects could not occur, here via alpha receptors.

What is still speculation in the case of the pancreas has been investigated in the case of the liver. That Mg depletion goes via adrenergic receptors has been shown for α-adrenoceptors (Jakob et al, 1989) as well as β-adrenoceptors (Romani and Scarpa, 1990ab). If so, then the down-regulation of these receptors must be taken into consideration, especially during the 10 days of EPI secretion enhancement in combination with Mg deficiency. Indeed, Mg deficient animals did not respond to a greater degree to EPI-induced Mg depletion of the liver because of a significantly lower number of β-adrenoceptors. In adrenal medullectomized Mg deficient animals, there was a normal response because EPI hypersecretion from the medulla during the prefeeding time was blocked. For more extensive details, please refer to the chapter by Rauter and co-workers in this volume. High Mg pretreatment significantly reduces stress-induced liver glycogen depletion. This is also important for the continuation of EPI secretion, since we have shown (Porta et al, 1989) that too pronounced a fall in liver glycogen triggers adrenal medullary EPI secretion.

Corticosterone increases, usually triggered by EPI actions (Porta et al, 1987; Shimizu et al, 1984) were also completely abolished during a 24 hour EPI treatment in rats maintained on high Mg diets. This especially clear effect could not only be demonstrated under laboratory conditions in rats, but also during a field experiment with young army officers. Stress-induced cortisol increases were virtually abolished in Mg pretreated officier trainees, as suggested previously by Golf et al, (1983, 1984). A very similar reaction was observed concerning total plasma NE, which was significantly increased after stress in the placebo, but not in the Mg treated group. Also of interest were the differences in mitogen stimulation of lymphocytes, which, according to preliminary results in the case of the placebo group but not of the Mg pretreated group was always significantly suppressed during stress.

We have also examined the plasma EPI suppressing action of long-term feeding in pigs, since electrocution and slaughterhouse conditions do lead to significant elevations in plasma EPI values. Since most pigs are treated in this manner and this very treatment leads to enhanced plasma EPI concentrations and to watery and unsavory meat, a 30% reduction in EPI values, especially when measured under such conditions, is remarkable. Moreover, mitochondrial numbers and their shape as well as a slight increase in MAO activity in skeletal muscle during decreases in EPI plasma values are encouraging (Porta *et al*, 1985). The significant negative correlation between MAO activity in skeletal muscle and plasma EPI values suggests an important role of increased numbers of mitochondria in skeletal muscle in degradation of catecholamines. In conclusion, Mg pretreatment clearly and significantly curbs catecholamine and corticoid secretion during stress in rats and man. We have pointed to a probable direct mechanism in the case of Mg induced protection of liver glycogen reserves during hypersecretion of EPI. Down-regulation of catecholamine receptors which are responsible for tissue loss of Mg during stress creates a time-limiting factor for stress-induced tissue Mg turnover. A paradoxical increase in pancreatic insulin production during stress in Mg deficient rats may also be blamed for receptor down-regulation. Mg effects upon mitogen stimulation of lymphocytes and upon muscle mitochondria in connection with catecholamine degradation invite further investigations.

ACKNOWLEGEMENTS

This work was supported by VERLA-PHARM, Tutzing, Germany, and partly supported by a grant from the Austrian Federal Minister of Defense (105.982/1/91), as well as by grant number 7509 med of the Austrian Fonds zur Förderung der wissenschaftlichen Forschung. The authors are indebted to Mrs. M. Werner and Ms. H. Schunko for excellent technical assistance.

REFERENCES

Emsenhuber, W., Porta, S., Felsner, P., Supanz, S., Helbig, J., and Classen, H.G. (1990). Nur übernormale Mg Aufnahme kann die Katecholamin -sekretion während Stress vermindern. *Magnesium Bulletin* 12, 57-60.

Golf, S.W., Happel, O., and Graef, V. (1984). Plasma aldosterone, cortisol and electrolyte concentrations in physical exercise after magnesium supplementation. *Journal of Clinical Chemistry and Clinical Biochemistry* 22, 717-721.

Golf, V.S., Graef, V., Gerlach, J.J., and Seim, K.E. (1983). Veränderungen der Serum CK und Serum CK-MB Aktivitäten in Abhängigkeit von einer Magnesiumsubstitution bei Leistungssportlerinnen. *Magnesium Bulletin* 2, 43-46.

Jakob, A., Becher, J., Scotti, G., and Fritzsch, G. (1989). Alpha-1 adrenergic stimulation causes Mg^{2+} release from perfused rat liver. *FEBS Letters* **246**, 127-130.

Mangge, H., Beaufort, F., Neubauer, M., Samitz, M., and Schauenstein, K. (1990). Peripheral blood lymphocytes of nonleukemic lymphoma patients exhibit aberrant expression of T cell activation markers after polyclonal stimulation *in vivo*. *Cancer* **66**, 677-683.

Porta, S., Emsenhuber, W., Felsner, P., Schauenstein, K., and Supanz, S. (1989). Norepinephrine triggers medullar epinephrine depletion during normoglycemia. (1989). *Life Sciences* **45**, 1763-1769.

Porta, S., Emsenhuber, W., Helbig, C., Helbig, J., Waltersdorfer, G., and Classen, H.G. (1991). Reversed stress response of the endocrine rat pancreas in Mg depletion. In: B. Lasserre and J. Durlach (Eds.), "Magnesium, A Relevant Io," pp. 331-336. London: John Libbey and Company.

Porta, S., Egger, G., Sadjak, A., Supanz, S., Pürstner, P., and Höfler, H. (1984). Peculiar effects during controlled release adrenaline application in rats on insulin, glucose and pancreas. *Frontiers in Hormone Research* **12**, 26.

Porta, S., Egger, G., Kubat, R., Sattelberger, R., and Seewann, S. (1978). Der Einfluβ von Glucose und Fructose auf chronischen, adrenalin-induzierten Stress bei Ratten. In: O. Mayrhofer-Krammel (Ed.), "Kohlenhydrate in der Infusionstherapie," New York: Springer Verlag.

Porta, S., Rinner, I., Egger, G., Rangetiner, B., and Sadjak, A. (1985). Enhancement of adrenaline plasma levels shortens adrenaline half life. *Hormone and Metabolic Research* **17**, 264-265.

Rapado, A., Rovira, A., Grant, C., and Casla, A. (1990). Insulin resistance in patients with hypomagnesemia. *Magnesium Research* **3**, 56.

Rayssiguier, Y., Guezennec, C.Y., and Durlach, J. (1990). New experimental and clinical data on the relationship between magnesium and sport. *Magnesium Research* **3**, 93-102.

Ross, E.M., Maquire, M.E., Sturgill, T.W., Biltonen, R.L., and Gilman, A.G. (1977). The relationship of the β-adrenergic receptor and adenylate cyclase: studies of ligand binding and enzyme activity in purified membranes of S 49 lymphoma cells. *Journal of Biological Chemistry* **252**, 5761-5772.

Romani, A., and Scarpa, A. (1990a). Norepinephrine evokes marked Mg efflux from liver cells. *FEBS Letters* **269**, 37-40.

Romani, A., and Scarpa, A. (1990b). Hormonal control of Mg transport in the heart. *Nature* **346**, 841-844.

Shimizu, K. (1984). Effect of alpha$_1$ and alpha$_2$ adrenoceptor agonists and antagonists on ACTH secretion in intact and in hypothalamic deafferentated rats. *Japanese Journal of Pharmacology* **36**, 23-33.

Ziegler, B., Hahn, H.-J., and Ziegler, M. (1985). Insulin recovery in pancreas and host organs of islet grafts. *Experimental and Clinical Endocrinology* **85**, 53-60.

Stress: Neuroendocrine and Molecular Approaches
Edited by R. Kvetnansky, R. McCarty and J. Axelrod

1992 Gordon and Breach Science
Publishers S.A., New York, USA.
Photocopying permitted by license only.

PERMANENT RESTORATION OF PITUITARY-ADRENOCORTICAL FUNCTION IN RATS WITH INHERITED STRESS-INDUCED ARTERIAL HYPERTENSION (ISIAH RATS)

E. V. Naumenko, L. N. Maslova, G. T. Shishkina
and A. L. Markel

Institute of Cytology and Genetics, Siberian Branch of the
Academy of Science of the USSR, Novosibirsk, USSR

INTRODUCTION

It is well known that the development of arterial hypertension in humans (Kaplan, 1986) and animals (Sowers *et al*, 1981; Hashimo *et al*, 1985; Hattori *et al*, 1986) is often accompanied by a change in function of the hypothalamic-pituitary-adrenocortical (HPA) system. This phenomenon depends partly on alterations of brain catecholamine (CA) mechanisms regulating both the level of arterial blood pressure and HPA function.

Recently, central adrenergic mechanisms involved in HPA regulation were studied in a newly developed strain of rats with inherited emotional stress-induced arterial hypertension (ISIAH rats). It appeared that selection for arterial hypertension did not affect basal corticosteroid levels. At the same time, the HPA response to emotional stress or to icv norepinephrine (NE) administration was decreased in ISIAH rats compared to normotensive Wistar controls (Naumenko *et al*, 1989a). The suppression of these responses was mediated through hereditary changes brought about by selection for specific adrenoceptor brain mechanisms. Alpha$_1$-adrenoceptors seem to play a major role in these alterations (Naumenko *et al*, 1989a).

Recent research showed that a brief increase in synthesis of NE and perhaps epinephrine (EPI) provoked by L-DOPA administration at 21-25 days of age was associated with a normalization in arterial blood pressure persisting over a long period of time in ISIAH rats (Naumenko *et al*, 1989b, 1990). Therefore, it may be that the effect of CA treatment in early life results in a permanent restoration of the mechanisms regulating HPA function in hypertensive rats, as well.

Here we describe our search for possible ways for permanent normalization of HPA function in ISIAH rats by activation of CA metabolism during a critical period of early life.

METHODS

The ISIAH strain was produced from Wistar rats.   A protocol describing the selection procedures was published in detail elsewhere (Markel *et al*, 1987).   In adult 4-5 month old ISIAH males used in these experiments, the basal level of arterial blood pressure was 171±3 mm Hg, and in Wistar males of the same age it was 120±3 mm Hg. The levels induced by emotional stress (1 hour of restriction in a small cage) were 200±3 and 136±3 mm Hg, respectively.

In each experimental series some of the litters were left intact and the rest were subjected to the experimental procedure.   At 21-25 days of age, the animals of some litters received daily injections of an aqueous suspension of L-DOPA (Reanal, Hungary, 1 mg/0.1 ml/10 g body weight, i.p.).   In the second series of experiments, some of the litters were treated with an aqueous suspension of Nakom (Lek, Yugoslavia) consisting of a combination (10:1) of L-DOPA and a peripheral L-amino acid decarboxylase inhibitor Carbidopa (1 mg and 100μg, respectively/0.1 ml/10 g body weight, i.p.).   Age-matched control rats received distilled water (0.1 ml/10 g body weight).

Young rats were weaned on day 30.   At 45 days of life they were separated by sex and housed in groups of 6 until 4-5 months of age.   Only adult males were used.   A week before the experiments, rats were removed from their home cages and placed into individual cages.

The HPA response to central adrenoceptor stimulation was studied in conscious rats 1 hour after NE bitartrate (Koch-Light,

England) administration into the lateral brain ventricle ($15\mu g/10\mu l$ of saline at pH 5.9).

At the time of decapitation, brains were removed and placed immediately on ice. The frontal cortex, hypothalamus, midbrain, pons and medulla were isolated. Dopamine (DA) and NE concentrations were determined flourimetrically (Schlumpf et al, 1974). The binding assays for alpha- and beta- adrenoceptors were performed using ^{3}H-prazosin (24.4 Ci/mmol, Amersham, England), an alpha$_1$-adrenoceptor antagonist (Greengrass and Bremmer, 1979), ^{3}H-clonodine (23.3 Ci/mmol, Amersham, England) an alpha$_2$-adrenoceptor agonist (U'Prichard et al, 1979), and ^{3}H-dihydroalprenolol (75.0 Ci/mmol, Amersham, England) a nonspecific beta-adrenoceptor antagonist (U'Prichard et al, 1987).

Determination of protein was performed by the method of Lowry et al (1951). Corticosterone was determined in peripheral blood plasma by a fluorometric method (Stahl and Dorner, 1966) or a competitive protein-binding radioassay (Murphy, 1967).

Student's t-test was used in the statistical treatment of the data.

RESULTS

Basal levels of plasma corticosterone were similar in adult ISIAH and Wistar rats. However, the response of the HPA axis induced by emotional stress or icv NE administration was significantly lower in hypertensive animals than in normotensive controls (Table 1).

Daily injection of L-DOPA on days 21-25 of postnatal life was associated with a significant increase of emotional stress responses and of the reponse of the HPA system to icv administration of NE in adult ISIAH rats up to the level of control normotensive Wistar strain (Table 1). At the same time, the responses of the HPA system to emotional stress or to icv NE injection were not changed in adult Wistar rats after L-DOPA treatment on days 21-25 of life (Table 1).

Administration of L-DOPA in combination with an inhibitor of peripheral aromatic l-amino acid decarboxylase (Carbidopa) on the same days did not change the restorative effect of the CA precursor in adult hypertensive animals. In these rats the response of the HPA system to both emotional stress and icv NE administration was similar to that in Wistar rats (Table 2).

The administration of L-DOPA in combination with Carbidopa

TABLE 1. Levels of plasma corticosterone (μg/dl, Mean ± SEM) in adult ISIAH rats treated with L-DOPA on days 21-25 of life.

Series of experiments	ISIAH	Wistar	P value ISIAH versus Wistar
	Resting level		
Intact	12.6±1.49(10)	11.8±1.31(9)	NS
Water	10.4±1.62(21)	8.2±1.07(10)	NS
L-DOPA	12.4±2.05(20)	13.3±1.04(10)	NS
	Emotional stress		
Intact	18.2±2.42(8)	33.8±2.57(10)[b]	0.001
Water	19.7±2.94(9)[a]	36.1±4.21(10)[b]	<0.01
L-DOPA	36.7±3.37(11)[b]	26.8±3.63(10)[b]	NS
	Saline, icv		
Water	8.9±1.04(5)	7.6±0.80(3)	NS
L-DOPA	9.8±1.11(5)	13.1±4.19(4)	NS
	Norepinephrine, icv		
Water	7.5±1.73(4)	38.9±1.74(5)[d]	<0.001
L-DOPA	41.6±3.33(6)[cd]	36.4±5.45(6)[e]	NS

Number of rats is given in parentheses. NS, nonsignificant; [a]P<0.05, [b]P<0.01 versus resting level; [c]P<0.001 versus water-treated rats; [d]P<0.001, [e]P<0.01 versus saline control.

TABLE 2. Levels of plasma corticosterone (μg/dl, Mean ± SEM) in adult ISIAH rats treated with L-DOPA + Carbidopa on days 21-25 of life and in intact Wistar rats.

Series of experiments	Water	L-DOPA+ Carbidopa	Wistar Intact
Resting level	2.4 ± 1.13 (10)	2.8 ± 1.04 (10)	6.6 ± 1.19 (10)[a]
Emotional stress	25.8 ± 2.52 (12)[b]	47.1 ± 4.13 (14)[b]	42.7 ± 5.54 (10)[b]
Saline, icv	4.2 ± 1.35 (4)	6.7 ± 1.85 (4)	4.8 ± 1.22 (4)
Norepinephrine icv	9.3 ± 1.52 (5)[c]	27.4 ± 2.97 (5)[eh]	32.2 ± 6.19 (5)[dg]

Number of rats is given in parentheses. [a]P<0.05 versus ISIAH rats; [b]P<0.001 versus resting level; [c]P<0.05, [d]P<0.01, [e]P<0.001 versus saline control; [g]P<0.01, [h]P<0.001 versus water-treated ISIAH rats.

was followed by marked changes of the brain CA system in ISIAH rats. A significant decrease in concentrations of DA in the midbrain, pons and medulla, and an increase in NE content in the hypothalamus and medulla were obtained in adults who had received the CA precursor and Carbidopa on days 21-25 of life (Table 3).

After treatments in early ontogeny, there were no significant changes in either density or affinity of adrenoceptors in the frontal cortex, hypothalamus and pons in adult ISIAH rats. In the medulla, B_{max} and K_d for alpha$_2$- and beta- adrenoceptors were not changed as well, whereas the number of alpha$_1$- adrenoceptors was significantly decreased in L-DOPA and Carbidopa-treated rats (Table 4).

DISCUSSION

The present data indicate that genetic alterations of HPA function in ISIAH rats can be corrected by drugs influencing CA metabolism during a short period in early postnatal ontogeny. The decreased response of the HPA system to emotional stress or to icv NE injection was restored completely in adult hypertensive animals by injection of the CA precursor L-DOPA at 21-25 days of age. Similar normalizing effects were obtained after L-DOPA treatment at the same age period on arterial blood pressure (Naumenko et al, 1989b, 1990). However, when L-DOPA was given at an age earlier than the 16th day of life, it did not provide any improving effect on arterial blood pressure (Naumenko et al, 1989b) or HPA function (Maslova et al, 1991).

TABLE 3.   Dopamine and norepinephrine content (µg/g tissue, Mean ± SEM) in brain regions of adult rats treated with L-DOPA + Carbidopa on days 21-25 of life.

Brain region	Dopamine		Norepinephrine	
	Water	L-DOPA + Carbidopa	Water	L-DOPA + Carbidopa
Cortex	0.316±0.046	0.239±0.028	0.245±0.028	0.258±0.012
Hypothalamus	0.800±0.087	0.753±0.054	1.170±0.120	1.748±0.059[c]
Midbrain	0.724±0.047	0.517±0.042[b]	0.427±0.029	0.422±0.017
Pons	0.442±0.055	0.275±0.018[a]	0.539±0.016	0.508±0.018
Medulla	0.339±0.028	0.249±0.022[a]	0.484±0.011	0.519±0.012[a]

Each group consisted of 10 rats. [a]P<0.05, [b]P<0.01, [c]P<0.001 versus water-treated rats.

TABLE 4.   Specific binding of the alpha $_1$-adrenoceptor ligand ^{3}H-prazosin in brain regions of adult ISIAH rats treated with L-DOPA + Carbidopa on days 21-25 of life (Mean ± SEM).

| Brain region | B_{max}, fmol/mg protein | | K_d, nM | |
	Water	L-DOPA + Carbidopa	Water	L-DOPA + Carbidopa
Cortex	107.95±6.54	101.47±4.07	0.104±0.014	0.105±0.010
Hypothalamus	74.95±6.11	81.55±4.52	0.085±0.018	0.092±0.011
Pons	37.36±1.68	37.15±2.35	0.086±0.009	0.084±0.015
Medulla	53.20±1.62	32.42±2.89[a]	0.101±0.007	0.074±0.018

Each group consisted of 5 rats. [a]$P < 0.001$ versus water-treated animals.

L-DOPA was administered on days 21-25 of life at a dose that affected both peripheral and central CA systems. Therefore, no exact conclusions regarding the role of peripheral or central CAs in the restoration of HPA function may be drawn from the results of these experiments. Meanwhile, the experiments with injections of the same dose of L-DOPA in combination with Carbidopa, the peripheral decarboxylase inhibitor, which did not change the effect of L-DOPA, clearly indicate that the permanent normalization of HPA function in ISIAH rats depends on brain CA systems.

Previously, it was found that injection of FLA-57, an inhibitor of dopamine-beta-hydroxylase, abolished the restorative effect of L-DOPA on the HPA system of adult rats when given 3 hours before L-DOPA (Maslova *et al*, 1991). These results suggest that the effect of L-DOPA on the HPA axis is produced through an enhancement of NE and, possibly EPI synthesis. DA increases at early ages do not seem to be involved in this effect.

The question arises, which of the links of brain noradrenergic mechanisms are involved in the effects of L-DOPA? Previously, it was shown that, in contrast to Wistar rats, icv NE administration to ISIAH rats did not result in an increase of plasma corticosterone levels. Moreover, icv phenylephrine administration, activating the HPA system in Wistar rats, caused a suppression of plasma corticosterone levels in ISIAH rats (Naumenko *et al*, 1989a). Besides, the density of alpha$_1$-adrenoceptors increased in the medulla of ISIAH rats as compared to Wistar rats (Shishkina *et al*, 1991).

In the present work, it was revealed that treatments with L-DOPA alone or in combination with Carbidopa at an early age were associated with a significant increase of HPA responses to emotional stress as well as to icv NE administration in adult hypertensive rats up to levels of the Wistar control strain. In addition, the density of alpha$_1$- adrenoceptors was decreased in the medulla of adult ISIAH rats treated with L-DOPA and Carbidopa during early ontogeny. Therefore, one of the effects of L-DOPA may be a normalization of the number of alpha$_1$- adrenoceptors in the medulla.

Recently in our laboratory, we found that concentrations of NE were decreased in the hypothalamus, midbrain, hindbrain and in areas of the locus coeruleus and nucleus tractus solitarius in ISIAH rats compared with Wistar controls (Gordienko, 1990). In the present experiments, we found a marked increase of NE concentrations in the hypothalamus and medulla, i.e., in the brain regions which participate in regulation of both arterial blood pressure and HPA function (Plotsky et al, 1989) in adult ISIAH rats treated with L-DOPA and Carbidopa at 21-25 days of age. These data are consistent with our previous investigation of tyrosine hydroxylase activity in brain regions of L-DOPA-treated ISIAH rats. The level of tyrosine hydroxylase activity was found to be higher in the hindbrain and frontal cortex of adult ISIAH rats treated with CA precursors in early ontogeny compared to control hypertensive animals (Naumenko et al, 1989b). Therefore, the second cause of the restorative effect of CA precursors may be changes in brain NE metabolism.

In contrast to NE, the role of changes in brain DA concentrations is more difficult to discern from available data. On the one hand, the experiments with an inhibitor of dopamine-beta-hydroxylase showed that the effect of L-DOPA was produced through an enhancement of synthesis of brain NE and perhaps EPI. DA did not seem to be involved in this effect (Maslova et al, 1991). On the other hand, in adulthood, a clear-cut decrease in DA content was found in the midbrain, pons and medulla in L-DOPA and Carbidopa treated ISIAH rats. It may be assumed that brain DA changes found in adult hypertensive animals after L-DOPA administration during early ontogeny are associated not with the mechanism of HPA regulation but with a mechanism regulating another neuroendocrine system. For example, it is well known that a brain DAergic mechanism is involved in control of the hypothalamic-pituitary-gonadal complex (Weiner and Ganong, 1978; Naumenko, 1985), and alterations in pituitary-testicular function have been reported in spontaneously hypertensive rats (Rodriguez-Padilla et al, 1987).

Therefore, the results demonstrate that the long-lasting restorative effect of L-DOPA on HPA function of genetically hypertensive rats was caused by an increase of brain CA synthesis in early ontogeny. In adults, this restorative effect was associated with (i) a normalization of brain adrenoceptor responses to NE, (ii) a decrease of medulla alpha$_1$-adrenoceptor number, and (iii) an elevation of NE content in the hypothalamic and medulla.

REFERENCES

Gordienko, N.I. (1990). Brain catecholamine levels in rats with inherited stress-induced arterial hypertension. *Patologicheskaya Fiziologiya Experimental'naya Terapia* N3, 38-40 (in Russian).

Greengrass, P., and Bremmer, R. (1971). Binding characteristics of ^{3}H-prazosin to rat brain adrenergic receptors. *European Journal of Pharmacology* 55, 323-326.

Hashimoto, K., Hattori,T., Murakami, K., Suemaru,S., Kawada, Y., Kageyama, J., and Ota, Z. (1985). Reduction in brain immunoreactive corticotropin-releasing factor (CRF) in spontaneously hypertensive rats. *Life Sciences* 36, 643-647.

Hattori, T., Hashimoto, K., and Ota, Z. (1986). Brain corticotropin releasing factor in the spontaneously hypertensive rat. *Hypertension* 8, 1027-1031 .

Kaplan, N.M. (1986). "Clinical Hypertension". Baltimore, Williams & Wilkins.

Lowry, O.H., Rosenbrough, M.J., Parr, A., and Randall, R.J. (1951). Protein measurement with the Folin phenol reagent. *Journal of Biological Chemistry* 193, 265-275.

Markel, A.L., Amstislavski, S.Y., and Naumenko, E.V. (1987). The central adrenergic mechanisms of blood pressure regulation in rats with inherited arterial hypertension. *Biogenic Amines* 4, 339-338.

Maslova, L.N., Markel, A.L., and Naumenko, E.V. (1991). Treatment with L-DOPA in early life restored pituitary-adrenocortical response to emotional stress in adult rats with inherited arterial hypertension. *Brain Research* 546, 55-60.

Murphy, B.E.P. (1967). Some studies of the protein-binding of steroids and their application to the routine micro and ultramicro measurement of various steroids in body fluids by competitive protein binding radioassays. *Journal of Clinical Endocrinology and Metabolism* 27, 973-990.

Naumenko, E.V. (1985). Role of catecholamines and serotonin in the control of the hypothalamic-pituitary-testicular complex - A review. *Biogenic Amines* 2, 173-190.

Naumenko, E.V., Markel, A.L., Amstislavsky, S.Y., and Dygalo, N.N. (1989a). Brain adrenergic mechanisms of adrenocortical regulation in rats with inherited stress-induced arterial hypertension. In: G.R. Van Loon, R. Kvetnansky, R. McCarty and J. Axelrod (Eds.), "Stress: Neurochemical and Humoral Mechanisms," pp.453-460. New York: Gordon and Breach.

Naumenko, E.V., Maslova, L.N., Gordienko, N.I., Amstislavski, S.Y., Dygalo, N.N., and Markel, A.L. (1989b). Persistent hypotensive effect of L-DOPA given early during development to rats with inherited stress-induced arterial hypertension. *Developmental Brain Research* **46**, 205-212.

Naumenko, E.V., Maslova, L.N., and Markel, A.L. (1990). Correction of arterial blood pressure in adult rats with inherited stress-induced arterial hypertension by enhancement of catecholamine metabolism in early postnatal period. *Endocrinologia Experimentalis* **24**, 421-428.

Plotsky, P.M., Bruhn, T.O., and Vale, W. (1985). Evidence for multifactor regulation of the adrenocorticotropin secretory response to hemodynamic stimuli. *Endocrinology* **116**, 633-639.

Rodriguez-Padilla, M., Bellido, C., Pinilla, L., and Aguilar, E. (1987). Secretion of LH in spontaneously hypertensive rats. *Journal of Endocrinology* **113**, 225-260.

Schlumpf, M., Lichtensteiger, W., Langemann, H., Waser, P.Y., and Hefti, F. (1974). A flourimetric micromethod for the simultaneous determination of serotonin, noradrenaline and dopamine in milligram amounts of brain tissue. *Biochemical Pharmacology* **23**, 2437-2446.

Shishkina, G.T., Markel, A.L., and Naumenko, E.V. (1991). Brain adrenoceptors in rats with inherited stress-sensitive arterial hypertension. *Genetika* **27**, 279-284 (in Russian).

Sowers, J., Tuck, M., Asp, N.D., and Sollars, E. (1981). Plasma aldosterone and corticosterone response to adrenocorticotropin, angiotensin, potassium and stress in spontaneously hypertensive rats. *Endocrinology* **108**, 1216-1221.

Stahl, F., and Dorner, G. (1966). Eine einfache spezifische routine Mehtode zur fluorometrischen Bestimmung von unkonjugierten 11-hydroxycorticosteroiden in Korperflussigkeiten. *Acta Endocrinologica* **51**, 175-185.

U'Prichard, D.C., Bylund, D.B., and Snyder, S.H. (1987). $(+)$-^{3}H-epinephrine and $(-)$-^{3}H-dihydroalprenolol binding to beta$_1$- and beta$_2$-noradrenergic receptors in brain, heart and lung membranes. *Journal of Biological Chemistry* **253**, 5090-5102.

U'Prichard, D.C., Bechtel, W.D., Rouot, B.N., and Snyder, S.H. (1979). Multiple apparent alpha-noradrenergic receptor binding sites in rat brain: effect of 6-hydroxydopamine. *Molecular Pharmacology* **16**, 47-60.

Weiner, R.I., and Ganong, W.F. (1978). Role of brain monoamines and histamine in regulation of anterior pituitary secretion. *Physiological Review* **58**, 905-976.

Stress: Neuroendocrine and Molecular Approaches
Edited by R. Kvetnansky, R. McCarty and J. Axelrod

1992 Gordon and Breach Science
Publishers S.A., New York, USA.
Photocopying permitted by license only.

A NEW PERSPECTIVE FOR THE STUDY OF CENTRAL NEURONAL NETWORKS IMPLICATED IN STRESS

F. Menzaghi[1], A. Burlet[1], J. W. A. M. Van Oers[2],
G. Barnanel[3], F. J. H. Tilders[2], J. P. Nicolas[1] and C. Burlet[1]

[1]Laboratory of Cellular Biology, INSERM 308, Nancy, France
[2]Department of Pharmacology, Free University, Amsterdam
The Netherlands
[3]URA 1197 CNRS, Montpellier, France

INTRODUCTION

Various peptides have been implicated in the hypothalamic control of adrenocorticotropin (ACTH) responses to a stressor (Dallman *et al*, 1987). However, contrasting with the relative ease with which the biological effects of peptides are established, studies on their physiological roles are complicated and often hampered by the lack of appropriate tools. Thus, the neuronal networks which may control the ACTH responses to different stressors remain unclear. As an alternative, we propose an immunopharmacological strategy which could lead to long-term disturbances and even to selective lesions of some hypothalamic peptidergic neurons.

Recently, we demonstrated that local injection of ascites fluid containing an $IgG_{2\alpha}$ monoclonal antibody (MAb) to specific peptides such as vasopressin (VP) or corticotropin releasing factor (CRF) close to the paraventricular nuclei (PVN) of rats leads to specific antibody uptake, subsequent specific anterograde transport (Burlet *et al*, 1987; Leon-Henri *et al*, 1989; Burlet *et al*, 1990a) and short-lasting biological effects (less than 24 hours) (Burlet *et al*, 1990b; Menzaghi *et al*, 1991).

It is not fully clear by what mechanisms antibodies act and become internalized into specific cells (Burlet et al, 1987,1990a; Tilders et al, 1990). Whatever the mechanism(s) involved in the specific recognition of peptidergic neurons and internalization of MAb, it has been reported that the formation of an immune-complex in rat brain initiates the stimulation of endogenous complement pathway (Schupf et al, 1987) and results in the induction of membrane pores on the targeted cells (Howard et al, 1988; Podack et al, 1984) which subsequently allow passive passage of IgG across the membrane (Bhakdi et al, 1987). In this way, we hypothesized that toxins such as ricin A chain, which can not cross cell membranes, may passively penetrate some neurons via these porous membranes and produce some subsequent specific irreversible damage. Our preliminary observations that the duration of the biological effects of VP-MAb were extended when toxins were injected simultaneously with the MAb supported this hypothesis (unpublished data).

Based on these observations, it was deemed of interest to study whether administration of similar toxin mixtures could disturb one peptidergic pathway which plays an important role in the regulation of ACTH responses to a stressor. The present study describes the effects of a MAb directed against rat CRF (CRF-MAb) injected together with non-linked toxins such as ricin A chain and monensin - just above the PVN. To ensure that *in vivo* administration of such a toxic mix interacts specifically with CRF neurons of the PVN, we examined the effects on the ACTH response to ether stress and described the state of CRF immunoreactive neurons in such hypothalamic areas, by means of an electron microscopic immunocytochemical method.

METHODS

Toxin/CRF monoclonal antibody mixture

The toxin mixture consisted of a monoclonal antibody to rat/human CRF (Van Oers et al, 1989) injected together with purified ricin A chain and monensin. CRF-MAb was a cytotoxic $IgG_2\alpha$ injected as an ascites fluid (IgG concentration: 60 μM). Ricin A chain inactivates the 60S ribosomal subunit (Carrasco et al, 1975; Endo et al, 1987) and monensin, in particular, raises the pH of prelysosomal and lysosomal compartments and inhibits intracellular protein degradation (Grimde,

1983), thus enhancing the toxic effect of ricin. Ricin A chain and monensin were used, respectively, at a final concentration of 8 μM and 50 nM (dissolved in artificial cerebrospinal fluid). The concentration of monensin is insufficient to inhibit CRF-mediated ACTH release by itself (Sobel *et al*, 1988; Mollenhauer *et al*, 1990). A control mixture consisted of equivalent amounts of toxins mixed with non-specific rat IgG.

Intracerebral Injection and Blood Sampling

Central injections of these toxin mixtures (0.25 μl per nucleus) were made by means of intracerebral cannulae implanted stereotaxically above the PVN of male Long-Evans rats. Five minutes before and 90 minutes after an intracerebral injection, blood samples (0.5 ml) were withdrawn via a chronic jugular catheter. Two final blood samples were withdrawn 25 and 90 minutes after inhalation of ether vapors. Experiments were conducted on freely moving rats, between 0800-1400 hours (rats were in a 12/12 hour light-dark regimen in which lights were on from 0700 hours) for a period of 2 days to 15 days. Plasma ACTH concentration was measured with a radioimmunoassay kit using rabbit anti-ACTH (ACTH K-PR, CIS, France). The sensitivity of the assay was 10 pg/ml of plasma. Data are expressed as means $\pm$ SEM and were subjected to analysis of variance and multiple protected t tests.

Electron Microscopic Immunocytochemical Methods

Two weeks after the intracerebral injections, animals were treated with colchicine (40 μg, intracisternal) to improve the visualization of CRF-synthesizing neurons in the PVN. Under Nesdonal® (Thiopental sodium 100 mg/kg) anaesthesia, rats were perfused 24 hours later with a histological fixative (glutaraldehyde 1 % and paraformaldehyde 1 %, v/v in cacodylate buffer 0.01M, pH 7.2). Frontal sections (100-200 μm) of brain were made using a microtome and subsequently permeabilized with saponin (0.05 %) and processed for CRF immunoreactivity using the PAP method and DAB as enzymatic substrate. They were postfixed in 1% OsO_4, dehydrated and flat embedded in araldite. Before observations, ultrathin sections were contrasted with uranyl acetate. Staining controls included sections incubated with normal rabbit serum.

The polyclonal antibody to CRF was prepared by one of us (Dr. G. Barbanel) and used at a final dilution of 1:2000. This antibody showed 100 % cross-reactivity with conjugate rCRF-thyroglobulin and no cross reactivity (< 0.01 %) with α-helical CRF (4-19), vasopressin, ACTH, ß-endorphin, vasoactive intestinal peptide, glucagon or somatostatin.

RESULTS

Effects of PVN Injection of Toxin/CRF-MAb Mixture on Ether-Induced ACTH Release

Injection of a mixture containing non-specific rat immunoglobulin (control IgG), monensin and ricin A chain did not affect plasma ACTH levels; both resting and stimulated ACTH levels were not different from those found in untreated rats (Figure 1 A-B). Likewise, no differences were found 24 hours following the intracerebral injection.   On the other hand, intra-PVN injection of a mixture containing CRF-MAb, monensin and ricin A chain failed to modify basal levels but markedly altered plasma ACTH levels measured 25 minutes after ether stress.   Stress-induced ACTH release was decreased from 836 ± 90 pg/ml (toxin/control IgG group) to 253 ± 22 pg/ml (toxin/CRF-MAb group) (Figure 1 D).   Reexposure of rats to ether vapor 24 hours after administration of the toxin/CRF-MAb mix still resulted in a diminished ACTH response (Figure 1D) [253 ± 22 pg/ml (day 1) versus 254 ± 39 pg/ml (day 2)].   Likewise, ACTH release was still inhibited 15 days after single administration of the toxin/CRF-MAb [215 ± 26 pg/ml (day 2) versus 223 ± 38 pg/ml (day 15)] (Figure 2).

Injection of CRF-MAb without toxins also significantly reduced the ACTH response.   However, when the rats were stressed again 24 hours after the injection, the ACTH response returned to its normal high value [340 ± 32 pg/ml (day 1) vs 1101 ± 77 pg/ml (day 2)] (Figure 1 C).

Effects of PVN Injection of Toxin/CRF-MAb Mix on CRF Immunoreactive Neurons of the PVN

Chronic cannula induced the appearance of a dramatic gliosis, 1 or

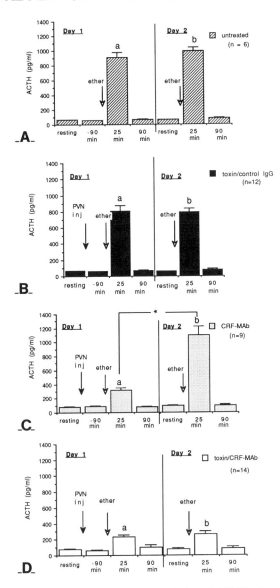

FIGURE 1. Acute inhibition of ether stress-induced ACTH release following a single bilateral injection of CRF-MAb alone or CRF-MAb added to toxins close to the PVN. Right panel (day 2) refers to same animals challenged with ether 24 hours after PVN injection. Values represent the means ± SEM. a,b:p <0.05 versus preinjection value (resting); *: p<0.0001.

2 mm above the PVN; but no significant lesion of brain tissue was observed in the PVN. CRF immunoreactivity was stained in the cytoplasm of numerous profiles. Most of them were close to macrophage-like cells, characterized by a dense chromatin and an abundant cytoplasmic population of lysosomcs and multivcsicular bodies (Figure 3A). CRF-immunoreactive neurons presented different morphological features. Some of them appeared with a non-disturbed organization : CRF immunoreactivity was simultaneously found in dense granules of 80-120 μm diameter scattered in the cytoplasm and in larger structures including 5-8 granular units ; both Golgi apparatus and rough endoplasmic reticulum (RER) were not obviously altered. On the other hand, numerous neuronal profiles were disturbed and different modifications could be found : in addition to different types of immunoreactive granular corpses, some neurons appeared with a large nucleus containing an abnormally dense chromatin and presenting a shape which varied from a profile with large indentations to a star-like profile. In other neurons, the nucleus disappeared, large accumulations of dense material were scattered in the cytoplasm, RER cisternae were disrupted but CRF immunoreactivity still was observed in dense granules of different diameter. Finally, large immunoreactive structures showing a total internal disorganization could be observed within the cytoplasm of macrophage-like cells (Figure 3B).

DISCUSSION

These data demonstrate that local bilateral administration of toxins added to an $IgG_{2\alpha}$ monoclonal antibody to CRF, close to the paraventricular nuclei, caused a marked and prolonged reduction of the ACTH response to ether vapors. This inhibitory effect was rapid in onset (less than 90 minutes) and lasted for at least 15 days. This long-term disturbance was unlikely to be a consequence of tissue damage or a non-specific action of toxins on neurons of the PVN, since normal high ACTH responses to stress were observed after a single and even after successive PVN injections of the same volume of the toxin/control IgG mix. Similarly, the long-term inhibition of ACTH release was not due to an adaptation to stress as we (Menzaghi *et al*, 1991) and others (Cook *et al*, 1973; Murakami *et al*, 1989) have observed that the daily repetition of ether anaesthesia does not induce modification of the ACTH response.

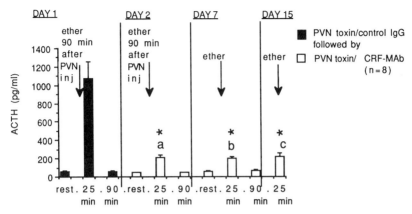

FIGURE 2. Acute and long-term inhibition of ether stress-induced ACTH release following a single bilateral injection of a toxin/CRF-MAb mix close to the PVN. Right panel refers to the same animals challenged with ether 7 days and 15 days after PVN injection. Animals were their own controls (control injection on day 1). Values represent the mean ± SEM. a,b,c:p <0.0001 versus preinjection value (resting); *:p<0.0001 versus 25 minute post-ether (control day 1).

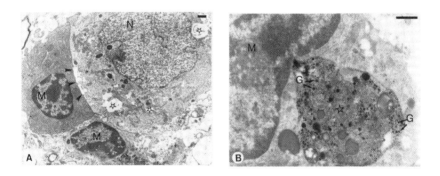

FIGURE 3. A--Electron micrograph of ultrathin section of the rat PVN after pre-embedding staining for CRF, 15 days after PVN administration of a toxin/CRF-MAb mix. Two macrophage-like cells (M) show close relationships with an immunoreactive neuron (N). Large vacuoles appear along the contact line (headarrow) and within the cytoplasm (star). Bar represents 0.5 μm. B--Electron micrograph of ultrathin section of the rat PVN after pre-embedding staining for CRF, 15 days after PVN administration of a toxin/CRF-MAb mix. A part of an immunoreactive neuron (star) is totally included in the cytoplasm of a macrophage-like cell (M). The total disorganization of the immunoreactive structures is noticeable. Numerous immunoreactive granules are still present (G). Bar represents 0.5 μm.

The use of a mixture rather than an immunoconjugate was justified by the concentration of IgG. It has been demonstrated that the quantity of IgG molecules is a major requirement for complement activation (Howard et al, 1988) and the concentration of IgG is 1000 times less concentrated in an immunoconjugate than in ascites fluid (Ghetie et al, 1986; Gros et al, 1985). Moreover, it has been demonstrated that the coupling of toxins to a monoclonal antibody decreases their cytotoxicity and diminishes the affinity of the antibody (Blakey et al, 1988).

Based on these observations, it is tempting to speculate that this long-term effect resulted from a specific central action of toxins on CRF neurons. PVN injections of CRF-MAb without toxins also induced an inhibitory effect on the ACTH response to ether. However, this effect was similar to the one obtained after intracerebroventricular injection of a polyclonal antibody to CRF (Rivier et al, 1982; Ono et al, 1985) and did not persist more than 24 hours. In our experiment, since the CRF-MAb was specifically internalized into CRF neurons of the PVN after intranuclear injection (Burlet et al, 1990a), we considered the possibility that not only the classical immunoneutralization mechanism, but also antibody uptake, interfaced with the control of ACTH secretion.

These findings suggest two mechanisms of action of the toxin/CRF-MAb mix:
- a short-term effect (less than 24 hours), probably not dependent on the presence of toxins and resulting from immunoneutralization of extracellular CRF and/or some membrane disturbances induced by the recognition of the cells
- a long-term effect (at least 15 days) dependent on the specific internalization of toxins via the complement pathway and subsequent cytotoxic effects. This long-term effect due to the presence of toxins is supported by our preliminary immunocytochemical results. At the electron microscopic level, we observed that numerous macrophagic cells were abnormally close to CRF immunoreactive neurons. Whether these results corresponded to the elimination of deficient CRF neurons damaged by the toxin mixture remains to be established; nevertheless, we could observe an intracellular disorganization of CRF neurons surrounded by these macrophagic cells. Moreover, we just observed (in preparation) that numerous CRF immunoreactive neurons appeared in the dorsal part of the supraoptic nuclei which did not include such cells under standard condition. An intense activity of

biosynthesis was observed at the Golgi level in these neurons. Similarly, other CRF neuronal populations located in extra-hypothalamic areas (central amygdala and bed nucleus of the stria terminalis) appeared to be stimulated.

These preliminary observations lead to the conclusion that the use of a toxin/antibody mixture will prove to be a valuable tool to compromise the function of specific peptidergic neurons of the hypothalamus and to dissect the contribution of different neuronal systems in determining the ACTH response to a stressor.

REFERENCES

Bhakdi, S., and Tranum-Jensen, J. (1987). Damage to mammalian cells by proteins that form transmembrane pores. *Reviews of Physiology, Biochemistry and Pharmacology* 107, 148-223.

Blakey, D.C., Wawrzynczak, E.J., Wallace, P.M., and Thorpe, P.E. (1988). Antibody toxin conjugates: a perspective. In: H. Waldmann (Ed.), "Monoclonal Antibody Therapy," pp. 50-90. Basel: Karger.

Burlet, A.J., Leon-Henri, B.P., Robert, F.R., Arahmani, A., Fernette, B.M.L., and Burlet, C.R. (1987). Monoclonal anti-vasopressin (VP) antibodies penetrate into VP neurons *in vivo*. *Experimental Brain Research* 65, 629-638.

Burlet, A.J., Menzaghi, F., Tilders, F.J.H., Van Oers, J.W.A.M., Nicolas, J.P., and Burlet, C.R. (1990a). Uptake of a monoclonal antibody to corticotropin-releasing factor (CRF) into rat hypothalamic neurons. *Brain Research* 517, 283-293.

Burlet, A.J., Haumont-Pellegri, B., Tankosic, P., Arahmani, A., Fernette, B., Giannangeli, F., Nicolas, J.P., and Burlet, C.R. (1990b). The monoclonal antibody to neuropeptide: a new tool to act *in vivo* on the peptidergic neuron activity. In: B. Greenstein (Ed.), "Neuroendocrine Research Methods," pp. 355-382. Harwood: Church.

Carrasco, L., Fernandez-Puentes, C., and Vasquez, D. (1975). Effects of ricin on ribosomal sites involved in the interaction of the elongation factors. *European Journal of Biochemistry* 54, 499-503.

Cook, D.M., Kendall, J.W., Greer, M.A., and Kramer, R.M. (1973). The effect of acute or chronic ether stress on plasma ACTH concentration in the rat. *Endocrinology* 93, 1019-1024.

Dallman, M.F., Akana, S.F., Cascio, C.S., Darlington, D.N., Jacobson, L., and Levin, N. (1987). Regulation of ACTH secretion: Variations on a theme B. *Recent Progress in Hormone Research* 43, 113-167.

Endo, Y., and Tsurughi, K. (1987). RNA N-glycosidase activity of ricin A-chain. Mechanism of action of the toxic lectin ricin on eukarystic ribosomes. *Journal of Biological Chemistry* 262, 8128-8130.

Ghetie, M.A., Laky, M., Moraru, I., and Ghetie, V. (1986). Protein A vectorized toxins. 1. Preparation and properties of protein A-ricin toxin conjugates.

Molecular Immunology **23**, 1373-1379.

Grimde, B. (1983). Effects of carboxylic ionophores on lysosomal protein degradation in rat hepatocytes. *Experimental Cell Research* **149**, 27-31.

Gros, O., Gros, P., Jansen, F.K., and Vidal, H. (1985). Biochemical aspects of immunotoxin preparation. *Journal of Immunological Methods* **81**, 283-297.

Howard, J., and Hugues-Jones, N. (1988). Complement mediated lysis with monoclonal antibodies. In: H. Waldmann (Ed.), "Monoclonal Antibody Therapy," pp.1-15. Basel: Karger.

Leon-Henri, B., Burlet, A., Chauvet, J., Nicolas, J.P., and Burlet, C. (1989). The vasopressin neuron is the target of the monoclonal antibodies raised against MSEL-neurophysin injected *in vivo*. *Neuroendocrinology* **49**, 125-133.

Menzaghi, F., Burlet, A., Van Oers, J.W.A.M., Tilders, F.J.H., Nicolas, J.P., and Burlet, C. (1991). Long-term inhibition of stress-induced ACTH release by intracerebral administration of a monoclonal antibody to rat corticotropin-releasing factor together with ricin A chain and monensin. *Journal of Neuroendocrinology* **3**, 469-475.

Mollenhauer, H.H., Moue, D.J., and Rowe, L.D. (1990). Alteration of intracellular traffic by monensin: mechanism, specificity, and relationship to toxicity. *Biochimica Biophysica Acta Reviews on Biomembranes* **1031**, 225-246.

Murakami, K., Akana, S., Dallman, M.F., and Ganong, W.F. (1989). Correlation between the stress-induced transient increase in corticotropin-releasing hormone content of the median eminence of the hypothalamus and adrenocorticotropic hormone secretion. *Neuroendocrinology* **49**, 233-241.

Ono, N., Samson, W.K., McDonald, J.K., Lumpkin, M.D., Bedran De Castro, J.C., and Mc Cann, S.M. (1985). Effects of intravenous and intraventricular injection of antisera directed against corticotropin-releasing factor on the secretion of anterior pituitary hormones. *Proceedings of the National Academy of Sciences* (USA) **82**, 7787-7790.

Podack, E.R., and Tschopp, J. (1984). Membrane attack by complement. *Molecular Immunology* **21**, 589-604.

Rivier, C., Rivier, J., and Vale W. (1982). Inhibition of adrenocorticotropic hormone secretion in the rat by immunoneutralization of corticotropin-releasing factor. *Science* **218**, 377-379.

Schupf, N., and Williams, C.A. (1987). Psychopharmacological activity of immune complexes in rat brain is complement dependent. *Journal of Neuroimmunology* **13**, 293-303.

Sobel, D.O., and Shakir, K.M. (1988). Monensin inhibition of corticotropin releasing factor mediated ACTH release. *Peptides* **9**, 1037-1042.

Tilders, F.J.H., Van Oers, J.W.A.M., White, A., Menzaghi F., and Burlet A. (1990). Antibodies to neuropeptides: biological effects and mechanisms of action. In: J.C. Porter and D. Jezova (Eds.), "Circulating Factors and Neuroendocrine Function," pp. 135-146. New York: Plenum Press.

Van Oers, J.W.A.M., Tilders, F.J.H., and Berkenbosch, F. (1989). Characterization and biological activity of a rat monoclonal antibody to rat/human corticotropin-releasing factor. *Endocrinology* **124**, 1239-1246.

Stress: Neuroendocrine and Molecular Approaches
Edited by R. Kvetnansky, R. McCarty and J. Axelrod

THE HYPOTHALAMO-PITUITARY-ADRENAL AXIS IN RODENTS: CORTICOTROPIN RELEASING HORMONE/VASOPRESSIN CO-EXISTENCE AND CYTOKINE EFFECTS

M. H. Whitnall[1], R. S. Perlstein[2],
E. H. Mougey[3], and R. Neta[2]

[1]Departments of Physiology and [2]Experimental Hematology
Armed Forces Radiobiology Research Institute,
Bethesda, MD; and [3]Neuroendocrinology and
Neurochemistry Branch, Department of Medical Neurosciences, Walter
Reed Army Institute of Research, Washington, DC USA

INTRODUCTION

Interleukin-1 alpha (IL-1α) and IL-1ß are peptide hormones that act on the same receptors and stimulate differentiation, activation and proliferation of immune cells (Durum *et al*, 1990). These hormones also have pleiotropic effects on many other tissues (Durum *et al*, 1990), and are referred to collectively as IL-1. IL-1 activates the hypothalamo-pituitary-adrenal axis (HPA), in part by stimulating hypothalamic neurosecretory cells that release corticotropin-releasing hormone (CRH) into the portal capillary system (Besedovsky *et al*, 1986; Sapolsky *et al*, 1987; Uehara *et al*, 1987; Berkenbosch *et al*, 1989; Weidenfeld *et al*, 1989) (Figure 1).

There are two subtypes of CRH neurosecretory cells in rats (Figure 1), distinguished by 1) the presence or absence of high levels of co-existent vasopressin (VP) in their axons, 2) different distributions

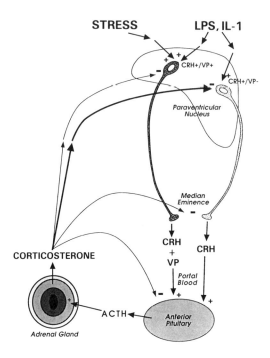

FIGURE 1.   Schematic diagram of regulation of CRH neurosecretory cells in rats.
Broken arrows represent indirect mechanisms.

of their perikarya in the hypothalamus, and 3) different sensitivities to
glucocorticoid inhibition (Whitnall, 1988ab, 1990; de Goeij *et al*, 1991).
In addition, several stressors selectively activate CRH+/VP+ axons,
as shown by the finding that neurosecretory vesicles were drastically
depleted from CRH+/VP+ terminals during stress, with no effect on
CRH+/VP- axons (Whitnall, 1989).

Recent results suggested that IL-1 may have an effect on the
hypothalamo-pituitary-adrenal axis (HPA) opposite to that of stress,
i.e., selective activation of the CRH+/VP- subtype (Berkenbosch *et al*,
1989). However, the only way to study directly the physiology of CRH
neurosecretory cell subtypes is to assay vesicle depletion using EM
immunocytochemistry (Whitnall, 1989). Therefore, we employed this
technique to answer the question of whether one or both CRH
neurosecretory cell subtypes are activated by IL-1 in rats. We also
initiated a study on CRH/VP co-existence in mice.

METHODS

Serial ultrathin sections were cut through the median eminence of mice, or 400 g male Sprague-Dawley rats that had received ip injections of IL-1α, IL-1β, endotoxin (bacterial lipopolysaccharide, LPS), or vehicle (0.01% human serum albumin in PBS) five hours before sacrifice. LPS administration is a model for studying the inflammatory response, and the HPA response to LPS is mediated at least partially by endogenous IL-1 (Rivier *et al*, 1989). Vesicle depletion from axon terminals in the external zone of the median eminence was quantified as described previously (Whitnall, 1989). Radioimmunoassays for PRL and ACTH in trunk blood were performed as described previously (Mougey *et al*, 1986; Perlstein *et al*, 1991). Reported values are means and standard errors, calculated using the number of animals in each group as N. Differences between groups were deemed significant if p values were less than 0.05, according to Fisher's PLSD test.

RESULTS

Ten μg of IL-1α or IL-1β, or 1.0 mg of LPS, produced significant increases in plasma ACTH (vehicle control: 58.1 ± 3.6 pg/ml, N=9; IL-1α: 127 ± 9.1 pg/ml, N=3; IL-1β: 174 ± 34 pg/ml, N=6; LPS: 154 ± 16 pg/ml, N=6). PRL levels were significantly decreased by all three treatments (vehicle control: 6.09 ± 1.2 ng/ml, N=10; IL-1α: 1.93 ± 0.24 ng/ml, N=3; IL-1β: 1.90 ± 0.62 ng/ml, N=6; LPS: 2.79 ± 0.65 ng/ml, N=7).

In vehicle-injected rats, the numbers of vesicle-containing axonal profiles per 500 μm² sample area were 21.8 ± 2.5 and 19.9 ± 2.4 for the CRH+/VP+ and CRH+/VP- subtypes, respectively (N=13). Five hours after an IL-1α injection, both subtypes exhibited significant depletion of neurosecretory vesicles, resulting in decreases in vesicle-containing profiles of 71% and 66% for the CRH+/VP+ and CRH+/VP- subtypes, respectively (N=3). Similarly, injection of 10 μg IL-1β caused significant decreases in vesicle-containing profiles of 78% and 64%, respectively (N=6). The corresponding significant decreases after LPS injection were 75% and 42% (N=6).

In view of the importance of the co-existence of VP and CRH to HPA function in rats, we were interested in determining whether the CRH+/VP+ and CRH+/VP- subtypes of neurosecretory cells exist in other species. We performed EM immunocytochemical studies on five female C3H/HeN mice, one female CD2F1 mouse and two male CD2F1 mice, and discovered to our surprise that all CRH neurosecretory axons in the external zone of the median eminence contained proVP-derived peptides. Hence, IL-1 induced ACTH responses in mice cannot depend on selective activation of a VP-deficient subtype of CRH neurosecretory cell, which is consistent with our results in rats.

DISCUSSION

There has been a recent suggestion that IL-1 may selectively activate the VP-deficient subtype of CRH neurosecretory cell (Berkenbosch *et al*, 1989). However, the present results directly demonstrate that IL-1α and IL-1β stimulate the HPA in rats by activating both CRH+/VP+ and the CRH+/VP- neurosecretory cell subtypes (Figure 1). These findings have some notable implications for the regulation of the HPA by cytokines:

1. IL-1 and stressors activate the HPA by different mechanisms, since several types of stressors selectively activate the VP-containing CRH neurosecretory cell subtype (Whitnall, 1989) (Figure 1).

2. Others have proposed that the synergistic effects of portal CRH and VP on ACTH release minimize adaptation of the HPA to repeated activation (Abou-Samra *et al*, 1986; Bilezikjian *et al*, 1987; Hashimoto *et al*, 1988; Hauger *et al*, 1990; Scaccianoce *et al*, 1991). In fact, corticosterone responses do not adapt to repeated administration of IL-1 in mice (Mengozzi and Ghezzi, 1991), which is consistent with the present finding of IL-1 induced release of both CRH and VP. Adaptation of HPA responses to IL-1 would be predicted by a selective effect of IL-1 on the VP-deficient subtype of CRH neurosecretory cell.

3. *In vitro* studies suggest that increasing portal CRH alone would activate a subset of pituitary corticotropes, while increases in CRH and VP would activate all corticotropes (Jia *et al*, 1991). Hence our results suggest that all corticotropes may be activated (indirectly) by IL-1.

4. It has been postulated that IL-1 stimulates the HPA by blocking glucocorticoid inhibition of this axis (Weidenfeld *et al*, 1989). Since both CRH neurosecretory cell subtypes are inhibited by glucocorticoids (Whitnall, 1988b), this hypothesis predicts that both

subtypes should be disinhibited by IL-1. Hence, the present finding of IL-1 induced vesicle depletion from both subtypes supports the disinhibition hypothesis (Weidenfeld et al, 1989).

Our finding of decreases in plasma PRL in response to LPS confirms a previous report (Hadzhikostova et al, 1979), and is consistent with IL-1 induced decreases in plasma PRL observed here. Decreases in PRL would serve as negative feedback to the immune system during inflammation (Russell, 1989). In addition to the effects of altered PRL, IL-1 induced activation of the HPA would also serve to keep the immune system in check (Munck et al, 1984).

The different organizations of the HPA in mice and rats, with respect to CRH/VP co-existence, may reflect the fact that these two ACTH secretagogues have a different relationship in mice and rats. In rats, VP by itself is a weak ACTH secretagogue, but it strongly potentiates CRH-induced ACTH secretion (Antoni, 1989). Hence, altering the ratio of VP to CRH in portal blood represents a powerful mechanism for altering the gain of the HPA in response to different physiological situations (Gillies and Lowry, 1979). The existence of VP-containing and VP-deficient subpopulations of CRH axons makes this independent regulation possible (Whitnall, 1988a, 1989). In mice, on the other hand, VP and CRH have similar potencies for inducing ACTH release, with no evidence of synergism (Castro et al, 1989). Therefore, independent regulation of portal VP and CRH would not be a powerful mechanism for altering the gain of the HPA in mice. Hence separate populations of neurosecretory cells with different concentrations of CRH and VP would offer less of an advantage in mice than in rats.

There is evidence in mice that the ACTH response to IL-1 is dependent on synergy with IL-6 induced by the IL-1 injections (Perlstein et al, 1991). The results of recent experiments combining injections of IL-1 with administration of antibodies to IL-6 support this hypothesis (unpublished data). It remains to be determined whether a similar mechanism operates in rats. We are currently investigating the mechanisms by which cytokines activate the HPA in both mice and rats.

ACKNOWLEDGMENTS

We thank Clint Wormley for helping with the RIAs, Dr. Peter Lomedico of Hoffman-LaRoche for donating the IL-1α, Dr. Patricia Kilian of Hoffman-LaRoche for donating the IL-1ß, and Dr. Wylie

Vale for donating the antiserum to CRH. This research was supported by the Armed Forces Radiobiology Research Institute, Defense Nuclear Agency, under work units 00105 and 00129. Views presented in this paper are those of the authors; no endorsement by the Defense Nuclear Agency or the Department of Defense has been given or should be inferred. Research was conducted according to the principles enunciated in the Guide for the Care and Use of Laboratory Animals prepared by the Institute of Laboratory Animal Resources, National Research Council.

REFERENCES

Abou-Samra, A.-B., Catt, K.J., and Aguilera, G. (1986). Biphasic inhibition of adrenocorticotropin release by corticosterone in cultured anterior pituitary cells. *Endocrinology* **119**, 972-977.

Antoni, F.A. (1989). Hypophysiotrophic neurones controlling the secretion of corticotropin: Is the hypothesis of a final common hypothalamic pathway correct? In: F.C. Rose (Ed.), "The Control of the Hypothalamo-Pituitary Adrenocortical Axis," pp. 317-329. Madison, CT: International University Press.

Berkenbosch, F., de Goeij, D.E.C., Del Rey, A., and Besedovsky, H.O. (1989). Neuroendocrine, sympathetic and metabolic responses induced by interleukin-1. *Neuroendocrinology* **50**, 570-576.

Besedovsky, H., del Rey, A., Sorkin, E., and Dinarello, C.A. (1986). Immunoregulatory feedback between interleukin-1 and glucocorticoid hormones. *Science* **233**, 652-654.

Bilezikjian, L.M., Blount, A.M., and Vale, W.W. (1987). The cellular actions of vasopressin on corticotrophs of the anterior pituitary: Resistance to glucocorticoid action. *Molecular Endocrinology* **1**, 451-458.

Castro, M.G., Gusovsky, F., and Loh, Y.P. (1989). Transmembrane signals mediating adrenocorticotropin release from mouse anterior pituitary cells. *Molecular and Cellular Endocrinology* **65**, 165-173.

de Goeij, D.C.E., Kvetnansky, R., Whitnall, M.H., Jezova, D., Berkenbosch, F., and Tilders, F.J.H. (1991). Repeated stress enhances AVP stores and colocalization with CRF in the median eminence of rats. *Neuroendocrinology* **53**, 150-159.

Durum, S.K., Oppenheim, J.J., and Neta, R. (1990). Immunophysiologic role of interleukin 1. In: J.J. Oppenheim, and E.M. Shevach (Eds.), "Immunophysiology. The Role of Cells and Cytokines in Immunity and Inflammation," pp. 210-225. New York: Oxford University Press.

Gillies, G., and Lowry, P. (1979). Corticotrophin releasing factor may be modulated by vasopressin. *Nature* **278**, 463-464.

Hadzhikostova, H., Visheva, N., Milanov, S., and Nikolov, N. (1979). Changes of ACTH, STH, TSH and prolactin levels in endotoxic shock in rats. *Agressologie* **20**, 203-206.

Hashimoto, K., Suemaru, S., Takao, T., Sugawara, M., Makino, S., and Fensuka, O. (1988). Corticotropin-releasing hormone and pituitary-adrenocortical responses in chronically stressed rats. *Regulatory Peptides* **23**, 117-126.

Hauger, R.L., Lorang, M., Irwin, M., and Aguilera, G. (1990). CRF receptor regulation and sensitization of ACTH responses to acute ether stress during chronic intermittent immobilization stress. *Brain Research* **532**, 34-40.

Jia, L.-G., Canny, B.J., Orth, D.N., and Leong, D.A. (1991). Distinct classes of corticotropes mediate corticotropin-releasing hormone- and arginine-vasopressin- stimulated adrenocorticotropin release. *Endocrinology* **128**, 197-203.

Mengozzi, M., and Ghezzi, P. (1991). Defective tolerance to the toxic and metabolic effects of interleukin 1. *Endocrinology* **128**, 1668-1672.

Mougey, E.H., Meyerhoff, J.L., Pennington, L.L., Kenion, C.C., and Kant, G.J. (1986). Pituitary cyclic AMP and plasma hormone responses to epinephrine administration *in vivo*. *Life Sciences* **39**, 2305-2313.

Munck, A., Guyre, P.M., and Holbrook, N.J. (1984). Physiological functions of glucocorticoids in stress and their relation to pharmacological actions. *Endocrine Reviews* **5**, 25-44.

Perlstein, R.S., Mougey, E.H., Jackson, W.E., and Neta, R. (1991). Interleukin-1 and interleukin-6 act synergistically to stimulate the release of adrenocorticotropic hormone *in vivo*. *Lymphokine and Cytokine Research* **10**, 141-146.

Rivier, C., Chizzonite, R., and Vale, W. (1989). In the mouse, the activation of the hypothalamic-pituitary-adrenal axis by a lipopolysaccharide (endotoxin) is mediated through interleukin-1. *Endocrinology* **125**, 2800-2805.

Russell, D.H. (1989). New aspects of prolactin and immunity: a lymphocyte-derived prolactin-like product and nuclear protein kinase C activation. *Trends in Pharmacological Sciences* **10**, 40-44.

Sapolsky, R., Rivier, C., Yamamoto, G., Plotsky, P., and Vale, W. (1987). Interleukin-1 stimulates the secretion of hypothalamic corticotropin-releasing factor. *Science* **238**, 522-524.

Scaccianoce, S., Muscolo, L.A.A., Cigliana, G., Navarra, D., Nicolai, R., and Angelucci, L. (1991). Evidence for a specific role of vasopressin in sustaining pituitary-adrenocortical stress response in the rat. *Endocrinology* **128**, 3138-3143.

Uehara, A., Gottschall, P.E., Dahl, R.R., and Arimura, A. (1987). Interleukin-1 stimulates ACTH release by an indirect action which requires endogenous corticotropin releasing factor. *Endocrinology* **121**, 1580-1582.

Weidenfeld, J., Abramsky, O., and Ovadia, H. (1989). Effect of interleukin-1 on ACTH and corticosterone secretion in dexamethasone and adrenalectomized pretreated male rats. *Neuroendocrinology* **50**, 650-654.

Whitnall, M.H. (1988a). Is there a "final common pathway" in the regulation of ACTH release? In: G.P. Chrousos, D.L. Loriaux, and P.W. Gold (Eds.), "Mechanisms of Physical and Emotional Stress," pp. 143-156. New York:

Plenum Publishing Corporation.

Whitnall, M.H. (1988b).   Distributions of pro-vasopressin expressing and pro-vasopressin deficient CRH neurons in the paraventricular hypothalamic nucleus of colchicine-treated normal and adrenalectomized rats. *Journal of Comparative Neurology* **275**, 13-28.

Whitnall, M.H. (1989). Stress selectively activates the vasopressin-containing subset of corticotropin-releasing hormone neurosecretory cells. *Neuroendocrinology* **50**, 702-707.

Whitnall, M.H. (1990).   Subpopulations of corticotropin-releasing hormone neurosecretory cells distinguished by presence or absence of vasopressin: confirmation with multiple corticotropin-releasing hormone antisera. *Neuroscience* **36**, 201-205.

Stress: Neuroendocrine and Molecular Approaches
Edited by R. Kvetnansky, R. McCarty and J. Axelrod

1992 Gordon and Breach Science
Publishers S.A., New York, USA.
Photocopying permitted by license only.

EFFECTS OF DIFFERENT INTERVALS OF IMMOBILIZATION ON IN VITRO SECRETION OF CRH

G. Cizza[1,2], R. Kvetnansky[3], A. Moazzez[1], S.E. Taymans[2],
G. P. Chrousos[2] and P. W. Gold[1]

[1]Clinical Neuroendocrinology Branch
National Institute of Mental Health,
[2]Developmental Endocrinology Branch
National Institute of Child Health and Human Development
[3]Clinical Neuroscience Branch
National Institute of Neurological Disorders and Stroke
Bethesda, Maryland USA

INTRODUCTION

Corticotropin releasing hormone is a 41 amino acid hypothalamic hormone that is thought to be the principal central stimulus to the pituitary-adrenal axis (Vale *et al*, 1981; Spiess *et al*, 1981) and whose central administration sets into motion a series of behavioral and physiological events that are adaptive during stressful situations (Gold *et al*, 1988b). Hypothalamic hormones other than CRH are also thought to modulate pituitary-adrenal responsiveness, including arginine vasopressin (Gillies *et al*, 1982). The latter is thought to be a weak secretagogue of ACTH alone but works synergistically with CRH in the stimulation of pituitary corticotrophs (Gillies *et al*, 1982).

    In the present study, we wished to determine whether the duration of a stressor influences the magnitude of CRH release from the hypothalamus, either under basal conditions or in response to norepinephrine (NE) or potassium chloride (KCl) and whether pituitary-adrenal responses after varying degrees of stress correlated with indices of CRH release. To accomplish this task, we exposed rats to varying intervals of immobilization stress. Pituitary-adrenal function was assessed by measurement of ACTH, beta-endorphin, and corticosterone in trunk blood at the time of sacrifice, and hypothalamic CRH release was assessed by *in vitro* hypothalamic organ culture studies, in which CRH release was assessed under basal conditions and after stimulation by either NE or KCl.

METHODS

Animals

Male F344/N rats 3-4 months old were purchased from Harlan-Sprague Dawley, Inc. (Indianapolis, IN). These inbred rats are cesarean derived and maintained behind a pathogen-free barrier under rigorously controlled conditions. Immediately after their arrival, animals were housed two per cage in shoe-box cages supplied with air filters and allowed one week to recover from transportation and novelty of the laboratory environment. Animals had free access to food and water, and were under constant conditions of temperature and humidity with a 12-hour light-12-hour dark cycle (lights on at 0600 hours). They were periodically inspected for signs of sickness to insure that only animals showing optimal health status were used in the experiments. At autopsy, subjects with evident organ pathology were eliminated from the study. All animals were examined for presence of pituitary adenomas.

Stress Procedure

On the day of the experiment, animals (at least 6 per group) were immobilized in a prone position by inserting their heads through steel wire loops fixed on a board and by fastening their limbs to 4 metal strips with adhesive tape. In this way, animals were immobilized for increasing intervals of time from 5 to 120 minutes and decapitated immediately after the end of the immobilization period. All animals were placed in a room in which the experiment took place 16 hours before the experiments and left undisturbed. In each experiment, controls were used which were decapitated at about 0830 hours. These animals were randomly taken from different cages. All immobilization sessions started at 0900 hours.

In Vitro Studies

Rats were sacrificed by decapitation. Utilizing sterile technique, hypothalamic blocks were rapidly removed with fine pointed curved scissors. The boundaries of the hypothalamic blocks consisted of the posterior border of the optic chiasma anteriorly, the anterior border of the mammillary bodies posteriorly, and the lateral hypothalamic sulci laterally. Dorsally, the cut was performed at about 3 mm from the ventral surface. Immediately after the explantation, the hypothalami (one explant per well, 48 multiwell plates, Costar, Cambridge, MA) were placed in a water-jacketed incubator at 37°C under a 5% CO_2 atmosphere. During preincubation and incubation, medium M 199 with modified Earle's salt (Gibco, Grand Island) containing 0.1 % BSA (fraction V, Sigma, St. Louis, MO) and 20 μM bacitracin (zinc salt, Aldrich, Milwaukee, WI) was used.

Experimental Protocols

The experimental design consisted of passages of the tissues through different wells, at 20 minute intervals.

Experiment 1. After a 2 hour preincubation, the hypothalami were transferred, using a nylon mesh grid (3x3 mm, Small Parts, Miami, FL) to 5 consecutive wells containing control media to assess basal secretion of CRH. In the 6th well, the stimulated secretory capacity, as well as the tissues' viability, were evaluated by exposing the hypothalami to a depolarizing (60 mM) concentration of KCl.

Experiment 2. Hypothalami explanted from controls and animals immobilized for 30, 60, and 120 minutes were exposed to control media in the first of 3 consecutive wells, and to media containing NE 10^{-9} M in the next of 3 wells.

Plasma hormone measurements. Trunk blood was collected in prechilled tubes and centrifuged at 4° C for 20 minutes. Plasma was separated and stored at -20° C until assayed. CRH, ACTH, corticosterone and beta-endorphin were measured as previously described (Calogero et al, 1989).

Statistics. Results are expressed as means (±SEM). CRH data were analyzed by one-way ANOVA followed by Fisher's LSD (Experiment 1) or unpaired Student's t tests (Experiment 2). Significance was defined at a level lower than 0.05 unless indicated otherwise.

RESULTS

Immobilization caused a 2- to 6-fold increase in plasma ACTH concentrations within 5 minutes. ACTH was found to be maximally elevated (760-1400 pg/ml) between 30 and 60 minutes of immobilization and remained elevated throughout the duration of the immobilization session (Table 1). Beta-endorphin and corticosterone concentrations increased and remained elevated during the entire stress session (Table 1).

In order to study the effect of immobilization on *in vitro* hypothalamic secretion of CRH, we exposed hypothalami from controls and from stressed animals to a non-specific stimulus (KCl), and to a specific stimulus, NE, at concentrations that were shown previously to stimulate CRH secretion (Calogero et al, 1988a). In both experiments, basal CRH secretion was unaffected by stress; however, KCl-stimulated secretion at 30 and 60 minutes of immobilization was significantly reduced compared to controls ($p < 0.05$) (Figure 1). NE stimulated

TABLE 1.   Effects of different intervals of immobilization on circulating levels of ACTH, beta-endorphin, and corticosterone in male F344 rats.

Hormone	Immobilization Time (minutes)				
	0	5	30	60	120
ACTH (pg/ml)	118.8±12.2	554.1±184.3	1429.1±516.9	761.5±192.8	368.2±111.4
Beta-Endorphin (pg/ml)	32.3±3.7	126.6±24.1	314.3±76.7	176.1±46.0	80.2±17.3
Corticosterone (ng/ml)	42.8±13.7	156.9±33.8	493.3±113.6	516.2±82.6	336.1±49.5

Data represent mean±SEM (n=6-8).

CRH secretion significantly above baseline in the control group and after 30 minutes of immobilization (unpaired t test); conversely it did not affect CRH secretion in the group stressed for 60 and 120 minutes (Figure 2).

DISCUSSION

The immobilization stress paradigm utilized in the present study has been previously documented as a reliable and reproducible stimulus to stress-responsive neurotransmitter systems, as indicated by catecholamine responses in outbred rats (Makara *et al*, 1986). In the present study, we replicated this finding using pituitary-adrenal activation in F344/N rats as an index. Hence, the reproducibility of this paradigm holds for pituitary-adrenal as well as catecholamine responses in inbred rats whose CRH neurons are hyperactive compared to outbred strains.

CRH is well-known to play a critical role in stress responses, but may not be the principle stimulus for ACTH release under all circumstances (Scaccianoce *et al*, 1991). Vasopressin, angiotensin II, catecholamines, and other secretagogues are also influential in stimulating release of ACTH from the anterior pituitary, and may be differentially important under varying circumstances (Rivier and Vale, 1983). Because there are technical problems with *in vivo* assessments of hypothalamic CRH secretion, we decided to determine CRH secretion *in vitro* as an indirect index of CRH secretion *in vivo* for comparison with the degree of immobilization-induced pituitary-adrenal activation. To determine *in vitro* CRH secretion, we measured this parameter under basal conditions and after stimulation with either KCl or NE.

Our *in vitro* data show that after increasing intervals of immobilization, the secretion of CRH becomes less sensitive to either non-specific or specific stimuli. In fact, with KCl stimulation, the releasable pool of CRH is progressively decreased by stress for up to 60 minutes. We have previously shown that NE is capable of stimulating CRH release from hypothalamic organ cultures, with a maximal stimulatory dose of 10^{-9}M (Calogero *et al*, 1988a). We also reported here that this concentration of NE is a potent stimulus of *in vitro* CRH release in both animals that had been unstressed prior to sacrifice and in those that had been immobilized for 30 minutes.

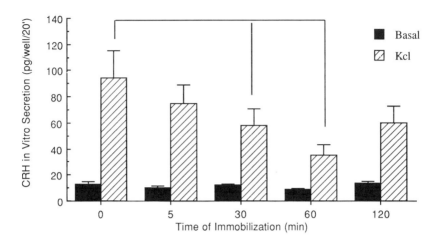

Figure 1. Effects of different intervals of immobilization on basal and KCl-stimulated hypothalamic CRH secretion *in vitro*. Lines connecting bars indicate a significant difference at the 0.05 level by ANOVA followed by Fisher's LSD.

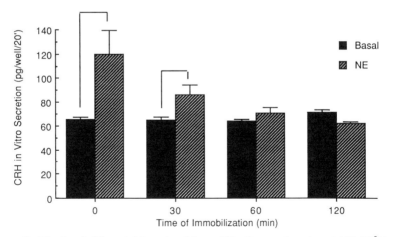

Figure 2. Effects of different intervals of immobilization on basal and NE 10^{-9}M stimulated hypothalamic CRH secretion *in vitro*. Lines connecting bars indicate a significant difference (p <0.004 by unpaired t test). Note that with stress, CRH secretion progressively decreases in the presence of both a non-specific (KCl) (Figure 1) and a specific stimulus (NE).

However, the magnitude of *in vitro* CRH release was less in animals sacrificed after thirty minutes of immobilization stress while NE-induced CRH release could not be demonstrated in animals that had been restrained for periods of 60 minutes or more prior to sacrifice. Hence, we surmise that as with KCl-induced CRH release, the pool of CRH that is releasable following incubation with NE diminishes after a progressively longer duration of immobilization stress. A desensitization of adrenergic receptors to NE released *in vivo* during prolonged stress may also play a role in the progressive decline in the response of CRH neurons in animals exposed to relatively long periods of immobilization stress. Thus, in animals stressed for 60 minutes prior to sacrifice, a CRH response to NE was no longer evident, while at this time point CRH was still released in response to KCl.

In our previous studies exploring CRH release from hypothalamic organ cultures utilizing outbred rats, we reported CRH release only after an overnight pre-incubation of hypothalami prior to stimulation with NE (Calogero *et al*, 1988a) or other neurotransmitters (Calogero *et al*, 1988b). We surmise that this preincubation allowed the neutralization of the effects of a variety of inhibitors of CRH release present in the media. In the present study, we could detect CRH release following exposure to NE after only two hours despite the fact that we used the same experimental design. We cannot definitely account for our capacity to cause the release of CRH after only two hours in rats that had been unstressed prior to sacrifice or stressed for 30 minutes. We speculate that CRH neurons in this strain are hyper-responsive to a variety of stimuli, including neurotransmitters such as NE and 5-hydroxytryptamine, inflammatory mediators, and environmental stimuli (Sternberg *et al*, 1988a, 1988b).

Despite the *in vitro* decrease of CRH secretion in animals exposed to progressively longer periods of immobilization stress prior to sacrifice, plasma ACTH and corticosterone remained substantially elevated over the whole immobilization period. Assuming that *in vitro* secretion of CRH is, in some way, related to *in vivo* secretion of CRH immediately before explantation, we speculate that, under these conditions, an ACTH secretagogue other than CRH plays a major role in the sustained stimulation of the HPA axis during prolonged stress. These may include vasopressin and angiotensin II.

In summary, we have shown that the responsiveness of CRH neurons decreases with increasing duration of a stressor, presumably because of a decrease in the releasable pool of CRH. Moreover, we also surmise that after prolonged exposure to a stressful stimulus, there is a desensitization of CRH neurons to neurotransmitters released during stress, including NE. We also suggest that as the duration of the stressor increases, factors other than CRH may play a progressively greater role in sustaining pituitary-adrenal activation. Finally, we have presented data that the immobilization paradigm utilized here is capable of eliciting reproducible responses using pituitary-adrenal measures in an inbred rat strain shown to have a hyper-responsive HPA axis.

The potential clinical implications of the present data are unclear. However, in light of our suggestion that the response of CRH neurons to NE released during stressful stimuli falls progressively with increasing duration of the stressor, we speculate that the sustained hypercortisolism and hyperarousal in melanchonic depression (Gold *et al*, 1988a) may reflect a failure for the development of this noradrenergic desensitization in depressed patients. Recently, the genes for a variety of proteins that promote the desensitization of the beta-noradrenergic receptor have been cloned, including beta-adrenergic receptor kinase, and ß-arrestin (Lefkowitz R.J.; personal communication), and it is likely that other protein factors influence the regulation of other subtypes of adrenergic receptors as well.

ACKNOWLEDGMENTS

We wish to thank Mrs. Darina Kvetnanska for excellent technical assistance.

REFERENCES

Calogero, A.E., Gallucci, W.T., Chrousos, G.P., and Gold, P.W. (1988a). Catecholamine effects upon rat hypothalamic corticotropin-releasing hormone secretion. Clinical implications. *Journal of Clinical Investigation* **82**, 839-846.

Calogero, A.E., Gallucci, W.T., Chrousos, G.P., and Gold, P.W. (1988b). Interaction between GABAergic neurotransmission and rat hypothalamic corticotropin-releasing hormone secretion *in vitro*: Theoretical and clinical implications. *Brain Research* **463**, 28-36.

Calogero, A.E., Kamilaris, T.C., Gomez, M.T., Johnson, E.O., Tartaglia, M.E., Gold, P.W., and Chrousos, G.P. (1989). The muscarinic cholinergic agonist arecoline stimulates the rat hypothalamic-pituitary-adrenal axis through a centrally-mediated corticotropin releasing hormone-dependent mechanism. *Endocrinology* 125, 2445-2453.

Gillies, G. E., Linton, A., and Lowry, P. J. (1982). Corticotropin releasing activity of new CRH is potentiated several times by vasopressin. *Nature* 299, 355-357.

Gold, P.W., Goodwin, F.K., and Chrousos, G.P. (1988a). Clinical and biochemical manifestation of depression (Part 1). *New England Journal of Medicine* 319, 348-353.

Gold, P.W., Goodwin, F.K., and Chrousos, G.P. (1988b). Clinical and biochemical manifestations of depression (Part 2). *New England Journal of Medicine* 319, 413-420.

Makara, G.B., Kvetnansky, R., Jezova, D., Jindra, A., Kakucska, I., and Oprsalova, Z. (1986). Plasma catecholamines do not participate in pituitary-adrenal activation by immobilization stress in rats with transection of nerve fibers to the median eminence. *Endocrinology* 119, 1757-1762.

Rivier, C., and Vale, W. (1983). Modulation of stress-induced ACTH release by corticotropin releasing factor, cathecholamines and vasopressin. *Nature* 305, 325-327.

Scaccianoce, S., Muscolo, L.A., Cigliana, G., Navarra, D., Nicolai, R., and Angelucci, L. (1991). Evidence for a specific role of vasopressin in sustaining pituitary-adrenocortical stress response in the rat. *Endocrinology* 128, 3138-3143.

Spiess, J., Rivier, J., Rivier, C., and Vale, W. (1981). Primary structure of corticotropin-releasing factor from ovine hypothalamus. *Proceedings of the National Academy Sciences (USA)* 78, 6517-6521.

Sternberg, E.M., Hill, J.M., Chrousos, G.P., Kamilaris, T., Listwak, S.J., Gold, P.W., and Wilder, R.L. (1989a). Inflammatory mediator-induced hypothalamic-pituitary-adrenal axis activation is defective in streptococcal cell wall arthritis-susceptible Lewis rats. *Proceedings of the National Academy of Sciences (USA)* 86, 2374-2378.

Sternberg, E.M., Young, W.S. III, Bernardini, R., Calogero, A.E., Chrousos, G.P., Gold, P.W., and Wilder, R.L. (1989b). A central nervous system defect in biosynthesis of corticotropin-releasing hormone is associated with susceptibility to streptococcal cell wall-induced arthritis in Lewis rats. *Proceedings of the National Academy of Sciences (USA)* 86, 4771-4775.

Vale, W., Spiess, J., Rivier, C., and Rivier, J. (1981). Characterization of a 41 -residue ovine hypothalamic peptide that stimulates secretion of corticotropin and ß-endorphin. *Science* 213, 1394-1397.

Stress: Neuroendocrine and Molecular Approaches
Edited by R. Kvetnansky, R. McCarty and J. Axelrod

1992 Gordon and Breach Science
Publishers S.A., New York, USA.
Photocopying permitted by license only.

NEUROENDOCRINE RESPONSES TO FOOTSHOCK STRESS IN ANESTRUS EWES

J. Polkowska and F. Przekop

Institute of Animal Physiology and Nutrition
Polish Academy of Sciences
05-110 Jablonna, Poland

INTRODUCTION

In previous work, we have found that brief periods of electric footshock (3 days) modified CRH/ACTH (Polkowska and Przekop, 1988) and LHRH (Przekop *et al*, 1988) secretion in anestrous ewes. Accumulation of LHRH in the ME coincides with concomitant depletion of irCRH in this structure and this observation suggests that the increase of CRH under stress conditions might be an important factor in suppression of LHRH and gonadotropins in stressed ewes. To elucidate the role of hypothalamic-pituitary-adrenal hormones in gonadotropin secretion during stress conditions in anestrous ewes, we investigated in the present study interrelationships between CRH/ACTH/cortisol and LHRH/LH secretion with different periods of long-term intermittent footshock stress.

MATERIALS AND METHODS

The experiment was performed on three-year-old Polish Merino ewes during the non-breeding season in 3 series (6 experimental and 3 control animals in each series). Stress was induced by applying electric

467

pulses of current intensity 3 mA during 20 minutes/hour for 9 days according to a procedure described by Przekop and co-workers (1985). In the first series, ewes were subjected to footshock stress for 8 or 12 days and sacrificed immediately after the end of stimulation. Hypothalamic and pituitary hormones were analyzed by immunocyto-chemistry. Procedures, antibody used and statistical analysis of LH_B and ACTH cells were described previously (Polkowska and Przekop, 1988). In the second series, blood samples for LH and cortisol assays were taken from jugular vein catheters at 2 hours intervals, on the day before stress (control, 0 day) and the 1st, 4th, 8th and 12th days of stress. LH and cortisol were determined according to the methods of Stupnicki and Madej (1976) and Stupnicki (1985). In the third series, stress lasted for 4 days and blood samples for analysis of LH pulse frequency were taken at 15 minute intervals for 6 hours on days 0, 1 and 4. An LH pulse episode was defined according to criteria of Goodman and Karsch (1980).

RESULTS

In control animals, irLHRH stored in nerve terminals of the ME was present in moderate amounts (Figure 1-1a). irLHRH perikarya situated mainly in the preoptic area (AP) were rare, 2-5 cells per slide (Figure 1-3a). In animals subjected to 8 days of intermittent stress, some visible increase of the density of irLHRH stored in the ME was observed. This phenomenon was even more pronounced in the ME of ewes stressed for 12 days (Figure 1-1b). In this group, the number of visible LHRH perikarya in the AP increased significantly up to 10-15 cells per slide (Figure 1-3b). LH_B cells in the pituitary gland of anestrous ewes (control) represented about 10% of the total cell population (Figure 1-2a, Table 2). The stress procedure resulted in an increase in the cell population up to 14.8% (P≤0.001) (Table 2). Some of them were hypertrophied and there was a shift of ir material (Figure 1-b). The daily mean LH concentration in plasma exhibited similar levels for all intervals of stimulation. In addition, LH pulse frequency was within the range observed in unstressed animals (Table 1).

In control ewes, a high density of irCRH material was seen in the ME (Figure 1-4a). After 8 or 12 days of stress, no differences were observed in the content of irCRH between control and experimental animals (Figure 1-4b). No differences were seen in the number

TABLE 1.   Plasma cortisol and LH concentrations and episodic frequency of LH for control and stressed ewes.

| Group of Animals | | Stressed | | | |
	Control	1st Day	4th Day	8th Day	12th Day
Mean of 24hour cortisol (ng/ml)	8.60 ±0.58	16.04[a] ±0.63	13.71[a] ±0.74	7.80 ±1.39	8.32 ±1.10
Mean of 24hour LH (ng/ml)	2.03 ±0.62	1.97 ±0.59	2.26 ±0.47	2.04 ±0.47	1.89 ±0.64
Episodic frequency	1.8 (0-3)	1.5 (0-3)	1.5 (0-3)		

Values are means ± SEM. [a] significantly different at P≤0.01.

and features of ACTH cells in the pituitary glands of control and experimental ewes (Table 2, Figures 1-5ab). A significant increase of cortisol concentration in plasma was observed only on the 1st and 4th days of stress. On the 8th and 12th days of stress, plasma cortisol concentrations remained at control levels (Table 1).

DISCUSSION

These results together with our previous data (Polkowska and Przekop, 1988; Przekop *et al*, 1988) concern the influences of exposure to acute and prolonged stressful stimuli on two hormonal systems, CRH/

TABLE 2.   Percentage of LH_B and ACTH cells in the pars distalis.

Group Number	Number of animals	LH %	ACTH %
0 Control	3	10. 1±0.8	11.0±0.6
1 Day Stress	3		3.8±0.8[a]
3 Day Stress	3		15.8±0.4[a]
8 Day Stress	3	12.0±0.9	10.9±0.6
12 Day Stress	3	14.8±1.1	11.8±0.7

Values are means ± SEM. [a] significantly different at P≤0.001.

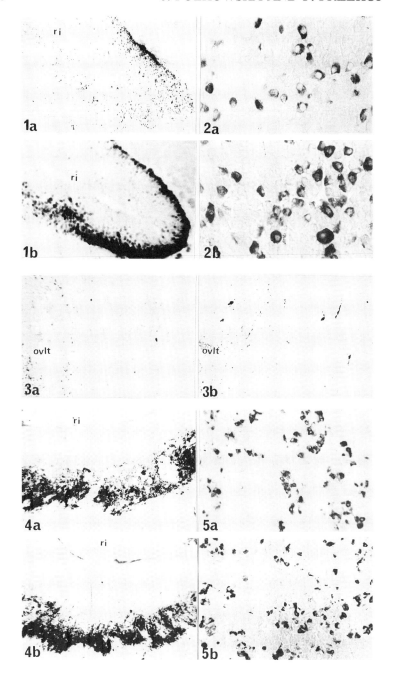

FIGURE 1.   Part 1. irLHRH in the ME of a representative sheep from groups: (a) control and (b) 12 days of intermittent stress. Note the accumulation of irLHRH in the ME of stressed ewes. x 28. ri -- infundibular recess. PART 2. irLH$_B$ cells in the pituitary glands of a representative sheep from groups: (a) control and (b) 12 days of intermittent stress.   Note the accumulation of irLH$_B$ material and hypertrophy of LH$_B$ cells in stressed ewes. x224. PART 3. irLHRH perikarya and axons in the AP and OVLT of brains of representative sheep from groups (a) control and (b) 12 days of intermittent stress. Note the higher number of LH perikarya in stressed sheep. x 28. OVLT -- organum vasculosum of the lamina terminalis. PART 4. irCRH in the ME of representative sheep from groups: (a) control and (b) 12 days of intermittent stress. Note the lack of differences in irCRH between groups. x28. ri -- infundibular recess. PART 5. irACTH cells in the pituitary glands of representative sheep from groups: (a) control and (b) 12 days of stress. Note the lack of differences in the number and morphology of cells between groups. x 70.

ACTH/cortisol and LHRH/LH, in anestrous ewes.   Our findings indicate that these hormones react in different ways to stressful stimulation and the responses depend on the time course of stressful stimulation.   The highest activity of the CRH/ACTH/cortisol axis appears immediately after exposure to the stressor, whereas the depression of LHRH activity was not seen until day 3 of stress (Przekop et al, 1988; Polkowska and Przekop, 1988). With long-term stress, there was a return of activity of the CRH/ACTH/cortisol axis to control levels but a suppression of the LHRH neuronal system probably by inhibition of hormone transport along neuronal fibers and its release from the ME. The changes observed in LH$_B$ cells after long-term stimulation without changes in LH plasma concentrations may be interpreted only as an augmentation of hormone synthesis but not suppression of its release.   The main changes in the LHRH/LH axis appear at a time when CRH/ACTH/cortisol secretion returns to control values.   In the light of these results, it seems reasonable to suggest that elevated activity of the CRH/ACTH/cortisol axis during brief stressful stimulation can initiate changes in the LHRH/LH axis but can not be considered as a major factor affecting LHRH/LH secretory activity in anestrous ewes subjected to long-term stress.   It may be that long-term stressful stimulation suppresses the secretion of hypothalamic LHRH by inhibition of its transport along neuronal fibers and its release from the ME and CRH/ACTH/ cortisol activity during long-term stress is not responsible for these changes.

REFERENCES

Goodman, R.L., and Karsch, F.J. (1980). Pulsatile secretion of luteinizing hormone: differential suppression by ovarian steroids. *Endocrinology* **107**, 1286-1290.

Polkowska, J., and Przekop, F. (1988). Immunocytochemical changes in hypothalamic and pituitary hormones after acute and prolonged stressful stimuli in the anestrous ewe. *Acta Endocrinologica* (Copenhagen) **118**, 269-276.

Przekop, F., Stupnicka, E., Wolinska-Witort, E., Mateusiak, K., Sadowski, B., and Domanski, E. (1985). Changes in circadian rhythm and suppression of plasma cortisol level after prolonged stress in sheep. *Acta Endocrinologica* (Copenhagen) **110**, 540-545.

Przekop, F., Polkowska, J., Mateusiak, K. (1988). The effect of prolonged stress on the hypothalamic luteinizing hormone releasing hormone (LHRH) in the anestrous ewe. *Experimental and Clinical Endocrinology* **91**, 334-340.

Stupnicki, R. (1985). Radioimmunoassay of cortisol. In: F. Kobot and R. Stupnicki (Eds.), "Radio-Immunoassays and Radiocompetition Methods: Applied in Clinics," pp. 235-244. Warsaw: P.Z.W.L.

Stupnicki, R., and Madej, A. (1976). Radioimmunoassay of LH in the blood plasma of farm animals. *Endokrinologie* **68**, 6-13.

Stress: Neuroendocrine and Molecular Approaches
Edited by R. Kvetnansky, R. McCarty and J. Axelrod

1992 Gordon and Breach Science
Publishers S.A., New York, USA.
Photocopying permitted by license only.

LONG-LASTING EFFECTS OF A SINGLE FOOTSHOCK SESSION ON BEHAVIORAL AND ENDOCRINE RESPONSIVENESS IN MALE RATS

H. H. Van Dijken[1], J. Mos[2], M. Th. M. Tulp[2]
and F. J. H. Tilders[1]

[1]Department of Pharmacology, Medical Faculty, Free University
Van der Boechorststraat 7, 1081 BT, Amsterdam, The Netherlands
[2]Department of Pharmacology, Duphar B.V., P.O. Box 900, 1380 DA
Weesp, The Netherlands

INTRODUCTION

Chronic intermittent stress has been considered as an important factor in the etiology of psychopathology in humans (Anisman and Zacharko, 1982). However, surprisingly little is known about the frequency and number of stress sessions needed to induce behavioral changes in animals.

Recently, we reported that rats exposed to a brief session of inescapable footshock (15 minutes) showed a decrease in behavioral activity in an open field. This effect developed with time following shocks, reaching a maximum 7 days after the shock session. Moreover, these rats showed an exaggerated immobility response to a sudden change in environmental conditions. Behavioral changes induced by shock exposure persisted for at least 28 days after the shock session. We conclude that these changes in behavioral response reflect an increased sensitivity to stressful environmental stimuli. These observations prompted us to study the long-lasting effects of a short period of inescapable footshocks on neuroendocrine systems known to

473

be involved in the stress response of animals.

MATERIALS AND METHODS

Male Wistar rats (Harlan CPB, Zeist, The Netherlands) weighing 300-350 g were housed individually in macrolon cages under a reversed day-night cycle (lights off from 0700 to 1900 hours). After two weeks of acclimatization, rats were randomly assigned to 4 groups (n = 6). Two groups of rats (S) were subjected to ten 1 mA footshocks of 6 seconds duration each spread over a period of 15 minutes (mean shock interval of 90 seconds, range of 24-244 seconds). Control rats (C) were placed in the shockbox for 15 minutes without receiving shocks. Subsequently, rats were handled daily for 10 days and twice daily for the next 3 days. On day 14 postshock, 6 C and S rats were rapidly decapitated and trunk blood was collected in heparinized tubes. Cerebral cortical tissue was prepared and used in a ß-adrenoceptor binding assay as described earlier (Nahorski, 1978). For each individual animal, the K_d and B_{max} of $[^3H]$-dihydroalprenolol ($[^3H]DHA$) binding was determined. The remaining 6 C and S rats were tested in a noise test as described earlier (Van Dijken *et al*, 1991). Briefly, rats were placed in an observation cage (80x60x50 cm) under dim light, while an 80-85 dB "white" noise was produced. Three minutes after the start of the test, the noise was switched off (reduction to 55-60 dB). Locomotion and immobility duration were scored during the 3 minutes preceding and the 3 minutes following switching off the noise. Thereafter, rats were immediately decapitated and trunk blood was collected.

Blood was centrifuged (1000 g, 10 minutes) and plasma was stored at -20°C until radioimmunoassays of ACTH, corticosterone (CS), and prolactin (PRL) were conducted according to methods described earlier (Van Oers and Tilders, 1991; Van Oers *et al*, 1989). Statistical analysis of behavioral data was conducted with Mann-Whitney U test or Wilcoxon-Matched pairs test and neuroendocrine data were analyzed with Student's t-test (p < 0.05 is significant).

RESULTS AND DISCUSSION

Rats subjected to footshock showed reduced locomotion during the first 3 minutes of the noise test as compared to C rats 14 days after the shock session. After switching off the noise, both C and S rats showed a decrease in locomotion, which did not influence the difference between the two treatment groups (Figure 1). Furthermore, S rats tended to show more immobility (p=0.06) than C rats during the first 3 minutes of the test. Switching off the noise induced a marked increase in immobility in S rats, whereas only a tendency towards increased immobility was observed in C rats (p=0.06). This resulted in a marked difference in immobility between C and S rats during the last 3 minutes of the noise test (Figure 1). These long-lasting behavioral effects of a brief shock session are in accord with our earlier observations (Van Dijken *et al*, 1991ab). Immobility is a species-specific defensive behavioral response to threatening or stressful stimuli in rodents (Bolles, 1970), which indicates that rats subjected to the brief shock session are sensitized to environmental stimuli for a long period of time. This interpretation is corroborated by the effects of shocks on behavioral response to open field exposure as earlier reported, which includes a decrease in behavioral activity (e.g., locomotion and rearing) and an increase in defecation (Van Dijken *et al*, 1991ab).

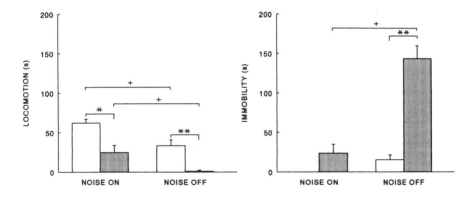

FIGURE 1.   Effect of a brief period of inescapable footshock on behavioral response to noise test exposure 14 days after the shock session. The open bars represent C rats (n=6) and the hatched bars represent S rats (n=6). The "noise on" and the "noise off" observation periods in the noise test were 3 minutes each. Data are expressed as means (SEM). *p<0.05, **p<0.01 versus C rats. +p<0.05 versus noise on period.

Furthermore, the present results of the hormone measurements also support this interpretation. As can be seen in Figure 2, exposure to footshock did not affect basal plasma levels of ACTH, CS, and PRL measured at 14 days post-shock. Exposure to the noise test resulted in increased ACTH and CS levels in both C and S rats. However, C rats did not show a PRL response, which is probably due to the relatively high basal levels of PRL (Kant *et al*, 1989). Nevertheless, a marked PRL response was found in S rats exposed to the noise test. Also the CS response in S rats was greater in magnitude than in C rats, whereas the ACTH response only tended to be increased in S rats (p = 0.06) relative to C rats (Figure 2). Thus, a short period of inescapable footshock induced long-lasting behavioral, pituitary-adrenal, and PRL hyperresponsiveness to a mild stressor. The fact that these three different output systems are affected in a parallel manner suggests a common underlying mechanism for these alterations. According to the dual-process theory of habituation (Groves and Thompson, 1970), sensitization of responses occurs in a "state" system. Others have argued that in animals the hypothalamic-pituitary-adrenal (HPA) axis is part of the arousal system, which is a "state" system, and that the activity of the HPA-axis is a reliable and sensitive measure of arousal or "stress" (Hennessey and Levine, 1979). Therefore, the arousal system is a likely candidate for the mediation of the long-lasting behavioral and endocrine effects induced by footshock.

In this arousal system, which is supposed to be located in the reticular formation and limbic areas, the norepinephrine (NE) ascending projections from the locus coeruleus play an important role in mediating general alertness or vigilance (Smelik, 1987). Moreover, NE neurotransmission is involved in stress-induced behavioral impairment such as depression and anxiety (Anisman and Zacharko, 1982; Glavin, 1985; Gray, 1982; Weiss *et al*, 1985). Therefore, we measured ß-adrenoceptor binding in cerebral cortex 14 days following the shock session as a first step in elucidating the neuronal substrate of shock-induced alterations. However, C and S rats did not differ in the B_{max} (0.32 ± 0.02 versus 0.32 ± 0.02, respectively) and K_d (107.4 ± 6.2 versus 109.7 ± 5.8, respectively) of [^{3}H]DHA binding. Thus, cortical ß-adrenoceptor binding characteristics do not explain the shock-induced alterations in behavioral and hormonal responses. Nevertheless, this finding does not exclude changes in signal transfer via these receptors, since it is known that stressful experiences may

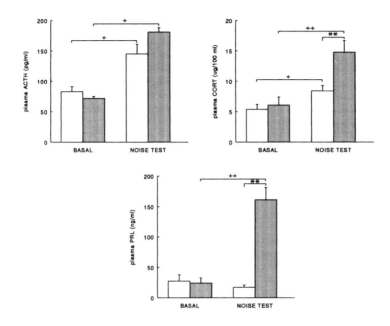

FIGURE 2. Effects of a brief period of inescapable footshock on basal levels and noise test-induced responses of ACTH, CORT, and PRL in plasma 14 days after the shock session. Open bars represent C rats (n=6) and hatched bars represent S rats (n=6). Data are expressed as means (SEM). **p<0.01 versus C rats. +p<0.05, ++p<0.01, +++p<0.001 versus basal levels.

alter the functional response of these receptors in cortex without alterations in binding properties (Stone, 1987). Possibly, shock-induced alterations in NE neuro-transmission should be investigated at other loci in brain such as the locus coeruleus (Weiss *et al*, 1985) or the hypothalamus (Irwin *et al*, 1986). Further, serotonin and dopamine may also play a role in the mediation of shock-induced changes (Gray, 1982; Smelik, 1987).

The fact that a single and brief session of inescapable footshock induces long-lasting changes in behavioral and endocrine responses is striking, since such effects are generally reported as a consequence of chronic intermittent stress. For instance, it has been reported that chronic intermittent stress increases HPA responses to a subsequent novel stressor (Sakellaris and Vernikos-Danellis, 1975; Vernikos et al, 1982). A similar phenomenon has been described for sympathetic-adrenal medullary responses to a novel stimulus following chronic stress (Konarska et al, 1989). In a different chronic stress paradigm (chronic unpredictable stress) originally developed by Katz and colleagues (Katz et al, 1981; Soblosky and Thurmond, 1986), animals were subjected to a variety of different stressors for 14 to 21 days. This procedure resulted in decreased behavioral activity in an open field and an impaired behavioral "activation response" to a novel stressor. In addition, the CS responses to the open field or a novel stressor were increased without changes in basal levels of this hormone (Katz et al, 1981, Soblosky and Thurmond, 1986). The behavioral and hormonal changes induced by Katz's procedure bear a striking resemblance to our shock-induced changes in these parameters. This suggests that time following the first exposure to a stressor in chronic intermittent stress paradigms may be more important than the number of "stress" sessions. As reported earlier, the effects of a single shock session on behavior in an open field increased with time reaching a maximum at 7 days (Van Dijken et al, 1991ab). In addition, S rats did not show an altered response as compared to C rats during the first 3 minutes of the noise test on day 1, and the immobility response to switching off the noise was smaller than we usually observe at 14 days after footshock (unpublished observation). It is interesting to note that similar time-dependent effects of different stressors have also been recognized by others (see Antelman, 1988 for a review). This emphasizes the need to reconsider the aspect of the variable time between stressor and response measurement in studies on the effects of chronic stress.

In conclusion, exposure to a brief period of inescapable footshock induces a process of sensitization of behavioral and hormonal responses to mildly stressful environmental stimuli in male rats on a long-term basis. We speculate that these effects may be useful in the search for new antidepressants and/or anxiolytics.

ACKNOWLEDGMENTS

The reagents for the determination of rPRL were kindly supplied by

NIDDK-NIH. We thank Dr. T.J. Benraad (Nijmegen, The Netherlands) for his gift of the corticosterone antiserum and Dr. G.B. Makara (Budapest, Hungary) for the ACTH antiserum. R. Binnenkade and P.A.J. Jonkergouw are acknowledged for their help with the hormone determinations.

REFERENCES

Anisman, H., and Zacharko, R.M. (1982). Depression: The predisposing influence of stress. *Behavioral and Brain Sciences* 5, 89-137.

Antelman, S.M. (1988). Time-dependent sensitization as the cornerstone for a new approach to pharmacotherapy: Drugs as foreign/stressful stimuli. *Drug Development Research* 14, 1-30.

Bolles, R.C. (1970). Species-specific defense reactions and avoidance learning. *Psychological Review* 77, 32-48.

Glavin, G.B. (1985). Stress and brain noradrenaline: A review. *Neuroscience and Biobehavioral Reviews* 9, 233-243.

Gray, J.A. (1982). Précis of the neuropsychology of anxiety: An enquiry into the functions of the septo-hippocampal system. *Behavioral and Brain Sciences* 5, 469-534.

Groves, P.M., and Thompson, R.F. (1970). Habituation: A dual-process theory. *Psychological Review* 77, 419-450.

Hennessey, J.W., and Levine, S. (1979). Stress, arousal and the pituitary-adrenal system: A psychoendocrine hypothesis. *Progress in Psychobiology and Physiological Psychology* 8, 133-178.

Irwin, J., Ahluwalia, P., and Anisman, H. (1986). Sensitization of norepinephrine activity following acute and chronic footshock. *Brain Research* 379, 98-103.

Kant, G.J., Mougey, E.H., and Meyerhoff, J.L. (1989). ACTH, prolactin, corticosterone and pituitary cyclic AMP responses to repeated stress. *Pharmacology, Biochemistry and Behavior* 32, 557-561.

Katz, R.J., Roth, K.A., and Carroll, B.J. (1981). Acute and chronic stress effects on open field activity in the rat: Implications for a model of depression. *Neuroscience and Biobehavioral Reviews* 5, 247-251.

Konarska, M., Stewart, R.E., and McCarty, R. (1989). Sensitization of sympathetic-adrenal medullary responses to a novel stressor in chronically stressed laboratory rats. *Physiology and Behavior* 46, 129-135.

Nahorski, S.R. (1978). Heterogeneity of cerebral ß-adrenoceptor binding sites in various vertebrate species. *European Journal of Pharmacology* 51, 199-209.

Sakellaris, P.C., and Vernikos-Danellis, J. (1975). Increased rate of response of the pituitary-adrenal system in rats adapted to chronic stress. *Endocrinology* 97, 597-602.

Smelik, P.G. (1987). Adaptation and brain function. In: E.R. de Kloet, V.M. Wiegant and D. de Wied (Eds.), Progress in Brain Research, Volume 72, pp. 3-9. Amsterdam: Elsevier Science Publishers B.V.

Soblosky, J.S, and Thurmond, J.B. (1986). Biochemical and behavioral correlates of chronic stress: Effects of tricyclic antidepressants. *Pharmacology, Biochemistry and Behavior* **24**, 1361-1368.

Stone, E.A. (1987). Central cyclic-AMP-linked noradrenergic receptors: New findings on properties as related to the actions of stress. *Neuroscience and Biobehavioral Reviews* **11**, 391-398.

Van Dijken, H.H., Van der Heyden, J.A.M., Mos, J., and Tilders, F.J.H. (1991). Long-lasting behavioral changes after a single footshock stress session. A model of depression? In: B. Olivier, J. Mos and J.L. Slangen (Eds.), "Animal Models in Psychopharmacology," pp. 231-236. Basel: Birkhduser Verlag.

Van Dijken, H.H., Van der Heyden, J.A.M., Mos, J., and Tilders, F.J.H. (1991). Inescapable footshocks induce progressive and long-lasting behavioral changes in male rats. Submitted.

Van Oers, J.W.A.M., and Tilders, F.J.H. (1991). Non-adrenocorticotropin mediated effects of endogenous corticotropin-releasing factor on the adrenocortical activity in the rats. *Journal of Neuroendocrinology* **3**, 119-121.

Van Oers, J.W.A.M., Tilders, F.J.H., and Berkenbosch, F. (1989). Characterization and biological activity of a rat monoclonal antibody for rat/human corticotropin-releasing factor. *Endocrinology* **124**, 1239-1246.

Vernikos, J., Dallman, M.F., Bonner, C., Katzen, A., and Shinsako, J. (1982). Pituitary-adrenal function in rats chronically exposed to cold. *Endocrinology* **110**, 413-420.

Weiss, J.M., and Goodman-Simson, P. Neurochemical basis of stress-induced depression. *Psychopharmacology Bulletin* **21**, 447-457.

Stress: Neuroendocrine and Molecular Approaches
Edited by R. Kvetnansky, R. McCarty and J. Axelrod

1992 Gordon and Breach Science
Publishers S.A., New York, USA.
Photocopying permitted by license only.

ADRENOCEPTOR REGULATION OF PARAVENTRICULAR NUCLEUS NEURONAL ACTIVITY AS RELATED TO HYPOTHALAMO-PITUITARY ADRENOCORTICAL RESPONSES

D. Saphier

Department of Pharmacology and Therapeutics, Louisiana State
University Medical Center, Shreveport, Louisiana USA

INTRODUCTION

The majority of neurons in the dorsal medial parvocellular components of the hypothalamic paraventricular nucleus (PVN) synthesize corticotropin releasing factor (CRF) and send projections to the median eminence to regulate adrenocorticotropic hormone (ACTH) secretion (Swanson *et al*, 1983). Advances have been made in the identification of the afferent projections and the chemical nature of the pathways innervating the PVN and regulating HPA responses (Antoni, 1986; Feldman and Saphier, 1989; Ganong, 1980; Jones *et al*, 1987; Plotsky, 1987; Saphier, 1989; Saphier and Feldman, 1989ab, 1991; Swanson and Sawchenko, 1983).

There is incomplete agreement as to the role of NE in the regulation of HPA activity, with some studies indicating an inhibitory component, although the consensus indicates a facilitatory role (Al-Damluji, 1988; Feldman and Saphier, 1989; Ganong, 1980; Jones *et al*, 1987). Facilitatory actions of NE are probably mediated via α_1- and α_2-adrenergic mechanisms, while ß-adrenoceptors may be inhibitory (Plotsky, 1987; Saphier and Feldman, 1989b, 1991). PVN innervation arises from the A1, A2, and A6 norepinephrine (NE)-containing, and the C1-3 epinephrine (EPI)-containing brain stem cell groups (Cunningham and Sawchenko, 1988), with their projections ascending via the ventral noradrenergic bundle (VNAB), where many fibers also stain for the co-transmitter, neuropeptide Y (NPY) (Sawchenko et al, 1985). These brainstem projections are believed to play a primary role in the regulation of PVN CRF-secreting neurons in response to hemodynamic stimuli (Plotsky, 1987). Some 70% of NE innervation of the PVN arises from the A2 cell group, with a further 20% arising from the A1 system in the caudal ventrolateral medulla (Cunningham and Sawchenko, 1988). The NE fibers terminate in proximity with CRF-immunoreactive perikarya and appear to regulate HPA activity (Swanson et al, 1983).

Injection of the catecholaminergic neurotoxin, 6-hydroxydopamine (6-OHDA), into the VNAB depletes hypothalamic NE, and abolishes the circadian rhythm of ACTH and corticosteroid (CS) secretion (Szafarczyk et al, 1985), and also reduces CS responses to neurogenic stimuli (Feldman and Saphier, 1989). Electrical stimulation of the VNAB elevates portal blood CRF concentrations (Plotsky 1987), and NE injections next to the PVN increase ACTH and CS secretion (Saphier and Feldman 1989b). These observations are also supported by in vitro studies (Jones et al, 1987), and demonstrate that NE stimulates the release of CRF.

METHODS

Electrophysiological recording experiments were performed using male rats under urethane anesthesia. Animals were secured in a stereotaxic frame, and stimulating electrodes were lowered into the A1, A2, A6 or C2 cell groups (Saphier, 1989), or into the VNAB. Three groups of animals were prepared for the VNAB experiments: (I) 7 control rats; (II) 9 animals given injections of α-methylparatyrosine (AMPT) methyl ester (315 mg/kg, i.p.) 27 and 3 hours prior to experiments; (III) 7 rats received an i.c.v. injection of 200 μg of 6-OHDA, 1-2 weeks preceding the experiment. These groups were used to determine the effects of NE depletion upon PVN responses, and the hypothalamus was removed for analysis of tissue NE content using HPLC (Saphier and Feldman, 1989a). A transpharyngeal approach was used to expose the median eminence and a non-concentric bipolar electrode was used for identification of PVN neurons, recorded using glass electrolyte-filled microelectrodes. Stimulation was achieved using bipolar square-wave pulses of 1 mA current strength and 1 msec total duration at frequencies of 0.5 and 5 Hz, and 50 Hz trains of stimuli. For iontophoresis, a multibarreled electrode was used, containing: (a) 0.0015 M ergotamine tartrate; (b) 0.025 M tolazoline HCl; (c) 0.025 M propranolol HCl; (d) 0.025 M l-phenylephrine HCl; (e) 0.025 M clonidine HCl; (f) 0.025 M NE bitartrate. One barrel was used for current balancing with drugs ejected as anions or cations, using 2-250 nA of current (Saphier and Feldman, 1991).

Neuroendocrine experiments were performed using rats with a stimulating electrode implanted into the VNAB, and chronic guide cannulae implanted bilaterally above each PVN. Drugs were injected using 29-gauge needles supplied by a microsyringe, and VNAB stimulation consisted of 1 msec pulses of 0.1 mA current strength, at a frequency of 20 Hz for 2 minutes. Rats were also prepared with PVN cannulae only. Experiments were carried out under pentobarbital anesthesia induced 15 minutes before testing. The following agonists were injected (1 μl/PVN), 25 minutes after anesthesia as 20 nmole doses, with NE administered at 10 nmoles also: α_1, l-phenylephrine HCl; α_2, clonidine HCl; $\beta_{1/2}$, isoproterenol HCl. NPY was also used at doses of 10 and 20 pmoles. Fifteen minutes following injection, blood was taken from a jugular vein and plasma frozen for estimation of plasma CS concentrations ([CS]) using fluorimetry. The antagonists were injected over a 2 minute period and 8 minutes later, VNAB stimulation was initiated, following which a blood sample was taken 13 minutes later. The antagonists (20 nmoles/PVN) were: α_1, ergotamine tartrate, prazosin HCl; α_2, yohimbine HCl, tolazoline HCl; $\beta_{1/2}$, propranolol HCl. Control injections were either 0.9% saline or dimethyl sulphoxide/saline (DMSO/saline; 0.32:0.68, V/V), the latter for dissolving ergotamine and yohimbine. In some animals, 6-OHDA (12.5 μg/PVN), was injected to examine the effect of PVN NE depletion upon responses to VNAB stimulation (Saphier and Feldman, 1989b).

RESULTS

Extracellular electrical activity was recorded from 203 PVN neurons antidromically identified as projecting to the median eminence. Cells were located 2.29±0.03 mm above the base of the brain, within the CRF-rich component of the PVN, their mean firing rate was 3.2±0.3 Hz and antidromic invasion latency was 9.9±0.3 msec. A1 or A2 stimulation evoked excitatory responses from most cells (76% and 85%, respectively), while A6 and C2 stimulation evoked more inhibition (43% and 59%, respectively). Most responses (56%) were not clearly defined in terms of latency and were only observed following delivery of 5-10 single shocks at 0.5 Hz. Response reversals were observed, after delivery of 50 Hz stimulation, for 20% of the cells tested.

VNAB stimulation at 0.5 or 5 Hz excited most of the cells (89/110, 81%) but only 43% showed clear, stimulus-locked activation with onset latency of 42±8 msec and offset at 89±12 msec; the remaining cells showed overall increases in firing after 5-10 stimuli. With 50 Hz trains of stimuli, the responses to inhibition for 21/68 of the excited cells were reversed. Inhibition of NE and EPI synthesis by AMPT did not alter firing of cells or responses to low frequency stimulation, but 50 Hz trains reversed the response direction of only 4/35 cells, (p < 0.05, χ^2-test). 6-OHDA treatment reduced the number of excited cells (10/47; p < 0.005), and both AMPT and 6-OHDA treatment reduced hypothalamic NE content to below detectable levels.

Iontophoresis of NE or of l-phenylephrine increased the activity of most cells tested (94% and 72%, respectively), while application of ergotamine reduced the activity of 44% of the cells tested and prevented the excitation following VNAB stimulation for 84% of the cells examined. Propranolol increased the activity of 77% of cells and prevented inhibitory responses following high frequency VNAB stimulation for 94% of the cells. Tolazoline evoked mixed responses from the cells examined, with a trend towards suppression of activity and potentiation of VNAB responses being observed. Clonidine elicited excitation from the majority of cells tested, with most cells exhibiting an after-inhibition.

Plasma CS following PVN saline injection was 116±3 ng/ml and VNAB stimulation increased this to 275±7 ng/ml with saline injection (p < 0.005, Student's t-test), and to 293±12 ng/ml (p < 0.005) with DMSO/saline. Ergotamine (89±14 ng/ml; p < 0.005) and prazosin (171±14 ng/ml; p < 0.005) prevented the rise in plasma CS, and for ergotamine the CS increase was lower than that in unstimulated rats (p < 0.05). Both yohimbine (128±5 ng/ml; p < 0.005) and tolazoline

(190±13 ng/ml; p<0.005) were also effective in attenuating the response. VNAB-stimulated increases in plasma CS were not prevented by propranolol (294±13 ng/ml) or by clonidine (272±8 ng/ml), while 6-OHDA treatment completely abolished the rise in plasma CS (111±5 ng/ml). NE at doses of 10 and 20 nmoles significantly increased plasma CS to 154±5 ng/ml (p<0.005) and 274±12 ng/ml (p<0.005), respectively. NPY at 10 and 20 pmoles elevated plasma CS to 189±19 ng/ml (p<0.005) and 278±11 ng/ml (p<0.005), respectively. A combination of 10 nmoles NE and 10 pmoles NPY elevated plasma CS to 247±14 ng/ml, more than NE (p<0.005) or NPY (p<0.01) alone at the same doses. Plasma CS following l-phenylephrine was 338±16 ng/ml (p<0.005), greater than basal CS (88±4 ng/ml), and clonidine also increased plasma CS to 255±7 ng/ml (p<0.005). Isoproterenol also caused an increase in plasma CS to 262±18 ng/ml (p<0.005).

DISCUSSION

Stimulation of the A1 and A2 brainstem cell groups and of the VNAB increased both HPA secretion and the activity of the majority of PVN neurons examined. The data are in line with those of other studies (Day *et al*, 1985), and confirm the functional inferences suggested by the presence of anatomically and histochemically identified projections to the PVN arising from the regions examined. The facilitory responses suggest a stimulatory role for NE, and the observations that α_1-antagonists prevented both HPA and PVN neuronal responses following VNAB stimulation, indicate that the effects of NE are mediated via α_1-adrenoceptors at the level of the PVN. Furthermore, the α_1-agonist l-phenylephrine also elicited increases in plasma CS and increased the activity of PVN neurons tested. Clonidine produced similar responses, suggesting the involvement of α_2-adrenoceptors.

C2 stimulation evoked inhibitory responses suggesting that EPI may inhibit PVN activity, and it seems likely that the lower basal firing rate of cells recorded from these rats may have been due to the sustained release of EPI during the course of the experiment. In addition, it seems likely that a greater release of NE (and EPI) following high frequency VNAB stimulation was responsible for the neuronal response reversals to inhibition that were observed. The idea that the reversals were due to ß-adrenoceptor activation was confirmed by the observation that such inhibition was prevented by propranolol. Propranolol alone also increased the firing of PVN neurons, and ergotamine reduced the firing, suggesting a tonic balance between α_1-adrenoceptor facilitation and ß-adrenoceptor inhibition of PVN activity under basal conditions. This was partly confirmed in the neuroendocrine experiment, when ergotamine alone reduced plasma CS, although the increases in plasma CS following the ß-agonist, isoproterenol, remains without explanation at the present time.

AMPT treatment did not alter PVN responses to single shock VNAB stimulation, despite the lack of hypothalamic NE content. Since the VNAB contains fibers co-staining for NPY (Sawchenko et al, 1985), it seems likely that this peptide was responsible for the continued responsiveness of the PVN neurons examined. High frequency stimulation of the VNAB in AMPT-treated animals did not evoke the response reversals, as seen in control rats. This indicates that NE/EPI were probably the transmitters responsible for the inhibitory response and suggests that the sole action of NPY is excitatory. The continued responsiveness of PVN cells following AMPT treatment raises the possibility that the remaining NPY may have been responsible for the excitatory responses. In support, NPY injected into the PVN increased plasma CS and this effect was additive with respect to NE (Saphier and Feldman, 1989b).

Perhaps one of the most significant findings in these studies was the observation that the majority (56%) of both excitatory and inhibitory responses were not clearly defined in terms of latency following stimulation. Thus, it was frequently necessary to deliver a number of single stimuli before a response could be observed, and responses were often sustained for many seconds following cessation of the stimulation. This lack of stimulus-locked synaptic responses suggests that a sustained afferent input may be necessary to evoke an activation of the HPA axis. In view of the direct innervation of the PVN by catecholaminergic fibers (Cunningham and Sawchenko, 1988), it seems likely that the reason for such responses may be a lack of close apposition of synaptic boutons with post-synaptic elements, requiring a slow diffusion of transmitter in order to evoke appropriate responses. Structural evidence for such organization exists in that 40-90% of NE terminals in the cerebral cortex lack synaptic specialization and, since NE is able to remain within the region of release for at least 10 seconds, there are likely to be long-term effects of the transmitter within the area (Vizi, 1981).

REFERENCES

Al-Damluji, S. (1988). Adrenergic mechanisms in the control of corticotrophin secretion. *Journal of Endocrinology* 119, 5-14.

Cunningham, Jr., E.T., and Sawchenko, P.E. (1988). Anatomical specificity of noradrenergic inputs to the paraventricular and supraoptic nuclei of the rat hypothalamus. *Journal of Comparative Neurology* 274, 60-76.

Day, T.A., Ferguson, A.V., and Renaud, L.P. (1985). Noradrenergic afferents facilitate the activity of tuberoinfundibular neurons of the hypothalamic paraventricular nucleus. *Neuroendocrinology* 41, 17-22.

Feldman, S., and Saphier, D. (1989). Extrahypothalamic neural afferents and the role of neurotransmitters in the regulation of adrenocortical secretion. In: F.C. Rose (Ed.), "The Control of the Hypothalamo-Pituitary-Adrenocortical Axis," pp. 297-316. Madison CT: International Universities Press.

Ganong, W.F. (1980). Neurotransmitters and pituitary function: regulation of ACTH secretion. *Federation Proceedings* 39, 2923-2930.

Jones, M.T., Gillham, B., Campbell, F.A., Al-Taher, A.R H., Chuang, T.T., and Di Sciuollo, A. (1987). Pharmacology of neural pathways affecting CRH secretion. *Annals of the New York Academy of Sciences* 512, 162-175.

Plotsky, P.M. (1987). Facilitation of immunoreactive corticotropin-releasing factor secretion into the hypophysial-portal circulation after activation of catecholaminergic pathways or central norepinephrine injection. *Endocrinology* 121, 924-930.

Saphier, D. (1989). Catecholaminergic projections to tuberoinfundibular neurones of the paraventricular nucleus: I. Effects of stimulation of A1, A2, A6 and C2 cell groups. *Brain Research Bulletin* 23, 389-395.

Saphier, D., and Feldman, S. (1989a). Catecholaminergic projections to tuberoinfundibular neurones of the paraventricular nucleus: II. Effects of stimulation of the ventral noradrenergic bundle: evidence for cotransmission. *Brain Research Bulletin* 23, 397-404.

Saphier, D., and Feldman, S. (1989b). Adrenoceptor specificity in the central regulation of adrenocortical secretion. *Neuropharmacology* 28, 1231-1237.

Saphier, D., and Feldman, S. (1991). Catecholaminergic projections to tuberoinfundibular neurones of the paraventricular nucleus: III. Effects of adrenoceptor agonists and antagonists. *Brain Research Bulletin* 26, 863-870.

Sawchenko, P.E., Swanson, L.W., Grzanna, R., Howe, P.R.C., Bloom, S.R., and Polak, J.M. (1985). Colocalization of neuropeptide Y immunoreactivity in brainstem catecholaminergic neurons that project to the paraventricular and supraoptic nuclei in the rat. *Journal of Comparative Neurology* 24, 138-153.

Swanson, L.W., and Sawchenko, P.E. (1983). Hypothalamic integration: organization of the paraventricular and supraoptic nuclei. *Annual Review of Neuroscience* 6, 269-324.

Swanson, L.W., Sawchenko, P.E., Rivier, C., and Vale, W. (1983). The organization of ovine corticotropin releasing factor (CRF)-immunoreactive cells and fibers in the rat brain: an immunohistochemical study. *Neuroendocrinology* 36, 165-186.

Szafarczyk, A., Alonso, G., Ixart, G., Malaval, F., and Assenmacher, I. (1985).

Diurnal-stimulated and stress-induced ACTH release in rats is mediated by ventral noradrenergic bundle. *American Journal of Physiology* **249**, E219-E226.

Vizi, E.S. (1981).   Release-modulating adrenoceptors.   In: G. Kunos (Ed.), "Neurotransmitter Receptors," (Volume 1), pp. 65-107. New York: Wiley Interscience.

Stress: Neuroendocrine and Molecular Approaches
Edited by R. Kvetnansky, R. McCarty and J. Axelrod

1992 Gordon and Breach Science
Publishers S.A., New York, USA.
Photocopying permitted by license only.

SOCIAL STRESS OF CROWDING IMPAIRS PITUITARY-ADRENOCORTICAL RESPONSIVENESS TO MONOAMINERGIC STIMULATION

J. Bugajski and E. Wieczorek

Institute of Pharmacology, Polish Academy of Sciences
Krakow, Poland

INTRODUCTION

It is generally accepted that psychological stressors like handling, crowding and novelty activate the hypothalamic-pituitary-adrenal (HPA) system. Daily exposure of animals to a stressor induces adaptation, as reflected by a decreased adrenocortical response to the stimulus (Armario *et al*, 1986; Natelson *et al*, 1988; Rivier and Vale, 1987). Acute and chronic stress alters the CRF content in brain regions that appear to play an essential role in regulating the pituitary-adrenal stress response and are involved in regulation of sympathomedullary system activity.

Numerous neuroendocrine and pharmacological findings support the hypothesis that adaptive changes induced by chronic stress include monoaminergic receptors. Repeated restraint causes a transient decrease in density of ß-adrenergic receptors in several regions of the brain (U'Pritchard and Kvetnansky, 1980; Stone and Platt, 1982; Stone *et al*, 1986). A variety of stressors induce short-term changes in sensitivity of central cholinergic and GABA-ergic receptors in discrete brain regions and the HPA axis in rats (Gabriel and Soliman, 1983; Biggio *et al*, 1990). Stress exposure also causes a rapid decrease in GABA receptor binding in the central nervous system (Biggio *et al*, 1981).

489

The purpose of the present study was to determine how chronic exposure of rats to the social stress of crowding influences the functional adaptability of some central neurotransmitter receptors known to be involved in mediation of HPA activity under basal and stress conditions.

METHODS

The experiments were carried out using male Wistar rats weighing 180-200 g at the beginning of the experiment. The rats were randomly assigned to one of two experimental groups: control and social stress of crowding. Control rats were housed 7 per cage (52x32x20 cm) and remained in their home cages until scheduled for treatment. Stressed rats were crowded in groups of 21 per cage (52x32x20 cm) for 7, 14 and 21 days. The effects of various receptor agonists were examined on the 7th, 14th and 21st day of crowding and were compared with the effects in control animals. All drugs were injected into the right lateral cerebral ventricle of conscious rats in a volume of 10 μl. Control rats were injected with the same volume of saline. One hour after drug injection rats were decapitated and their trunk blood was collected. Serum corticosterone concentration was determined fluorometrically.

To minimize circadian variability, all experiments were performed between 1000-1100 hours and animals were sacrificed between 1100-1200 hours, when plasma corticosterone is at a relatively low level.

The drugs used were: Clonidine (Boehringer), L-Phenylephrine hydrochloride, Carbamylcholine hydrochloride, r-Amino-n-Butyric acid (GABA), DL-Isoproterenol. Hcl, [D-Ala^2-D-Met^5]-Enkephalinamide-Acetate salt and [D-Ala^2-D-Leu^5]-Enkephalinamide Acetate salt (Sigma). Drugs were dissolved in 0.9% NaCl immediately before use.

The data are presented as means ± SEM. Statistical significance of differences between groups was estimated using t-tests.

RESULTS

Effects of Crowding on Corticosterone Response to Adrenergic Agonists

In crowded rats, basal serum corticosterone levels did not differ significantly from levels in control rats. Phenylephrine (30 μg icv), an α_1-adrenergic agonist, induced a considerably reduced corticosterone response in rats crowded for 7 days as compared with control rats. The corticosterone response to phenylephrine after 14 days of crowding did not differ from the control response, whereas after 21 days it was higher in crowded than in control rats (Table 1). A significant increase in serum corticosterone levels induced by clonidine, an α_2-adrenergic agonist (10 μg icv), was only moderately diminished after 1st and 2nd week of crowding. The corticosterone response to isoproterenol (10 μg icv), a ß-adrenergic agonist, was almost totally blocked in rats crowded for 7 days and also substantially diminished after 14 or 21 days of social stress.

TABLE 1.   Effects of crowding on corticosterone secretion stimulated by icv administration of monoaminergic receptor agonists in rats.

Treatment	Dose µg/rat	Crowding period (days)		
		7	14	21
Saline control	0	6.9 ± 0.6	9.7 ± 1.6	10.4 ± 1.7
Phenylephrine control	30	25.2 ± 5.5+	19.2 ± 2.5+	21.4 ± 2.6+
Phenylephrine stress	30	16.2 ± 1.7	22.7 ± 3.1	18.5 ± 2.3
Saline control	0	6.9 ± 0.6	7.6 ± 1.2	10.0 ± 2.4
Clonidine control	10	21.6 ± 3.1+	28.9 ± 3.0++	29.5 ± 2.8++
Clonidine stress	10	21.6 ± 3.7	23.1 ± 3.5	23.8 ± 3.7
Saline control	0	6.3 ± 0.7	7.0 ± 0.9	9.5 ± 2.8
Isoproterenol control	10	19.3 ± 2.6++	20.1 ± 5.6+	21.0 ± 3.4+
Isoproterenol stress	10	9.7 ± 1.3*	15.0 ± 2.1	16.5 ± 2.3
Saline control	0	7.4 ± 1.4	11.0 ± 2.1	10.4 ± 1.7
GABA control	10-50x	16.7 ± 4.3+	27.5 ± 6.3+	26.9 ± 2.7++
GABA stress	10-50x	7.6 ± 0.8*	8.8 ± 1.7*	10.3 ± 1.6*
Saline control	0	7.4 ± 1.4	11.0 ± 2.1	7.4 ± 1.7
Carbachol control	1	29.3 ± 3.4++	28.1 ± 4.6+	24.9 ± 3.2++
Carbachol stress	1	14.3 ± 1.4*	11.8 ± 1.3*	13.3 ± 1.0*

All drugs were administered intracerebroventricularly and 1 hour later rats were sacrificed. ˣ A dose of 50 µg was given to rats after 21 days of crowding. Values are means ± SEM. + p < 0.05 and ++ p>0.001 versus saline controls. * p < 0.05 versus agonist control group.

Effects of Crowding on Corticosterone Response to GABA and Carbachol

The increase in serum corticosterone levels induced in control rats by icv administration of GABA (10 - 50 μg), a GABA$_A$ receptor agonist, was completely absent in rats crowded for 1, 2 or 3 weeks. Also, a significant increase in serum corticosterone induced by carbachol (1 μg icv), a cholinergic muscarinic receptor agonist, was almost totally suppressed in rats crowded for 1, 2 or 3 weeks. These results indicate that responsiveness of central GABA and muscarinic receptors involved in stimulation of CRF and ACTH secretion is completely eliminated by crowding stress.

Effects of Crowding on Corticosterone Response to Enkephalinamide

In contrast to the considerable diminution by crowding stress of the HPA response to different monoaminergic receptor agonists, the corticosterone responses to two non-degradable enkephalins, met- and leu-enkephalinamide, were not affected significantly. This suggests that central opioidergic receptors are not affected by the social stress of crowding in rats (Table 2).

DISCUSSION

The results of the present experiments clearly demonstrate that in rats social stress of crowding causes a considerable decrement of the HPA response to different central monoaminergic receptor agonists.

TABLE 2. Effects of crowding on corticosterone secretion stimulated by icv administration of enkephalins.

Treatment	Dose	Crowding period (days)		
	μg/rat	7	14	21
Saline control	10 μl	11.0 ± 2.3	11.0 ± 2.3	12.8 ± 0.4
Met-Enk. control	1	25.7 ± 2.6+ +	31.2 ± 5.8+	32.6 ± 3.2+
Met-Enk. stress	1	27.6 ± 4.9	29.5 ± 3.7	26.3 ± 4.9
Saline control	10 μl	9.2 ± 1.5	9.7 ± 1.6	10.4 ± 1.7
Leu-Enk. control	1	27.1 ± 3.5+ +	27.3 ± 4.5+	24.7 ± 2.9+ +
Leu-Enk. stress	1	21.9 ± 4.1	24.4 ± 4.6	20.0 ± 3.7

All drugs were administered icv and 1 hour later rats were sacrificed. Enk.= Enkehalinamide. Values are means ± SEM. + $p < 0.05$ and + + $p < 0.001$ versus saline controls.

Crowding for 7 days considerably reduced the corticosterone response to phenylephrine, an α_1-adrenergic receptor agonist. This suggests that crowding stress considerably desensitizes α_1-adrenoceptors involved in the pituitary-adrenocortical response.

The sensitivity of α_1-adrenoceptors was reduced presumably due to high levels of catecholamines released during stress (Anisman *et al*, 1987). Our results indicate that crowding does not significantly influence the responsiveness of α_2-adrenoceptors involved in CRF and ACTH secretion since only a moderate decrease of the normal stimulatory effect of clonidine was observed after 2 or 3 weeks of social stress. Chronic immobilization stress lasting 14 days was shown to affect α_2-adrenoceptor density in different brain regions and decreased the binding affinity in the hypothalamus (Nukina *et al*, 1987; Stanford *et al*, 1984). The latter change might be responsible for the moderate diminution of corticosterone responses to clonidine stimulation in the present experiment. A stronger inhibition of the isoproterenol-induced corticosterone response in crowded rats is in agreement with other studies that have demonstrated a significant desensitization of ß-adrenoceptors following exposure of animals to a variety of stressors.

The most dramatic results of the present experiments were the demonstration that central GABA-ergic and muscarinic cholinergic systems involved in central stimulation of the HPA axis were most susceptible to stress of crowding in rats. In fact, these two types of receptors were dramatically desensitized during 1-3 weeks of crowding, as no corticosterone response was elicited by icv administration of GABA or carbamylcholine. The lack of responsiveness of GABA and muscarinic receptors in crowded rats may be related to the coexistence of GABA and cholinergic receptors on the same synaptic terminals (Beaulieu and Somogy, 1991) and a known influence of repeated stress on both GABA-ergic and cholinergic mechanisms in different brain regions (Gottesfeld *et al*, 1978). It is known that handling or footshock stress causes a rapid decrease in GABA in different brain areas (Biggio *et al*, 1990). Our data strongly suggest that $GABA_A$ and cholinergic receptors play major roles in the pharmacology and physiopathology of stress and anxiety.

In contrast to the reported effects of enkephalinergic systems on behavioral changes and norepinephrine metabolism in brain regions induced by stress, we did not find any significant influence of these systems on corticosterone responses to the chronic stress of crowding in rats (Katoh et al, 1990; Tanaka et al, 1989).

It is not clear at present by what mechanism chronic stress induces specific monoaminergic receptor desensitization. Both profound desensitization of adenyl cyclase by impaired coupling between the receptor and the nucleotide regulatory protein, G_S, with prolonged exposure to circulatory catecholamines and down-regulation of total cellular receptors (Hausdorff et al, 1990) appear to be responsible for the observed decrease of HPA responsiveness to monoaminergic agonists in crowded rats.

REFERENCES

Anisman, H., Irwin, J., Bowers, W., Ahluwalia, P., and Zacharko, M. (1987). Variations of norepinephrine concentrations following chronic stressor application. *Pharmacology, Biochemistry and Behavior* **26**, 653-659.

Armario, A., Lopez-Calderon, A., Jolin, T., and Balasch, J. (1986). Response of anterior pituitary hormones to chronic stress. *Neuroscience and Biobehavioral Reviews* **10**, 245-250.

Beaulieu, C., and Somogy, P. (1991). Enrichment of cholinergic synaptic terminal on GABAergic neurons and coexistence of immunoreactive GABA and choline acetyltransferase in the same synaptic terminals in the striate cortex of the cat. *Journal of Comparative Neurology* **304**, 666-680.

Biggio, G., Concas, A., Corda, M., Giorgi, O., Sanna, E., and Serra, M. (1990). GABAergic and dopaminergic transmission in the rat cerebral cortex: effect of stress, anxiolytic and anxiogenic drugs. *Pharmacology and Therapeutics* **48**, 121-142.

Biggio, G., Corda, M., Concas, A., Demontis, G., Rossetti, Z., and Gessa, G. (1981). Rapid changes in GABA binding induced by stress in different areas of the rat brain. *Brain Research* **229**, 363-369.

Gabriel, N., and Soliman, K.F.A. (1983). Effect of stress on the acetylcholinesterase activity of the hypothalamus-pituitary-adrenal axis in the rat. *Hormone Research* **17**, 43-48.

Gottesfeld, Z., Kvetnansky, R., Kopin, I.J., and Jacobowitz, D.M. (1978). Effects of repeated immobilization stress on glutamate decarboxylase and choline acetyltransferase in discrete brain regions. *Brain Research* **152**, 374-378.

Hausdorff, W.P., Caron, M.G., and Lefkowitz, R.J. (1990). Turning off the signal: desensitization of ß-adrenergic receptor function. *FASEB Journal* **4**, 2881-2889.

Katoh, A., Nabeshima, T., and Kameyama, T. (1990). Behavioral changes induced by stressful situations: Effects of enkephalins, dynorphin, and their interactions.

Journal of Pharmacology and Experimental Therapeutics **253**, 600-607.

Natelson, B.H., Ottenweller, J.E., Cook, J.A., Pitman, D., McCarty, R., and Tapp, W. (1988). Effect of stressor intensity on habituation of the adrenocortical stress response. *Physiology and Behavior* **43**, 41-46.

Nukina, I., Glavin, G.B., and LaBella, F.S. (1987). Chronic stress affects alpha$_2$-adrenoceptors in brain regions of the rat. *Research Communications in Psychology, Psychiatry and Behavior* **12**, 53-60.

Rivier, C., and Vale, W. (1987). Diminished responsiveness of the hypothalamic-pituitary-adrenal axis of the rat during exposure to prolonged stress: A pituitary-mediated mechanism. *Endocrinology* **121**, 1320-1328.

Stanford, C., Fillenz, M., and Ryan, E. (1984). The effect of repeated mild stress on cerebral cortical adrenoceptors and noradrenaline synthesis in the rat. *Neuroscience Letters* **45**, 163-167.

Stone, E.A., and Platt, J.E. (1982). Brain adrenergic receptors and resistance to stress. *Brain Research* **237**, 405-414.

Stone, E.A., Platt, J.E., Herrera, A.S., and Kirk, K.L. (1986). Effect of repeated restraint stress, desmethylimipramine or adrenocorticotropin on the alpha and beta adrenergic components of the cAMP response to norepinephrine in rat brain slices. *Journal of Pharmacology and Experimental Therapeutics* **237**, 702-707.

Tanaka, M., Ida, Y., Tsuda, A., Tsujimaru, S., Shirao, I, and Oguchi, M. (1989). Met-enkephalin, injected during the early phase of stress, attenuates stress-induced increase in noradrenaline release in rat brain regions. *Pharmacology, Biochemistry and Behavior* **32**, 791-795.

U'Pritchard, D.C., and Kvetnansky, R. (1980). Central and peripheral adrenergic receptors in acute and repeated immobilization stress. In: E. Usdin, R. Kvetnansky and I.J. Kopin (Eds.), "Catecholamines and Stress: Recent Advances," pp. 299-308. New York: Elsevier.

Stress: Neuroendocrine and Molecular Approaches
Edited by R. Kvetnansky, R. McCarty and J. Axelrod

1992 Gordon and Breach Science
Publishers S.A., New York, USA.
Photocopying permitted by license only.

INTERACTION BETWEEN ADRENERGIC AND OPIOIDERGIC SYSTEMS IN THE REGULATION OF PITUITARY HORMONE SECRETION

D. T. Kiem[1], G. Moskovin[1], K. Yarygin[2] and G. B. Makara[1]

[1]Institute of Experimental Medicine, Hungarian Academy of Sciences
[2]Institute of Experimental Cardiology of Cardiology Center of USSR
Academy of Medical Sciences, USSR

INTRODUCTION

It is generally accepted that morphine and various opioid peptides act via different central opioid receptors to stimulate pituitary hormone secretion (Pechnich *et al*, 1985). Studies using alpha$_2$-adrenergic agonists and antagonists have also suggested that alpha$_2$-adrenoceptors have an important role in the regulation of pituitary hormone secretion (Cella *et al*, 1983; Krulich *et al*, 1982). Since the alpha$_2$-antagonist yohimbine as well as norepinephrine depleting compounds abolished the growth hormone(GH)-releasing effect of morphine, it has been suggested that alpha$_2$-mechanisms probably mediate the opiate action, and the GH-releasing effect of morphine may require an intact central noradrenergic system (Koenig *et al*, 1980). However, discrepant findings have also been found. In newborn rats, morphine is effective while clonidine is ineffective in stimulating GH secretion (Kuhn *et al*, 1981). In stressed animals, the effect of morphine on GH secretion remains, whereas the effect of clonidine on this hormone is abolished (Mounier *et al*, 1983). The GH releasing effect of morphine and clonidine is also inhibited by kappa opiate agonists (Krulich *et al*,

1986). In addition, we have also observed that the GH releasing effect of clonidine is lost in long-term castrated rats, while the stimulatory effect of morphine on GH secretion remained unchanged in these animals (Kiem et al, 1991). In this study, we used the desensitization phenomenon of receptors following chronic treatment with yohimbine or clonidine or the kappa opiate agonist U50-488H to evaluate the relationship between opioidergic and alpha$_2$-adrenergic system in the regulation of GH and prolactin (PRL) secretion.

MATERIALS AND METHODS

Male Wistar rats, weighing 250-300, g were used throughout these experiments.   Drugs included morphine sulphate (Alkaloida, Tiszavavari, Hungary); clonidine (C.H. Boehriger, Ingelheim, FRG); yohimbine (Chinoin, Hungary); and U50-488H, a selective kappa opiate receptor agonist (kindly provided by Vonvoiglander P.F.). All experiments were performed in randomized groups of conscious unrestrained animals.   Blood samples were collected by quick decapitation (within 1 minute of opening the cage) at 30 minutes after morphine, and 45 minutes after clonidine, U50-488H and yohimbine injections. Blood samples were collected between 0900-1000 hours.

Experiments

Experiment 1: Interaction between yohimbine, morphine and clonidine on GH secretion in intact rats.   Yohimbine (1 and 3 mg/kg i.p.) was injected, followed 15 minutes later by either morphine (3 mg/kg s.c.) or clonidine (31 μg/kg i.p.).

Experiment 2: The effect of morphine and clonidine on GH secretion in chronic clonidine- treated rats.   Morphine (3 mg/kg s.c.) or clonidine (250 μ/kg i.p.) was injected into rats chronically treated with clonidine in drinking water (2 μg/ml for 14 days until sacrifice).

Experiment 3: Effect of morphine and clonidine on GH secretion in chronic yohimbine- treated rats.   Animals were treated with yohimbine (2 mg/kg i.p. twice a day in the first week and three times a day in the second week) until sacrifice. These animals were acutely injected with either morphine (5 mg/kg s.c.) or clonidine (0.5 mg/kg i.p.).

Experiment 4: Interaction between U50-488H, yohimbine and morphine in chronic yohimbine-treated rats. Animals were injected first with either yohimbine (3 mg/kg) or with U50-488H (3 mg/kg). Morphine (3 mg/kg) was injected 15 minutes later.

Experiment 5: Effect of yohimbine on PRL secretion in rats made tolerant to the kappa opiate agonist U50-488H. Animals were injected with U50-488H at a dose of 2 mg/kg s.c. twice a day in the first week and 3 mg/kg three times a day in the second week. These animals were injected acutely with 5 mg/kg i.p. yohimbine on the 14th day.

Experiment 6: Kappa opiate receptor binding in rats chronically treated with yohimbine. Kappa opiate binding was studied with ^{3}H-ethylketocyclazocine (^{3}H-EKC) obtained from Amersham (UK). In all experiments, the incubation mixture consisted of 0.1 ml homogenate, 0.1 ml labelled ligand (or plus MR-2266). For B_{max} and K_D determinations, ^{3}H-EKC binding data were subjected to Scatchard analysis. Protein was determined by the method of Lowry *et al* (1951).

Plasma GH and PRL were measured by radioimmunoassay (RIA). Statistical analysis was done by two-way ANOVAs followed with Dunnett's test for multiple comparison, using logarithmic transformation of data.

RESULTS

Interactions between adrenergic and opioidergic systems occur in the regulation of GH secretion. Morphine and clonidine injections in intact male Wistar rats induced significant increases in plasma GH secretion. Pretreatment with yohimbine effectively prevented the GH releasing effect of morphine and clonidine (Figure 1).

In rats chronically treated with clonidine (2 μg/ml for 14 days; this administration leads to a desensitization of alpha$_2$-adrenoceptors), the GH response to an acute injection of clonidine was blocked, while the effects of morphine on this hormone remained unchanged (Figure 2).

In rats chronically treated with yohimbine (2 mg/kg i.p. 2-3 times daily for 14 days until sacrifice), the GH response to acute clonidine injection was blocked, while the effect of morphine was significantly enhanced (Figure 3). Interestingly, the inhibitory effect of

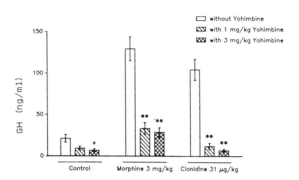

FIGURE 1:   The inhibitory action of yohimbine on the GH-releasing action of morphine and clonidine. Values are means ± SE. **<0.01; + p<0.05, compared to controls (n=5-6).

an acute injection of the kappa opiate agonist U50-488H and yohimbine on basal as well as morphine-induced GH secretion was lost in rats chronically treated with yohimbine (Figure 4), suggesting that kappa opiate receptors were either blocked or desensitized by long term alpha$_2$-adrenoceptor blockade.

Interactions between the adrenergic and opioidergic systems also occur in the regulation of PRL secretion. The stimulatory action of U50-488H and morphine on PRL secretion was lost in rats chronically treated with yohimbine (Figure 5).

Acute administration of yohimbine failed to stimulate PRL secretion in rats chronically treated with yohimbine (Figure 5). Yohimbine also failed to stimulate PRL secretion in rats made tolerant to the kappa opiate agonist U50-488H (Figure 6).

Alpha$_2$-adrenoceptor   antagonist-kappa   opiate   receptor interactions were evident. Chronic yohimbine treatment for 14 days changed kappa binding in some regions of rat brain. The B$_{max}$ of ^{3}H-EKC to cortical sites of control rats was 360±30 fmol/mg protein versus 180 ± 21 fmol/mg protein for yohimbine - treated   animals

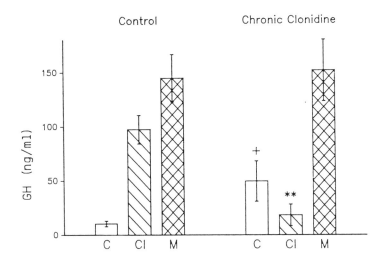

FIGURE 2: The effect of morphine (M) and clonidine (CL) on GH secretion in rats chronically treated with clonidine. *p<0.05; **p<0.01 compared to controls (n=5-6).

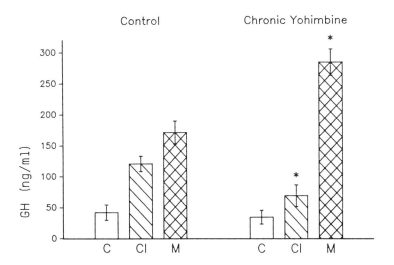

FIGURE 3: The effect of morphine (M) and clonidine (CL) on GH secretion in rats chronically treated with yohimbine. *p<0.05 compared to controls (n=5-6).

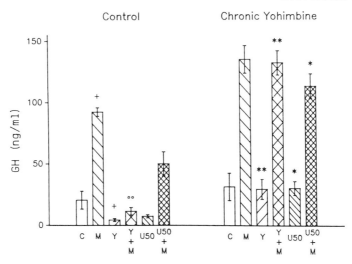

FIGURE 4:   The effect of acute yohimbine (Y) and U50-488H (U50) on the GH-releasing effects of morphine (M) in rats chronically treated with yohimbine. *p<0.05; **p<0.01 compared to saline-treated group, oo p<0.01 compared to morphine group without yohimbine treatment (N=5-6).

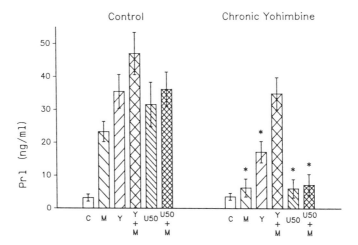

FIGURE 5:   The effect of U50-488H (U50), yohimbine (Y) and morphine (M) on plasma PRL levels in vehicle-and chronically yohimbine-treated rats.   The PRL-releasing effect of these compounds was abolished in rats chronically treated with yohimbine.   *p<0.01 compared to corresponding values of the vehicle-treated animals (n=5).

(p < 0.01). A change in the K_D for ^{3}H-EKC binding in cortex tissue was evident in controls (7.6 nM) versus yohimbine- treated animals (3.9 nM). Consistent with the results obtained in frontal cortex, a decreased number of kappa binding sites was also found in hippocampus (Figure 7).

DISCUSSION

The present data support a new view of the interrelationships between alpha-adrenergic and opioidergic systems in the modulation of GH and PRL release. We suggest that the GH-releasing effects of opiates is probably not mediated by, or is not dependent on alpha$_2$-adrenoceptors. The GH inhibitory as well as the PRL stimulatory actions of the alpha$_2$-adrenergic antagonist yohimbine might be mediated by kappa opiate receptors.

The concept of central alpha$_2$-adrenergic mechanisms mediating opiate actions on GH secretion was derived from the observation that the effect of morphine on GH secretion was effectively prevented by acute pretreatment with either yohimbine (Koenig et al, 1980), or phenoxybenzamine, a mixed alpha$_1$ and alpha$_2$-antagonist (Aden et al, 1976). However, several discrepant findings argue against an obligatory role for alpha$_2$-adrenoceptor exitation in mediating the GH stimulatory role of morphine. Morphine was able to elevate plasma GH even after stress-induced inhibition of GH secretion in rats (Mounier et al, 1983; Briski et al, 1984), while the GH releasing effect of clonidine was antagonized in the stressed animals (Mounier et al, 1983). Also, there is a difference in the maturation of opiate and alpha$_2$-adrenoceptor stimulation of GH secretion. The opiate system is already functional in the first week of life, while the alpha-adrenergic system matures only after puberty (Kuhn et al, 1981). In this study, we demonstrated that morphine's stimulatory effect on GH secretion remained unchanged when alpha$_2$-adrenergic receptors were desensitized by chronic clonidine treatment. Thus, the pathway for opiate-induced GH secretion is active when the alpha$_2$-adrenoceptor system is compromised. That alpha$_2$-adrenoceptors had been desensitized is indicated by the finding of no further GH elevation after an acute injection of a large dose of clonidine.

It is of interest to determine the possible mechanisms by which acutely administered yohimbine inhibits morphine - induced GH

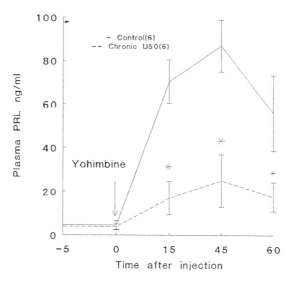

FIGURE 6: The PRL-releasing effect of yohimbine was blunted in the rats made tolerant to the kappa opiate agonist U50-488H. *$p < 0.05$ compared to vehicle-treated animals ($n = 5$).

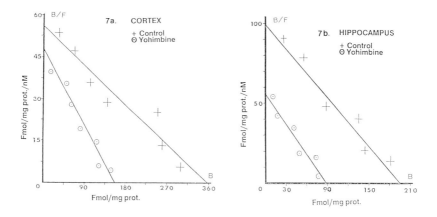

FIGURE 7: Kappa binding sites were significantly ($p < 0.01$) decreased in the frontal cortex (Figure 7a) and hippocampus (Figure 7b) of rats chronically treated with yohimbine ($n = 3$ per group).

responses since this mechanism might have misled previous authors to believe that morphine stimulates GH secretion via alpha$_2$-adrenoceptors. The following considerations may shed some light on this mechanism.

First, we have found that yohimbine's inhibitory action on morphine-induced GH secretion is lost in rats chronically pretreated with yohimbine. At the same time clonidine's stimulatory effect on GH remained inhibited. This suggests that the main inhibitory action of yohimbine on morphine-induced GH secretion doesn't result from blockade of alpha$_2$-adrenoceptor stimulation of GH secretion, but rather may be due to other pharmacological mechanisms (side effects).

Second, the loss of the inhibitory action of yohimbine on morphine-induced GH secretion was parallel with the failure of both yohimbine and U50-488H to inhibit basal and morphine-induced GH secretion in rats chronically treated with yohimbine. This suggests that yohimbine desensitization interferes with kappa opiate receptors and that the acute inhibitory action of yohimbine on the GH-releasing effect of morphine might also be due to activation of kappa opiate binding sites. Presynaptic kappa opiate receptors may become operative only after alpha$_2$-adrenoceptors have been blocked, similar to the case of cortical noradrenergic axons (Ramme *et al*, 1986). In the present study, we found the kappa opiate binding sites in rat brain were significantly decreased in rats chronically treated with yohimbine suggesting a heterologous desensitization of kappa opiate receptors.

There is *in vivo* evidence that kappa opiate receptor agonists may act as mu opiate receptor antagonists. Pretreatment with kappa opiate agonists effectively inhibited morphine-induced GH secretion (Krulich *et al*, 1986b). Studies on opiate receptor binding showed that kappa opiate agonists have a high affinity for mu opiate receptors (Magnan *et al*, 1982; Wood *et al*, 1982) and that they are antagonists at the mu binding sites (Gillan *et al*, 1981).

Yohimbine is able to stimulate PRL secretion (Gold *et al*, 1979; Lien *et al*, 1986), possibly by inhibiting dopaminergic receptors in the pituitary gland (Scatton *et al*, 1980). However, since yohimbine does not prevent the inhibitory action of dopamine on PRL secretion and failed to displace ^{3}H-spiperone in bovine pituitary membranes *in vitro* (Meltzer *et al*, 1983), dopaminergic mechanisms might not be important in yohimbine-induced PRL secretion. 5-HT receptors may not be involved in yohimbine-induced PRL secretion because the 5-HT receptor blockers minanserin and cyproheptadine did not reverse the

PRL releasing effect of yohimbine (Meltzer *et al*, 1983). Finally, since histamine H_1- and H_2-receptor blockers (chlorpheniramine, metiamide) do not prevent the PRL-releasing action of yohimbine, histaminergic mechanisms may not be involved in yohimbine-induced PRL secretion (Meltzer *et al*, 1983). In our study, the PRL releasing effect of acutely administered yohimbine was significantly attenuated in rats chronically treated with this drug. This finding together with the loss of PRL release to U50-488H in these animals, indicates that kappa opiate receptors rather than dopaminergic mechanisms mediate the PRL-releasing effect of yohimbine. Finding desensitized kappa-opiate receptors in rats chronically treated with yohimbine strongly supports our hypothesis.

In conclusion, these findings suggest that there is an interaction between kappa opiate receptors and alpha$_2$-adrenoceptors at a central level; the GH releasing effect of opiates does not require mediation by the alpha$_2$-adrenoceptor system. The inhibitory action of yohimbine on the GH releasing effect of morphine and the PRL releasing effect of yohimbine might be mediated via its side effect(s), possibly on kappa-opiate receptors.

REFERENCES

Aden, N.E., Grabowska, M., and Strombom, U. (1976). Different alpha-adrenoceptors in central nervous system mediating biochemical and functional effects of clonidine and receptor blocking agents. *Naunyn Schmiedeberg's Archives of Pharmacology* **292**, 43-52.

Briski, K.P., Quigley, K., and Meites, J. (1984). Counteraction by morphine of stress-induced inhibition of growth hormone release in the rat. *Proceedings of the Society for Experimental Biology and Medicine* **177**, 137-142.

Cella, S.G., Picotti, G.B., Morgese, M., Mantegazza, P., and Muller, E.E. (1983). Presynaptic alpha-receptor stimulation leads to GH release in the dog. *Life Sciences* **31**, 447-454.

Kiem, D.T., Bartha, L., Harsing, L.G., Jr., and Makara, G.B. (1991). Revaluation of the role of alpha$_2$-adrenoceptors in morphine-stimulated release of growth hormone. *Neuroendocrinology* **53**, 516-522.

Koenig, J., Mafield, M.A., Coppigs, R.J., McCann, S.M., and Krulich, L. (1980). Role of central nervous system neurotransmitters in mediating the effects of morphine on growth hormone and prolactin secretion in the rat. *Brain Research* **197**, 453-468.

Krulich, L., Koenig, J.L., Conway, S., McCann, S.M., and Mafield, M.A. (1986). Opioid kappa receptors and the secretion of prolactin (PRL) and growth hormone (GH) in the rat. II. GH and PRL release-inhibiting effects of the opioid kappa receptor agonists bremazocine and U50-488H. *Neuroendocrinology* **42**, 82-87.

Krulich, L., Mafield, M.A., Steele, M.K., McMillen, B.A., McCann, S.M., and Koenig, J.L. (1982). Differential effects of pharmacological manipulations of central alpha$_1$ and alpha$_2$-adrenergic receptors on the secretion of thyrotropin and growth hormone in male rats. *Endocrinology* **110**, 796-804.

Kuhn, C.M., and Schanberg, S.M. (1981). Maturation of central nervous system control of growth hormone secretion in the rat. *Journal of Pharmacology and Experimental Therapeutics* **217**, 152-156.

Magnan, J., Paterson, S.J., Tavani, A., and Kosterlitz, H.W. (1982). The binding spectrum of narcotic analgesic drugs with different agonist and antagonist properties. *Archives of Pharmacology* **21**, 487-497.

Mounier, F., Bluet-Pajot, M.T., Durant, D., and Schaub, C. (1983). Effect of acute stress on growth hormone release induced by morphine and clonidine. *Neuroendocrinology Letters* **5**, 221-226.

Pechnich, R., George R., and Poland, R.E. (1985). Identification of multiple opiate receptors through neuroendocrine responses. II. Antagonism of mu, kappa and sigma agonists by naloxone and Win 44, 441-3. *Journal of Pharmacology and Experimental Therapeutics* **232**, 170-177.

Ramme, D., Illes, P., Spath, L., and Starke, K. (1986). Blockade of alpha$_2$-adrenoceptors permits the operation of otherwise silent opioid kappa-receptors at the sympathetic axons of rabbit jejunal arteries. *Naunyn Schmiedeberg's Archives of Pharmacology* **334**, 48-55.

Wood, P.L. (1982). Multiple opiate receptors: Support for unique mu, delta and kappa sites. *Neuropharmacology* **3**, 105-107.